Progress in Molecular and Subcellular Biology

20

Springer

Berlin
Heidelberg
New York
Barcelona
Budapest
Hong Kong
London
Milan
Paris
Singapore
Tokyo

Alvaro Macieira-Coelho (Ed.)

Inhibitors of Cell Growth

With 22 Figures

Springer

Professor Dr. A. Macieira-Coelho
INSERM
73 bis, Rue du Maréchal Foch
78000 Versailles
France

ISBN 978-3-642-72151-9

Library of Congress Cataloging-in-Publication Data
Inhibitors of cell growth / Alvaro Macieira-Coelho (ed.) .
p. cm. -- (Progress in molecular and subcellular biology ; 20)
Includes bibliographical references and index.
ISBN 978-3-642-72151-9 ISBN 978-3-642-72149-6 (eBook)
DOI 10.1007/978-3-642-72149-6
1. Cells--Growth--Regulation. 2. Cancer cells--Growth--Regulation. 3. Growth factors. 4. Interferon. 5. Oncogenes. 6. Antioncogenes. I. Macieira-Coelho, Alvaro, 1932- II. Series.
QH604.I53 1998
571 . 8`49--dc21

Softcover reprint of the hardcover 1st edition 1998

Production: PRO EDIT GmbH, Heidelberg
Cover design: Meta Design, Berlin
Typesetting: Best set Typesetter Ltd., Hong Kong
SPIN 10636853 39/3137 5 4 3 2 1 0 - Printed on acid-free paper

Preface

The containment of cell growth is at the core of the homeostatic regulation of metazoans, and considerable progress has been made in the understanding of how this is achieved. Most knowledge comes from the isolation of molecules with positive and negative regulatory effects on cell proliferation, and most emphasis so far has been on these molecules.

Some of these molecules are already available for therapeutic purposes, and others look promising in this respect. This volume gives examples of such approaches.

The understanding of the control of cell growth is also fundamental to grasp phylogenic and ontogenic development. Why organisms have developed increasingly sophisticated mechanisms that control their size and that of their organs, how different cells originate, some destined for renewal and repair, others for specialized functions in a postmitotic state or evolving through division, others like the germinal cells waiting for the signal to start another organism.

There is one mechanism of growth containment, however, about which we know very little. It concerns the structural characteristics of the cell, i.e. the relationship between structure and function. How structure can change the response to identical signals. The positive and negative growth regulators may be conserved, but the structure and organization of the genetic material and of other cell components differ widely and are responsible to a great extent for the differences in cell proliferative behaviour.

It is not only through the response to molecules, however, that structure plays a part in the control of cell growth. There are intrinsic properties created by structural constraints that constitute in themselves a limit to proliferation. From simple unicellular organisms where proliferation may depend only upon the availability of nutrients, the complexity of the structural organization of the genetic apparatus evolved phylogenetically, and is certainly a determinant of how many times a cell should and can divide to fulfill its role during development and in homeostasis. The increasing structural complexity of the genome created an additional potential for the diversification of function but, on the other hand, originated constraints on the limits of somatic cell proliferation.

The structural organization also evolves during ontogenesis and through the life span of an organism. This is an important feature of the manifestations of aging that is dealt with herein.

This aspect of the control of cell growth revealed by experimental systems is looming and for the moment although obvious, its mechanisms can only be conjectured.

So although we are just beginning to understand the different aspects of the whole shebang, it is time to appraise how far we have gone and in what directions we should look. This is what this volume is about.

Versailles, France
May 1998

A. Macieira-Coelho

Contents

Introduction

Alvaro Macieira-Coelho[1]

The understanding of the mechanisms of the inhibition of cell proliferation concerns developmental processes such as the size of organs and organisms, cell differentiation, cell renewal during the animal life span, aging, and abnormal cell growth. This explains why growth inhibitors can be so versatile (for instance, interferon), their mechanism of action being as complex as the processes they regulate.

Although the search for negative growth regulators started at the beginning of the century (Lozzio et al. 1975) the field took longer to develop than that of positive growth factors, due to the fact that it is easier to obtain nonspecific inhibition of cell division such as obtained through toxicity than nonspecific stimulation. On the other hand, the methods used to evaluate growth inhibition, such as the measurement of thymidine incorporation, are often prone to artifacts (Diatloff et al. 1978).

There is an almost infinite number of molecules or cell fractions that have been reported to impede cell division, and new ones are progressively being identified. Thus, a comprehensive description of inhibitors of cell proliferation would require a considerable amount of space that is not within the scope of this series. The goal of this volume is to give enough examples of different types of identified physiological negative growth regulators for the reader to understand the general mechanism of how cell proliferation can be checked to fulfil the needs of the organism.

Examples of growth inhibitors isolated from specific cell types are given because they were initially identified in those cells; it does not mean that they are always specific to these cell types. In many instances one can speak of cell preference.

The arrest of cell division is not due to the presence or predominance of negative growth factors alone, or the absence of positive growth regulators; it can also be due to changes in molecular and supramolecular configurations that require a higher mobilization of energy to activate a molecule. In other words, structural changes can decrease the conformational flexibility and lower the probability for the activation of an energy barrier. Hence, the mo-

[1] INSERM, 73 bis, rue du Maréchal Foch, 78000 Versailles, France

Progress in Molecular and Subcellular Biology, Vol. 20
A. Macieira-Coelho (Ed.)

lecular composition and structure of the cell surface, of the cytoskeleton, and of the nuclear components influence the initiation and progression of the division cycle. This is most likely one of the causes of the arrest of cell division in situations such as the terminal stage of the human fibroblasts' life cycle.

Moreover, initiation and progression of the division cycle does not depend only on the activation or repression of the cyclin-dependent kinases, transcription factors, and regulatory gene products that have been identified (Harper 1997; Jackman and Pines 1997; Reed 1997).

In order to understand the mechanism of the physiological inhibition of cell division, it is necessary to grasp the organization of a cell in resting phase and how it is activated to enter the division cycle, duplicate the DNA, and split into two new cells.

Mammalian DNA is a molecule more than 1 m long, confined in a sphere, the nucleus, with a diameter of about 5 μ. This implies an elaborate folding, which is regulated inter alia by the structure of DNA (Crothers et al. 1990), by DNA-bound proteins, by enzymes, and by the anchorage of this makeup to a protein matrix. In addition, there are glycoproteins present whose role is yet to be determined (Zardi et al. 1976); since these molecules are highly charged, it is reasonable to assume that they must also function as regulators of chromatin conformation.

The scaffold upon which chromatin is anchored plays a crucial role in the organization of the high-order structure of DNA. There is indeed a protein framework, called the nuclear protein matrix, with which DNA is associated. DNA synthesis-initiating sites are preferentially located at the borders between condensed chromatin and interchromatin areas (Berezney and Coffey 1975), suggesting an important role for the nuclear matrix, in particular the peripheral nuclear region, in the initiation of DNA replication. The anchorage of DNA is crucial not only for replication but also for transcription, since nascent RNA is associated with the nuclear cage (Jackson et al. 1984).

The nuclear lamina, a filamentous protein meshwork lining the nucleoplasmic surface of the nuclear envelope, probably provides an anchoring site at the nuclear periphery for interphase chromatin (Gerace 1985). When the nuclear shell is isolated, it contains chromatin structures made of packed nucleosomes 28–32 nm thick (identical to the high-order solenoid DNA structure) that are associated with the three nuclear lamins (Bouvier et al. 1985).

The lamina is composed of proteins called lamins which seem to be intermediary structures between DNA-binding proteins and the cytoskeleton. Indeed, on the one hand, the lamina is tightly bound to chromatin since it can be dissociated from chromatin only by high salt solution which also extracts the tightly bound histones in the nucleosome cores (Bouvier et al. 1985). On the other hand, lamins have a striking sequence homology with intermediate filaments, a component of the cytoskeleton (Gerace 1985).

Thus, the anchorage of chromatin seems to be fulfilled with the preservation of the continuity with the cytoplasmic scaffold. In this way, DNA is linked to the cytoskeleton through its anchorage to the nuclear cage and via the former to the cell membrane and the extracellular matrix. The cytoskeleton and the nuclear matrix act as integrators not only of space but also of function within the cell. This whole structure can be seen as a three-dimensional manifold (Thurston and Weeks 1984) constituted by linked cranks where information flows to a great extent through changes in molecular configurations. The function of this network is regulated through the synthesis of molecules that have to assume the right steric configuration and through energy turnover. Cell behavior is determined by the way in which the network of structures is connected, i.e., its topology. Through the cell division cycle new topological constraints are created.

The first modifications originate at the cell surface and are then transmitted through the cytoskeleton to the nuclear cage. DNA has to unravel for gene transcription and replication. At the end of the G2 period, before mitosis, there is complete disruption of the initial topology with the disassembly of the nuclear envelope, chromosome condensation, and the reorganization of the cytoskeleton.

Experimental evidence favors this view. Stretching of the skin in vivo is enough to trigger cell division (Lorber and Milobsky 1968). Moreover, varying the adhesion of cells on surfaces whose physicochemical properties could be modified in a controlled fashion allowed the modulation of cell proliferation, differentiation, and malignancy (Macieira-Coelho and Avrameas 1972, 1973; Wahrman et al. 1981; Macieira-Coelho 1988). Modulation of cell proliferation by cell-substratum interactions leading to gene expression is mediated by physical forces (Macieira-Coelho et al. 1974).

From the cell periphery to the nucleus the information flows through the activation of a series of energy barriers. The tools used to switch from low-energy to high-energy molecular configurations are: the binding of ligands to their receptors, gradients of electrochemical potential created by ion pumps, Ca^{2+} mobilization, and phosphorylation and dephosphorylation. Ordering substrates in an advantageous configuration with respect to each other and thus lowering the energy needed to activate the barrier is another way to pass on information. This is what catalysts do (Harold 1986).

Changes in conformation through molecular binding take place, for instance, when the epidermal growth factor (EGF) binds to its receptor, which becomes a kinase. Growth inhibitors can act at this level, blocking the binding of growth factors (Bryckaert et al. 1988) or downregulating growth factor receptors (Paulsson et al. 1993). They can also block the division cycle, inducing changes in cell contraction (Montesano and Orci 1988) through the regulation, for instance, of the levels of fibronectin isoforms (Balza et al. 1988). It is also through variations in the anchoring of the cell that molecules of the extracellular matrix such as gangliosides (Ohsawa and Senshu 1987) and glycosaminoglycans (Wever et al. 1980; Westergreen-Thorsson et al. 1993) can act as inhibitors. Obviously, all molecules of the cell periphery intervene in this way in the transmission of information at this stage.

Variation in molecular configuration is in itself sufficient to trigger ion pumps. This is why cation channels can be activated in human fibroblasts by stretching (Stockbridge and French 1988), and growth factors, the cell substratum, and integrins can activate the Na^+/H^+ antiporter and increase intracellular pH (Moolenar et al. 1983; Margolis et al. 1988; Schwartz et al. 1991), a necessary step to trigger the division cycle. Inhibition of the antiporter impedes the transduction of the information and inhibits DNA synthesis (L'Allemain et al. 1984).

Once the signal has crossed the cell membrane, it progresses across the cytoplasm through the cytoskeleton to the nuclear matrix and to DNA. Indeed, it has been shown that microtubule depolymerization is sufficient to initiate DNA synthesis (Crossin and Carney 1981), that growth factors cause modifications in the higher-order chromatin structure (Ortega and Meneghini 1989), that cell contractility is coupled with proliferation (Macieira-Coelho and Azzarone 1990), and that $p34^{cdc2}$, a kinase necessary for cell-cycle progression which is associated with the cell scaffold, induces changes in cell shape, cytoskeletal organization, and chromatin structure (Lamb et al. 1990; Pockwinse et al. 1997). Moreover, mitogen-activated protein kinase (MAPK) regulates cytoskeletal organization via microtubule-specific interactions (Reszka et al. 1997).

Evidence in favor of the influence of intermediate filaments on chromatin conformation was reported (Hay and Deboni 1991); it was shown that the disruption of intermediate filaments induces chromatin motion. Thus, it makes sense that a growth arrest-specific gene codes for a protein that is a component and an organizer of the microfilament system (Brancolini et al. 1992). The cytoskeleton could play a role in the transmission to the nucleus of signal transducer activators of transcription (STATs).

It is interesting that positive and negative growth factors can use the same pathways to transmit information from the membrane to DNA. Platelet-derived growth factor (PDGF), EGF, growth hormone, and interferons all use the Janus kinases (JAK)-STAT pathway. The message could differ, depending on particular molecular configurations that create specific associations between molecules, which will lead to either the stimulation or inhibition of proliferation. Small molecular intermediates not yet identified could also confer specificity.

The information flows through the cytoplasm to the nuclear cage mainly through phosphorylation and dephosphorylation, a crucial tool used by leaving organisms to confer conformational flexibility. In this way, molecules may assume a configuration exposing specific domains favorable for molecular association or dissociation. Hence, many cell-cycle inducers and inhibitors and the products of genes transcribed during G1 are kinases and phosphatases or inhibitors of these enzymes (Martin-Castellanos and Moreno 1997). Through the creation of molecular configurations favorable for molecular binding, genes are transcribed or repressed, proteins are synthesized and degraded, etc. The different regulators interact within this web, thus it is not surprising that we still do not understand its normal regulation and even less its deregulation.

The paths initiated at the cell membrane, taken by the phosphorylation cascade, can be different. They can flow through the activation of adenylate cyclase, increase in cyclic AMP, and activation of protein kinase A (a serine and threonine kinase); the cascade can be initiated by the turnover of phosphatidylinositol and the activation of the phospholipid- and Ca^{2+}-dependent protein kinase C (another serine and threonine kinase); it can also flow through the JAK-STAT pathway after activation of a tyrosine kinase domain of a receptor following receptor occupancy, the goal being always the utilization of the same currency for energy transfer, i.e., transfer of inorganic phosphate (P_i), the ionized form of phosphoric acid, from ATP or GTP to another molecule.

Phosphorylation and dephosphorylation should not be regarded as a mechanism for turning an enzyme on or off but rather for tuning it across a range of forms, each of which responds in its own way to substrates and to regulatory molecules (Harold 1986). Hence, the same molecule can be used along the cell cycle depending upon the sites where it is phosphorylated. This is the case for $p34^{cdc2}$, which is maximally phosphorylated at Ser 277 during G1 and then dephosphorylated at this site. It becomes maximally phosphorylated at Thr 14 and Tyr 15 during G2 but dephosphorylated abruptly at the G2/M transition. During M phase it becomes phosphorylated on a threonine residue different from Thr 14 (Krek and Nigg 1991).

By binding Cdk2, cyclin A alters the orientation of ATP in the catalytic cleft, causing the T loop to move away from the catalytic cleft, thus neutralizing the conformation that keeps monomeric Cdk2 inactive (Harper 1997).

Protein degradation is another important tool to control cell-cycle progression. The levels of the cyclin-dependent kinase inhibitor p27 are regulated through its synthesis, phosphorylation, and degradation. In quiescent cells, p27 accumulates without an increase in mRNA or protein synthesis (Alessandrini et al. 1997). The regulation of p27 levels in this circumstance occurs via the ubiquitin-proteasome pathway, phosphorylation being a signal for ubiquitination. An ubiquitin-dependent proteolytic system also regulates cyclin B destruction and progression through mitosis (King et al. 1994).

Growth inhibitors can act through the induction of genes whose products have a negative regulatory role. This is the case with interferon alpha that, inter alia, induces the expression of the cyclin-dependent kinase inhibitors p21 (Hobeika et al. 1997) and p27 (Kuniyasu et al. 1997).

Mechanisms other than the mobilization of ions and P_i, responsible for the regulation of the division cycle during the G1 period, concern the synthesis of polyamines (for a review see Pegg 1986). These compounds are essential for cell growth in a fashion that is not well understood, and the induction of ornithine decarboxylase (ODC), a key enzyme involved in their synthesis, is a feature during the stimulatory action of different growth factors. Polyamines act at different levels necessary for the progression of the division cycle, such as protein synthesis or chromatin structure. Whatever their mechanism of action, the synthesis of ODC seems to constitute a check point in G1 since vitamin A, for instance, (Haddok and Russell 1979) arrests the cell at mid-G1 phase through the inhibition of ODC in a nontoxic reversible fashion. Vitamin A also induces G1 arrest and cell differentiation through hypophosphorylation of the retinoblastoma protein (Yen et al. 1996).

Physiological arrest during the S period has not been described. Many examples of G2 arrest have been reported in several organisms. In *Drosophila*, cells stop in G2 during embryogenesis due to the absence of positive growth regulators (Edgar and O'Farrell 1990). The G2 arrest of oocytes in some species is achieved through a mechanism that inactivates cyclin B-cdc2 kinase (Abrieu et al. 1997). In mammals, the presence of G2-arrested cells has been reported (Gelfant 1977) and endogenous glycopeptides were isolated from mammalian tissues, which induce a G2 arrest in a reversible manner (Charp et al. 1983). Glucocorticoids are also known to impose a G2 block (Das et al. 1983). Moreover, some gene products that regulate the transit in G1, such as p53 (Agarwal et al. 1995) and the retinoblastoma protein (Karantza et al. 1993), can stop cells during G2. The former seems to work through the stimulation of p21, an inhibitor of the G2/M-specific cdc2 kinase. The G2/M transition can also be checked by $p34^{cdc2}$, whose catalytic unit is subject to the action of a series of kinases and phosphatases (Epifanova and Brooks 1994).

After going through the G2 period, the nuclear envelope breaks down and the chromosomes prepare for separation into two new cells.

Obviously, one of the goals in understanding how the division cycle is controlled is to find how to correct it when it deviates from normalcy, in particular in relation to malignant growth.

A common denominator that became apparent in several cancers and transformed cells in vitro is the deregulation of the mechanisms of energy transfer through phosphorylation. The mechanism of action of protooncogenes and their derivative oncogenes is related to that of protein phosphorylation. Indeed, oncogene products are direct or indirect effectors of different corners of the web where energy mobilization and transduction takes place. Their products can intervene in the phosphorylation of various substrates and thus it should be expected that they play a role in such fundamental processes as development, growth, and differentiation. This also explains why, contrary to what is still found in much of the scientific literature, oncogenes cannot by themselves transform cells in vitro or trigger cancer, but only contribute to the deregulation of different metabolic pathways that create the conditions for the progressive escape from homeostasis.

Alterations in the expression of protooncogenes, through mutations, amplifications, transpositions, or recombinations must inevitably lead to deviations in homeostasis, since their products act at the most fundamental level of the life of a cell, the flow of energy transfer. Much of the future of cell biology depends on understanding this information-processing system.

Several works suggest a deregulation of phosphorylation during in-vitro cell transformation. The transforming src oncogene product, for instance, is a more active kinase than that of the src protooncogene and can accelerate oocyte maturation induced by progesterone (Spivack et al. 1984). Moreover, in v-src-transformed fibroblasts there is increased tyrosyl phosphorylation of JAKl, leading to constitutive activation of the kinase (Campbell et al. 1997). Deregulation of phosphorylation is probably one of the events responsible for increased conformational flexibility in the transformed cell (Macieira-Coelho 1990). The exact role of oncogenes will only be known when the regulation of the pathways of energy transduction and the information processing are elucidated.

Isolation of physiological growth inhibitors and understanding their mechanism of action will be crucial tools to elucidate these pathways and correct deviations from homeostasis.

References

Abrieu A, Dorée M, Picard A (1997) Mitogen-activated protein kinase activation down-regulates a mechanism that inactivates cyclin B-cdc2 kinase in G2-arrested ocytes. Mol Biol Cell 8:249–261

Agarwal ML, Agarwal A, Taylor WR, Stark GR (1995) p53 controls both the G2/M and the G1 cell-cycle checkpoints and mediates reversible growth arrest in human fibroblasts. Proc Natl Acad Sci USA 92:8493–8497

Alessandrini A, Chiaur DS, Pagano M (1997) Regulation of the cyclin-dependent kinase inhibitor p27 by degradation and phosphorylation. Leukemia 11:342–345

Balza E, Borsi L, Allemani G, Zardi L (1988) Transforming growth factor beta regulates the levels of different fibronectin isoforms in normal human cultured fibroblasts. FEBS Lett 228:42–44

Berezney R, Coffey DS (1975) Nuclear protein matrix: association with newly synthesized DNA. Science 189:291–293

Bouvier D, Hubert J, Seve AP, Bouteille M (1985) Characterization of lamina-bound chromatin in the nuclear shell isolated from HeLa cells. Exp Cell Res 156:500–512

Brancolini C, Bottega S, Schneider C (1992) Gas2, a growth arrest-specific protein, is a component of the microfilament nework system. J Cell Biol 117:1251–1261

Bryckaert MC, Lindroth M, Lönn A, Tobelem G, Wasteson A (1988) Transforming growth factor (TGF beta) decreases the proliferation of human bone marrow fibroblasts by inhibting the platelet-derived growth factor (PDGF) binding. Exp Cell Res 179:311–321

Campbell GS, Yu Cl, Jove R, Carter-Su C (1997) Constitutive activation of JAKl is src-transformed cells. J Biol Chem 272:2591–2594

Charp PA, Kinders RJ, Johnson JC (1983) G2 cell-cycle arrest induced by glycopeptides isolated from the bovine cerebral cortex. J Cell Biol 97:311–316

Crossin KL, Carney DH (1981) Evidence that microtubule depolymerization early in the cell cycle is sufficient to initiate DNA synthesis. Cell 23:61–71

Crothers DM, Haran TE, Nadeau JG (1990) Intrinsically bent DNA. J Biol Chem 265:7093–7095

Das HR, Lavin M, Sicuso A, Young DV (1983) The uncoupling of macromolecular synthesis from cell division in SV3T3 cells by glucocorticoids: the imposition of a G2 block. J Cell Phys 117:241–248

Diatloff C, Bengtson A, Billardon C, Macieira-Coelho A (1978) Lack of DNA synthesis inhibitory activity in an immunosuppressor obtained from spleen. Cell Tissue Kinet 11:317–322

Edgar BA, O'Farrell (1990) The three postblastoderm cell cycles of *Drosophila* embryogenesis are regulated in G2 by string. Cell 62:469–480

Epifanova OI, Brooks RF (1994) Negative control of cell proliferation in eukaryotes. Cell Prolif 27:373–394

Gelfant S (1977) A new concept of tissue and tumor cell proliferation. Cancer Res 37:3845–3862

Gerace L (1985) Structural proteins in the eukaryotic nucleus. Nature 318:508–509

Haddox MK, Russell DH (1979) Cell cycle-specific locus of vitamin A inhibition of growth. Cancer Res 39:2476–2480

Harold F (1986) A study of bioenergetics. WH Freeman, New York, 500 pp

Harper JW (1997) Cyclin-dependent kinase inhibitors. Cancer Surv 29:91–108

Hay M, Deboni U (1991) Chromatin motion in neuronal interphase nuclei – changes induced by disruption of intermediate filaments. Cell Motil Cytoskeleton 18:63–70

Hobeika AC, Subramaniam, Johnson HM (1997) IFN alpha induces the expression of the cyclin-dependent kinase inhibitor p21 in human prostate cancer cells. Oncogene 14:1165–1170

Jackman MR, Pines JN (1997) Cyclins and the G(2)/M transition. Cancer Surv 29:47–74

Jackson DA, McCready SJ, Cook PR (1984) Replication and transcription depend on attachment of DNA to the nuclear cage. J Cell Sci Suppl 1:59–79

Karantza V, Maroo A, Fay D, Sedivy JM (1993) Overproduction of Rb protein after G1/S boundary causes (G2 arrest. Mol Cell Biol 13:6640–6652

King RW, Jackson PK, Kirchner MW (1994) Mitosis in transition. Cell 79:563–571

Krek W, Nigg EA (1991) Differential phosphorylation of vertebrate $p34^{cdc2}$ kinase at the G1/S and G2/M transitions of the cell cycle: identification of major phosphorylation sites. EMBO J 10:305–316

Kuniyasu H, Yasui W, Ketahara K, Naka K, Yokosaki H, Akama Y, Hamamoto T, Tahara H, Tahara E (1997) Growth inhibitory effect of interferon beta is associated with the induction of cyclin-dependent kinase inhibitor p27 (Kipl) in a human gastric carcinoma cell line. Cell Growth Differ 8:47–52

L'Allemain G, Franchi A, Cragoe E Jr, Pouysségur J (1984) Blockade of the Na^+/H^+ antiport abolishes growth factor-induced DNA synthesis in fibroblasts. J Biol Chem 259:4313–4319

Lamb NJC, Fernandez A, Watrin A, Labbé JC, Cavadore JC (1990) Microinjection of $p34^{cdc2}$ kinase induces marked changes in cell shape, cytoskeletal organization, and chromatin structure in mammalian fibroblasts. Cell 60:151–165

Lorber M, Milobsky SA (1968) Stretching of the skin in vivo. A method of influencing cell division and migration in the rat epidermis. J Invest Dermatol 51:395–402

Lozzio BB, Lozzio CB, Bamberger EC, Lair SV (1975) Regulators of cell division: endogenous mitotic inhibitors of mammalian cells. Int Rev Cytol 42:1–47

Macieira-Coelho A (1988) Intracellular signal flow and genome activation. In: Bergener M, Ermini M, Stähelin HB (eds) Crossroads in aging. Academic Press, London, pp 3–23

Macieira-Coelho A (1990) Cancer and aging at the cellular level. In: Macieira-Coelho A, Nordenskjöld B (eds) Cancer and aging. CRC Press, Boca Raton, pp 11–37

Macieira-Coelho A, Avrameas S (1972) Modulation of cell behavior in vitro by the substratum in fibroblastic and leukemic mouse cell lines. Proc Natl Acad Sci USA 69:2469–2473

Macieira-Coelho A, Avrameas S (1973) Protein polymers as a substratum for the modulation of cell poliferation in vitro. In: Kruse PF, Patterson MK (eds) Tissue culture, methods and applications. Academic Press, New York, pp 379–383

Macieira-Coelho A, Azzarone B (1990) Correlation between contractility and proliferation in human fibroblasts. J Cell Phys 142:610–614

Macieira-Coelho A, Berumen L, Avrameas S (1974) Properties of protein polymers as substratum for cell growth in vitro. J Cell Phys 83:379–388

Margolis LB, Rozovskaja IA, Cragoe E (1988) Intracellular pH and cell adhesion to solid substrate. FEBS Lett 234:449–450

Martin-Castellanos C, Moreno S (1997) Recent advances on cyclins, CDKs and CDK inhibitors. Trends Cell Biol 7:95–98

Montesano R, Orci L (1988) Transforming growth factor beta stimulates collagen-matrix contraction by fibroblasts: implications for wound healing. Proc Natl Acad Sci USA 85:4894–4897

Moolenar WH, Tsien RY, Van der Saag PT, de Laat SW (1983) Na^{+}/H^{+} exchange and cytoplasmic pH in the action of growth factors in human fibroblasts. Nature 304:645–648

Ohsawa T, Senshu T (1987) Exogenous GM1 ganglioside caused G1-arrest of human diploid fibroblasts. Flow cytometric studies. Exp Cell Res 173:49–55

Ortega JM, Meneghini R (1989) Competence growth factors can cause modification in higher-order chromatin structure in mouse embryo 3T3 fibroblasts. J Cell Biochem 40:229–238

Paulsson Y, Karlsson C, Heldin CH, Westermark B (1993) Density-dependent inhibitory effect of transforming growth factor-beta-1 on human fibroblasts involves the down-regulation of platelet-derived growth facotr alpha-receptors. J Cell Phys 157:97–103

Pegg AE (1986) Recent advances in the biochemistry of polyamines in eukaryotes. Biochem J 234:249–262

Pockwinse SM, Krockmalnic G, Doxsey SJ, Nickerson J, Lian JB, Wijnen AJ van, Stein JL, Penman S (1997) Cell cycle-independent interaction of cdc2 with the centrosome which is associated with the nuclear matrix intermediate filament scaffold. Proc Natl Acad Sci USA 94:3022–3027

Reed SI (1997) Control of the G(1)/S transition. Cancer Surv 29:25–46

Reszka AA, Bulinski JC, Krebs EG, Fische EH (1997) Mitogen-activated protein kinase extracellular signal-regulated kinase 2 regulates cytoskeletal organization and chemotaxis via catalytic and microtubule-specific interactions. Mol Biol Cell 8:1219–1232

Schwartz MA, Ingber DE, Lawrence M, Springer TA, Lechene C (1991) Multiple integrins share the ability to induce elevation of intracellular pH. Exp Cell Res 195:533–535

Spivack JG, Prusoff WH, Tritton TR (1984) Microinjection of pp60V-src into *Xenopus* oocytes increases phosphorylation of ribosomal protein S6 and accelerates the rate of progesterone-induced meiotic recombination. Mol Cell Biol 4:1631–1634

Stockbridge LL, French AS (1988) Stretch-activated cation channels in human fibroblasts. Biophys J 54:187–190

Thurston PW, Weeks JR (1984) The mathematics of three-dimensional manifolds. Sci Am July:94–106

Wahrmann JP, Delain D, Bournoutian C, Macieira-Coelho A (1981) Modulation of differentiation in vitro. I Influence of the attachment surface on myogenesis. In Vitro 17:752–762

Westergreen-Thorsson G, Persson S, Isakson A, Onnervik PO, Malmström A, Fransson LA (1993) L-iduronate-rich glycosaminoglycans inhibit growth of normal fibroblasts independently of serum or added growth factors. Exp Cell Res 206:93–99

Wever J, Schachtschabel DO, Sluke G, Wever G (1980) Effect of short- or long-term treatment with exogenous glycosaminoglycans on growth and glycosaminoglycan synthesis of human fibroblasts (WI-38) in culture. Mech Ageing Dev 14:89–99

Yen A, Sturgill A, Varvayanis S, Chern R (1996) FMS (CSF-1 receptor) prolongs cell cycle and promotes retinoic acid-induced hypophosphorylation of retinoblastoma protein, G1 arrest, and cell differentiation. Exp Cell Res 229:111–125

Zardi L, Siri A, Santi L (1976) A serum protein associated with chromatin of cultured fibroblasts. Science 191:869–870

The Growth-Inhibitory Effects of TGFβ

Rafael E. Herrera[1]

1 Introduction

The transforming growth factor βs (TGF-β1, TGF-β2, and TGF-β3) are members of a large superfamily of peptides that control a variety of cellular processes including proliferation, differentiation, and development (Massague 1992; Massague et al. 1992; Kingsley 1994; Ravitz and Wenner 1997). In particular, TGF-β1 controls cellular differentiation, adhesion, cell cycle progression, and extracellular matrix production in a number of different cell types (Massague et al. 1991; Ravitz and Wenner 1997). TGFβ binding to high affinity cell surface receptors leads to initiation of signal transduction pathways that conclude with cell type-specific phenotypes. The TGFβ type I and II receptors are N-linked glycoproteins that interact with each other as well as with TGFβ itself (Wrana et al. 1992; Franzen et al. 1993; Chen and Weinberg 1995). Both the type I and II receptors contain cytoplasmic domains that show homology to serine-threonine kinases (Lin et al. 1992; Wrana et al. 1992; Franzen et al. 1993; Carcamo et al. 1994; Kingsley 1994; ten Dijke et al. 1994) and, indeed, the initiating signaling event is probably phosphorylation of the type I by the type II receptor (Wrana et al. 1992). In addition, the type II receptor can also phosphorylate itself (Chen and Weinberg 1995). The type III TGFβ receptor seems to be involved in regulation of ligand access to the type I and II receptors (Lopez-Casillas et al. 1993; Wang et al. 1991).

TGFβ effects on cellular proliferation are highly dependent on cell type as well as on growth conditions and the state of differentiation of the cell (Ravitz and Wenner 1997). Indeed, the same mouse embryonic cell line (C3H 10T1/2) is either growth-inhibited (Schwarz et al. 1988) or -stimulated (Kim et al. 1994a,b) by TGFβ, depending on growth conditions. In addition, mouse embryo fibroblasts are also either stimulated or inhibited by TGFβ, depending on their pRB status (Herrera et al. 1996b) or state of immortalization (Sorrentino and Bandyopadhyay 1989). It has been suggested that loss of pRb

[1] The University of Texas Health Science Center at San Antonio, Department of Medicine/Medical Oncology, 7703 Floyd Curl Dr., San Antonio, TX 78284-7884, USA

Progress in Molecular and Subcellular Biology, Vol. 20
A. Macieira-Coelho (Ed.)

function during tumorigenesis may frequently underlie the observed non-responsiveness of transformed cells to inhibition by TGFβ and may lead to stimulatory effects elicited by signals that are normally growth-inhibitory (Fynan and Reiss 1993; Herrera et al. 1996b). Indeed, TGFβ has been found to increase tumorgenicity of several tumor cell types both in vivo (Arrick et al. 1992; Steiner and Barrack 1992; Ueki et al. 1992; Arteaga et al. 1993a,b; Change et al. 1993; Fitzpatrick et al. 1994) and in vitro (Huggett et al. 1991; Arrick et al. 1992; Suardet et al. 1992; Fitzpatrick et al. 1994; Park et al. 1994; Rodeck et al. 1994). While the reasons for the growth stimulation by TGFβ remain obscure, it has recently become clear that TGFβ can have profound effects on components of the cell-cycle machinery, leading to an inhibition of cellular proliferation (see below). These inhibitory effects of TGFβ on cell-cycle progression and proliferation will be discussed in this chapter.

2 The Eukaryotic Cell Cycle

In order to understand TGFβs inhibitory effects on proliferation, a knowledge of the eukaryotic cell cycle is required. Figure 1 illustrates the phases of the cell cycle as well as several of its major regulators. The eukaryotic cell cycle is divided into four major phases termed G_1, S, G_2, and M. DNA replication during S phase is preceded by a growth phase, G_1. Similarly, mitosis or M phase

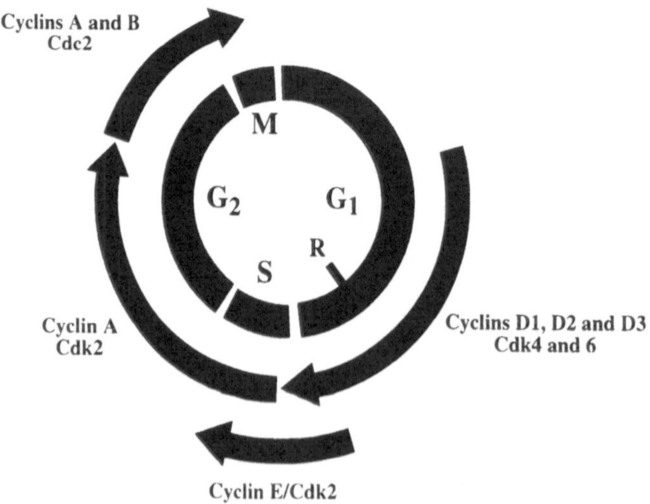

Fig. 1. The eukaryotic cell cycle. Further described in text

occurs after a second growth phase termed G_2. Actively proliferating cells will continue to traverse the cell cycle with kinetics that depend on cell type as well as on growth conditions. In the absence of mitogenic factors or in response to growth inhibitory signals, cells may exit the cycle and enter into a state of quiescence termed G_0. Cells in G_0 may reenter the cell cycle in response to proliferative signals. In addition, upon terminal differentiation the cell can also exit the cell cycle. However, in this case, reentry into the cycle is generally not observed.

Progress through the cell cycle is largely governed by the activity of a family of protein kinases, the cyclin-dependent kinases (CDKs) (Pines 1994, 1995; Grana and Reddy 1995). The activity of CDKs is modulated through association with their regulatory subunits, the cyclins. As depicted in Fig. 1, distinct cyclin/CDK complexes are expressed and active in the different phases of the cell cycle. Cyclin D/CDK4,6 complexes are activated early in G_1 in response to mitogenic stimuli and remain active until S phase entry (Baldin et al. 1993; Matsushime et al. 1992, 1994). Cyclin E/CDK2 is activated late in G_1 and remains active until the end of S phase (Dulic et al. 1992; Koff et al. 1992; Tsai et al. 1993). Cyclin A/CDK2 complexes become active in S phase and are present until mid-G_2 at which time cyclin A/Cdc2 and B cyclin complexes are produced and activated (Hunt 1989; Pagano et al. 1992). Passage through the cell cycle is dependent on the activity of each of these cyclin/CDK complexes in distinct phases of the cell cycle (Ravitz and Wenner 1997).

Occurring late in G_1 is the mammalian restriction (R) point (Pardee 1989). The R point was defined by Pardee and coworkers as the position in G_1 beyond which cells are committed to continue into S phase. Three criteria are used to define this point. After the restriction point cells (1) no longer need serum to continue into S, (2) are no longer sensitive to inhibition by TGFβ, and (3) are no longer sensitive to low levels of the protein synthesis inhibitor cycloheximide. Therefore, beyond the restriction point cells are no longer responsive to inhibitory signals and no longer need stimulatory signals for progression into S. It is before the R point in G_1 that cells commit to proceed through or to exit the cell cycle. In addition to exiting into G_0 or a state of differentiation as mentioned above, cells also may choose to apoptose during this part of G_1 as a consequence of response to apoptotic signals. Therefore, passage through G_1 and the R point is governed by cellular responses to growth-stimulatory and -inhibitory signals (Ducommun 1991; Brattain et al. 1994; Pines 1995).

Deregulation of G_1 progression is invariably observed in cancer cells as a consequence of the cells' inability to respond to inhibitory signals (Pardee 1989). Because of this, a major focus of cell cycle and cancer research is to determine the molecular events that enable cells to progress through the G_1 phase of the cell cycle. The survival and identity of the cell are critically

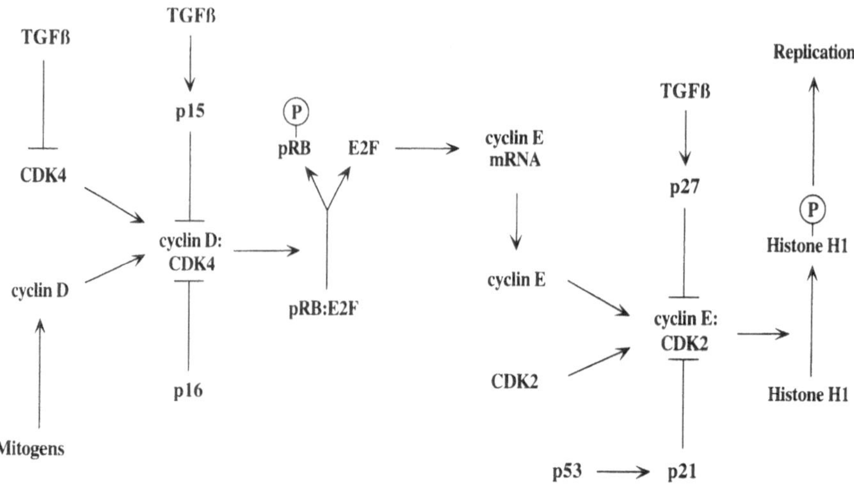

Fig. 2. A proposed molecular model defining G_1 progression. Important cell-cycle regulatory molecules are shown. However, this is a very incomplete diagram. Only those components relevant to this chapter are shown. Further described in text

dependent on the proper regulation of these events, deregulation of which can lead to cell death or to the uncontrolled proliferation that is characteristic of cancer cells. Therefore, a delineation of the events that prime the cellular genome for DNA replication may permit investigation into the possible disruption of these events in the progression towards tumorigenesis. Once derangements of these processes are identified in cancer cells, they may be used as experimental and clinical markers of the neoplastic phenotype. Furthermore, restoration of the normal processes becomes a possible therapeutic strategy for halting the cell's progression toward malignancy.

Recent advances have led to a molecular definition of the G_1 phase of the cell cycle as depicted in Fig. 2. As mentioned, G_1 progression is largely defined by the action of cyclin/CDK complexes as well as the product of the retinoblastoma gene, pRB, and several low molecular weight CDK inhibitors (CDKIs). The *Rb* tumor suppressor gene is inactivated in a wide range of human tumors. Its encoded protein, pRb, has been implicated in cell-cycle regulation and is known to regulate members of the E2F family of transcription factors. These, in turn, control the expression of a variety of genes expressed in the G_1 and S phases of the cell cycle (Riley et al. 1994). Therefore, loss of *Rb* function might be expected to result in deregulated gene expression and abnormal cell cycle progression.

pRb has been shown to be phosphorylated at a particular stage of the cell cycle (Buchkovich et al. 1989; Chen et al. 1989; De Caprio et al. 1989; Geng and Weinberg 1993). This phosphorylation occurs late in G_1 and is responsible for the inactivation of pRb. This event is thought to be critical for passage into S phase. Therefore, a characterization of the mediators of pRb phosphorylation and of the consequences of this event is central to an understanding of the molecular details of cellular progression through G_1 and into S.

Recently, it has been shown that CDKs, in concert with their regulatory cyclin subunits, can phosphorylate pRb (Hinds et al. 1992; Ewen et al. 1993a; Kato et al. 1993). The cyclin/CDKs of particular interest are cyclin E/CDK2 and cyclin D/CDK4,6. As mentioned, these molecules are active in G_1, the time of pRb phosphorylation, and indeed have been shown to phosphorylate pRb in vitro and in transfection studies. In addition, these regulatory proteins are essential for transit through G_1 and their premature activation leads to an accelerated G_1 (Ohtsubo and Roberts 1993; Quelle et al. 1993; Resnitzky et al. 1994) Thus, cyclin/CDKs are prime candidates for the pRb kinases. Indeed, it has been proposed that cyclin D/CDK4,6 complexes formed early in G_1 in response to mitogenic factors phosphorylate pRB complexed with the transcription factor E2F (Koh et al. 1995; Lukas et al. 1995; Medema et al. 1995). This critical G_1 event leads to activation of E2F due to its release from pRB (Weinberg 1995). E2F then activates transcription of a number of cellular genes involved in proliferation (Cobrinik 1996; La Thangue 1996; Sanchez and Dynlacht 1996; Slansky and Farnham 1996). Among these is another cyclin needed for G_1 progression, cyclin E (Koff et al. 1992; Tsai et al. 1993; DeGregori et al. 1995; Geng et al. 1996; Herrera et al. 1996c). While the substrates of cyclin E/CDK2 complexes remain largely unidentified, it has been suggested that the linker histone, H_1, is phosphorylated by cyclin/CDK complexes, in particular those containing CDK2, at the G_1/S border (Herrera et al. 1996c). It has been proposed that H_1 phosphorylation allows a relaxation in chromatin structure and enables the replication machinery to access the chromatin template (Roth and Allis 1992). Thus, a cascade of molecular events leading through G_1 is proposed (Fig. 2). However, a plethora of regulatory mechanisms modulate every step through this pathway. As depicted, TGFβ effects several of these mechanisms invariably leading to an inhibition of cellular proliferation.

3
Effects of TGFβ on Cell-Cycle Progression

As mentioned, treatment of various cell types with transforming growth factor β (TGFβ) results in a wide range of biological effects. These include growth inhibition, growth promotion, escape from contact inhibition, and induction of growth factor and extracellular matrix production (Massague et al. 1992). Growth inhibition by TGFβ invariably leads to cellular arrest in the G_1 phase of the cell cycle. Such cells express pRb in its underphosphorylated, growth-suppressing form (Laiho et al. 1990; Furukawa et al. 1992; Geng and Weinberg 1993). Thus, it has been suggested that TGFβ blocks progress through the G_1

phase by preventing phosphorylation and functional inactivation of pRb (Derynck 1994). Indeed, it has been shown that primary mouse embryo fibroblasts lacking a functional *Rb* gene are no longer growth-inhibited by TGFβ (Herrera et al. 1996b).

This inhibition of pRb phosphorylation has been explained at the biochemical level through the observation that TGFβ causes the synthesis or activation of several low molecular weight CDK inhibitors (CDKIs) (Hannon and Beach 1994; Polyak et al. 1994a; Toyoshima and Hunter 1994; Datto et al. 1995; C.Y. Li et al. 1995; J.M. Li et al. 1995; Reynisdottir et al. 1995). In addition, TGFβ has also been shown to inhibit the expression of various cyclin and CDK genes (Landesman et al. 1992; Barlat et al. 1993, 1995; Ewen et al. 1993b, 1995; Geng and Weinberg 1993). These various lines of research suggest a chain of events in which TGFβ induces or activates mechanisms which proceed to block pRb phosphorylation; pRb in turn continues to sequester a series of transcription factors, the activity of which is required for advancement into the late G_1 and S phases of the cell cycle.

3.1 TGFβ Inhibition of CDK and Cyclin Expression

TGFβ has been shown to repress CDK4 synthesis in mink lung epithelial cells (Mv1Lu) exiting contact inhibition without affecting cyclin D levels (Ewen et al. 1993b). This prevents the accumulation of active cyclin D/CDK4 complexes and therefore prevents pRb phosphorylation and G_1 progression. In addition, constitutive expression of CDK4 leads to TGFβ resistance in these cells (Ewen et al. 1993). In contrast, while TGFβ also can inhibit CDK2 activity, its constitutive expression failed to override TGFβ-induced cell cycle arrest (Ewen et al. 1993). Therefore, Ewen et al. suggest that the G_1 block induced by TGFβ in these cells is primarily mediated by a reduction in CDK4 expression. This then leads to an inhibition of CDK2 activity and pRb phosphorylation, secondary to inhibition of CDK4 synthesis. These workers then showed that TGFβ downregulates CDK4 synthesis by inhibiting its translation (Ewen et al. 1995). This inhibition was mapped to the 5′ untranslated region of the CDK4 mRNA (Ewen et al. 1995). In addition, it was found that mutant p53 confers TGFβ resistance in these cells by interfering with its repression of CDK4 synthesis (Ewen 1996; Ewen et al. 1995). In contrast, wild-type p53 represses CDK4 translation (Ewen 1996; Ewen et al. 1995). Similarly, TGFβ blocked a serum induction of CDK4 and CDK2 mRNA in human keratinocytes (HaCaT) (Geng and Weinberg 1993). However, in exponentially growing HaCaT cells (Hannon and Beach 1994; Reynisdottir et al. 1995) Mv1Lu cells (Reynisdottir et al. 1995), primary mouse embryo fibroblasts (Herrera et al. 1996b), and rat intestinal epithelial cells (Ko et al. 1994, 1995) TGFβ did not effect CDK4 or CDK6 protein levels. Therefore, TGFβ inhibits CDK4 synthesis in cells emerging from a G_0 state but does not display such effects in proliferating cells.

TGFβ has also been shown to inhibit the serum induction of cyclin E mRNA and protein observed late in G_1 (Geng and Weinberg 1993; Herrera et al. 1996b). In addition, the S phase induction of cyclin A was also lost after TGFβ treatment of HaCaT (Landesman et al. 1992; Geng and Weinberg 1993) and Chinese hamster lung fibroblasts (Landesman et al. 1992; Barlat et al. 1993). However, these effects may be secondary to inhibition of cyclin D/CDK4 activity as suggested in Fig. 2. In addition, Koff et al. found that in MvLul cells already expressing cyclin E, TGFβ inhibits the formation of active cyclin E/CDK2 without affecting protein levels (Koff et al. 1993). Therefore it has been suggested that TGFβ prevents cyclin E expression early in G_1 and active cyclin E/CDK2 complex formation late in G_1 (Geng and Weinberg 1993).

3.2 TGFβ Effects on p15 Activity

As mentioned above, TGFβ induces the synthesis or activation of several low molecular weight CDK inhibitors. In particular, TGFβ treatment of HaCaT cells leads to the induction of p15 or INK4B (Hannon and Beach 1994). p15 is a member of the p16/INK4 family of CDK inhibitors that preferentially inhibit cyclin D-associated kinase activity (Hannon and Beach 1994; Serrano et al. 1993). TGFβ treatment results in a 30-fold induction of p15 mRNA, increased binding of p15 to CDK4 and CDK6 complexes, and inhibition of their activities (Hannon and Beach 1994). In addition, TGFβ has also been shown to increase p15 mRNA accumulation as well as protein stability in human mammary epithelial cells (Sandhu et al. 1997). Interestingly, a TGFβ-resistant human mammary epithelial cell line displayed increased p15 mRNA in response to TGFβ but no stabilization of the protein nor inhibition of cyclin D associated kinase activity (Sandhu et al. 1997).

In contrast, it has been shown that DNA synthesis is inhibited by TGFβ in several p15-defective human tumor cell lines (Iavarone and Massague 1997). This inhibition is associated with reduced CDK4 and CDK6 kinase activities and an increase in inhibitory tyrosine phosphorylation of the proteins (Iavarone and Massague 1997). This study also shows that this effect is most likely due to reduced expression of the CDK-activating phosphatase Cdc25A in response to TGFβ. Indeed, TGFβ inhibits the accumulation of Cdc25A mRNA and protein levels in these cells. In addition, ectopic expression of Cdc25A in these cells leads to reduced TGFβ inhibition of CDK6 kinase activity and a reduction in the TGF-induced increase in G_1 arrested cells (Iavarone and Massague 1997). Finally, the same authors report that fibroblasts from p15 null mice are still responsive to TGFβ (Iavarone and Massague 1997).

3.3
TGFβ Effects on p27 Activity and CDKI Distribution

The CDK inhibitor p27 was originally identified as a heat-stable protein in inactive cyclin E/CDK2 complexes from cells arrested in G_1 by TGFβ or contact inhibition (Polyak et al. 1994) or in lovastatin-arrested cells (Hengst et al. 1994). Cloning of p27 (Polyak et al. 1994b; Toyoshima and Hunter 1994) revealed a highly conserved protein with similarity to p21, a previously identified CDKI (el-Deiry et al. 1993; Gu et al. 1993; Harper et al. 1993; Serrano et al. 1993; Xiong et al. 1993). Like p21, p27 inhibits cyclin E/CDK2, cyclin A/CDK2, and cyclin D/CDK4 kinase activities (Polyak et al. 1994b; Toyoshima and Hunter 1994). Inhibition by p27 is modulated by repression of cyclin/CDK catalytic function as well as interference of threonine-160 phosphorylation by the CDK-activating kinase (CAK) (Koff et al. 1993; Kato et al. 1994; Polyak et al. 1994; Slingerland et al. 1994). This is probably related to the fact that threonine-160 is located in the catalytic cleft of CDK2. Therefore, p27 binding to this region may interfere with catalytic function as well as CAK phosphorylation of threonine-160.

The levels of p27 do not fluctuate during the cell cycle or after treatment with TGFβ (Koff et al. 1993; Polyak et al. 1994b; Toyoshima and Hunter 1994; Reynisdottir et al. 1995). In addition, p27 can be found in heat-treated extracts from both proliferating and TGFβ or density-arrested cells (Polyak et al. 1994). Therefore, a mechanism distinct from increased p27 accumulation must be responsible for its proposed role in TGFβ-induced growth arrest. In proliferating cells, p27 can be found in association with complexes containing CDK2, CDK4, and CDK6 (Reynisdottir et al. 1995). It has been proposed that the stoichiometry of this association determines if it will be inhibitory. In accord with this notion, a stoichiometric requirement of two p21 molecules per molecule of CDK for inhibition has been suggested (Harper et al. 1995; Zhang et al. 1994). It has been shown that TGFβ treatment leads to an increase in levels and binding of p15 to cyclin D-dependent kinases, thus preventing p27 binding (Reynisdottir et al. 1995; Reynisdottir and Massague 1997; Sandhu et al. 1997) This newly liberated p27 is then able to associate with and inhibit cyclin E/CDK2 complexes by exceeding the stoichiometric requirement of p27 needed for inhibition. Indeed, ectopic expression of p15 in MvLu1 cells leads to an increase in cyclin E/CDK2-associated p27 (Reynisdottir et al. 1995). Therefore, a redistribution of p27 from cyclin D complexes to cyclin E complexes in response to TGFβ can explain its role in TGFβ-induced growth inhibition. Such redistribution may also be triggered by a decrease in CDK4 levels, as mentioned above. However, in vitro, p15 prevents p27 binding to cyclin D/CDK4 complexes only if it has access to the complex first (Reynisdottir and Massague 1997). This caveat to the model proposed above is resolved in the cell by distinct localizations of p15 and p27. In the cell, p15 is located mainly in the cytoplasm whereas p27 in predominately nuclear (Reynisdottir and Massague 1997). These properties have been proposed to explain the ability of p15 and p27 to inhibit CDK4 and CDK2 activities (Reynisdottir and Massague 1997).

4 Conclusions

The proliferative inhibition by TGFβ is manifested through the cytokine's barrage of the cell-cycle machinery. Much of the attack is performed by the action of cyclin-dependent kinase inhibitors. Activation of these molecules through regulation or redistribution leads to inhibition of various components of the cell-cycle machinery that insure passage through the G_1 phase of the cell cycle. While the focal point of many of these inhibitory mechanisms is pRb phosphorylation, TGFβ has also been shown to inhibit the proliferation of several cell lines that lack wild-type RB. However, the model of G_1 progression depicted in Fig. 2 suggests that even in cells lacking pRb, TGFβ would inhibit cyclin E/CDK2 activity and its subsequent phosphorylation of histone H1 and other substrates.

References

Arrick BA, Lopez AR, Elfman F, Ebner R, Damsky CH, Derynck R (1992) Altered metabolic and adhesive properties and increased tumorigenesis associated with increased expression of transforming growth factor beat 1. J Cell Biol 118:715–726

Arteaga CL, Dugger TC, Winnier AR, Forbes JT (1993a) Evidence for a positive role of transforming growth factor-beta in human breast cancer cell tumorigenesis. J Cell Biochem Suppl 17G:187–193

Arteaga CL, Hurd SD, Winnier AR, Johnson MD, Fendly BM, Forbes JT (1993b) Anti-transforming growth factor (TGF)-beta antibodies inhibit breast cancer cell tumorigenicity and increase mouse spleen natural killer cell activity. Implications for a possible role of tumor cell/host TGF-beta interactions in human breast cancer progression. J Clin Invest 92:2569–2576

Baldin V, Lukas J, Marcote MJ, Pagano M, Draetta G (1993) Cyclin D1 is a nuclear protein required for cell cycle progression in G_1. Genes Dev 7:812–821

Barlat I, Fesquet D, Brechot C, Henglein B, Dupuy d'Angeac A, Vie A, Blanchard JM (1993) Loss of the G_1-S control of cyclin A expression during tumoral progression of Chinese hamster lung fibroblasts. Cell Growth Differ 4:105–113

Barlat I, Henglein B, Plet A, Lamb N, Fernandez A, McKenzie F, Pouyssegur J, Vie A, Blanchard JM (1995) TGF-beta 1 and cAMP attenuate cyclin A gene transcription via a cAMP responsive element through independent pathways. Oncogene 11:1309–1318

Brattain MG, Howell G, Sun LZ, Willson JK (1994) Growth factor balance and tumor progression. Curr Opin Oncol 6:77–81

Buchkovich K, Duffy LA, Harlow E (1989) The retinoblastoma protein is phosphorylated during specific phases of the cell cycle. Cell 58:1097–1105

Carcamo J, Weis FM, Ventura F, Wieser R, Wrana JL, Attisano L, Massague J (1994) Type I receptors specify growth-inhibitory and transcriptional responses to transforming growth factor beta and activin. Mol Cell Biol 4:3810–3821

Chang HL, Gillett N, Figari I, Lopez AR, Palladino MA, Derynck R (1993) Increased transforming growth factor beta expression inhibits cell proliferation in vitro, yet increases tumorigenicity and tumor growth of Meth A sarcoma cells. Cancer Res 53:4391–4398

Chen F, Weinberg RA (1995) Biochemical evidence for the autophosphorylation and transphosphorylation of transforming growth factor beta receptor kinases. Proc Natl Acad Sci USA 92:1565–1569

Chen PL, Scully P, Shew JY, Wang JY, Lee WH (1989) Phosphorylation of the retinoblastoma gene product is modulated during the cell cycle and cellular differentiation. Cell 58:1193–1198

Cobrinik D (1996) Regulatory interactions among E2Fs and cell cycle control proteins. Curr Top Microbiol Immunol 208:31–61

Datto MB, Li Y, Panus JF, Howe DJ, Xiong Y, Wang XF (1995) Transforming growth factor beta induces the cyclin-dependent kinase inhibitor p21 through a p53-independent mechanism. Proc Natl Acad Sci USA 92:5545–5549

De Caprio JA, Ludlow JW, Lynch D, Furukawa Y, Griffin J, Piwnica-Worms H, Huang CM, Livingston DM (1989) The product of the retinoblastoma susceptibility gene has properties of a cell cycle regulatory element. Cell 58:1085–1095

DeGregori J, Leone G, Ohtani K, Miron A, Nevins JR (1995) E2F-1 accumulation bypasses a G_1 arrest resulting from the inhibition of G_1 cyclin-dependent kinase activity. Genes Dev 9:2873–2887

Derynck R (1994) TGF-beta-receptor-mediated signaling. Trends Biochem Sci 19:548–553

Ducommun B (1991) From growth to cell cycle control. Semin Cell Biol 2:233–241

Dulic V, Lees E, Reed SI (1992) Association of human cyclin E with a periodic G_1-S phase protein kinase. Science 257:1958–1961

el-Deiry WS, Tokino T, Velculescu VE, Levy DB, Parsons R, Trent JM, Lin D, Mercer WE, Kinzler KW, Vogelstein B (1993) WAF1, a potential mediator of p53 tumor suppression. Cell 75:817–825

Ewen ME (1996) p53-dependent repression of cdk4 synthesis in transforming growth factor-beta-induced G_1 cell cycle arrest. J Lab Clin Med 128:355–360

Ewen ME, Sluss HK, Sherr CJ, Matsushime H, Kato J, Livingston DM (1993a) Functional interactions of the retinoblastoma protein with mammalian D-type cyclins. Cell 73:487–497

Ewen ME, Sluss HK, Whitehouse LL, Livingston DM (1993b) TGF beta inhibition of Cdk4 synthesis is linked to cell cycle arrest. Cell 74:1009–1020

Ewen ME, Oliver CJ, Sluss HK, Miller SJ, Peeper DS (1995) p53-dependent repression of CDK4 translation in TGF-beta-induced G_1 cell-cycle arrest. Genes Dev 9:204–217

Fitzpatrick DR, Bielefeldt-Ohmann H, Himbeck RP, Jarnicki AG, Marzo AL, Robinson BW (1994) Transforming growth factor-beta: antisense RNA-mediated inhibition affects anchorage-independent growth, tumorigenicity and tumor-infiltrating T-cells in malignant mesothelioma. Growth Factors 11:29–44

Franzen P, Ichijo H, Miyazono K (1993) Different signals mediate transforming growth factor-beta 1-induced growth inhibition and extracellular matrix production in prostatic carcinoma cells. Exp Cell Res 207:1–7

Furukawa Y, Uenoyama S, Ohta M, Tsunoda A, Griffin JD, Saito M (1992) Transforming growth factor-beta inhibits phosphorylation of the retinoblastoma susceptibility gene product in human monocytic leukemia cell line JOSK-I. J Biol Chem 267:17121–17127

Fynan TM, Reiss M (1993) Resistance to inhibition of cell growth by transforming growth factor-beta and its role in oncogenesis. Crit Rev Oncog 4:493–540

Geng Y, Weinberg RA (1993) Transforming growth factor beta effects on expression of G_1 cyclins and cyclin-dependent protein kinases. Proc Natl Acad Sci USA 90:10315–10319

Geng Y, Eaton EN, Picon M, Roberts JM, Lundberg AS, Gifford A, Sardet C, Weinberg RA (1996) Regulation of cyclin E transcription by E2Fs and retinoblastoma protein. Oncogene 12:1173–1180

Grana X, Reddy EP (1995) Cell cycle control in mammalian cells: role of cyclins, cyclin dependent kinases (CDKs), growth suppressor genes and cyclin-dependent kinase inhibitors (CKIs). Oncogene 11:211–219

Gu Y, Turck CW, Morgan DO (1993) Inhibition of CDK2 activity in vivo by an associated 20K regulatory subunit. Nature 366:707–710

Hannon GJ, Beach D (1994) p15INK4B is a potential effector of TGF-beta-induced cell cycle arrest [see comments]. Nature 371:257–261

Harper JW, Adami GR, Wei N, Keyomarsi K, Elledge SJ (1993) The p21 Cdk-interacting protein Cip1 is a potent inhibitor of G_1 cyclin-dependent kinases. Cell 75:805–816

Harper JW, Elledge SJ, Keyomarsi K, Dynlacht B, Tsai LH, Zhang P, Dobrowolski S, Bai C, Connell-Crowley L, Swindell E, et al. (1995) Inhibition of cyclin-dependent kinases by p21. Mol Biol Cell 6:387–400

Hengst L, Dulic V, Slingerland JM, Lees E, Reed SI (1994) A cell cycle-regulated inhibitor of cyclin-dependent kinases. Proc Natl Acad Sci USA 91:5291–5295

Herrera RE, Chen F, Weinberg RA (1996a) Increased histone H1 phosphorylation and relaxed chromatin structure in Rb-deficient fibroblasts. Proc Natl Acad Sci USA 93:11510–11515

Herrera RE, Makela TP, Weinberg RA (1996b) TGF beta-induced growth inhibition in primary fibroblasts requires the retinoblastoma protein. Mol Biol Cell 7:1335–1342

Herrera RE, Sah VP, Williams BO, Makela TP, Weinberg RA, Jacks T (1996c) Altered cell cycle kinetics, gene expression, and G1 restriction point regulation in Rb-deficient fibroblasts. Mol Cell Biol 16:2402–2407

Hinds PW, Mittnacht S, Dulic V, Arnold A, Reed SI, Weinberg RA (1992) Regulation of retinoblastoma protein functions by ectopic expression of human cyclins. Cell 70:993–1006

Huggett AC, Ellis PA, Ford CP, Hampton LL, Rimoldi D, Thorgeirsson SS (1991) Development of resistance to the growth inhibitory effects of transforming growth factor beta 1 during the spontaneous transformation of rat liver epithelial cells. Cancer Res 51:5929–5936

Hunt T (1989) Maturation promoting factor, cyclin and the control of M-phase. Curr Opin Cell Biol 1:268–274

Iavarone A, Massague J (1997) Repression of the CDK activator Cdc25A and cell-cycle arrest by cytokine TGF-beta in cells lacking the CDK inhibitor p15. Nature 387:417–422

Kato J, Matsushime H, Hiebert SW, Ewen ME, Sherr CJ (1993) Direct binding of cyclin D to the retinoblastoma gene product (pRb) and pRb phosphorylation by the cyclin D-dependent kinase CDK4. Genes Dev 7:331–342

Kato JY, Matsuoka M, Polyak K, Massague J, Sherr CJ (1994) Cyclic AMP-induced G1 phase arrest mediated by an inhibitor (p27Kip1) of cyclin-dependent kinase 4 activation. Cell 79:487–496

Kim SJ, Romeo D, Yoo YD, Park K (1994a) Transforming growth factor-beta: expression in normal and pathological conditions. Horm Res 42:5–8

Kim TA, Ravitz MJ, Wenner CE (1994b) Transforming growth factor-beta regulation of retinoblastoma gene product and E2F transcription factor during cell cycle progression in mouse fibroblasts. J Cell Physiol 160:1–9

Kingsley DM (1994) The TGF-beta superfamily: new members, new receptors, and new genetic tests of function in different organisms. Genes Dev 8:133–146

Ko TC, Beauchamp RD, Townsend CM Jr, Thompson EA, Thompson JC (1994) Transforming growth factor-beta inhibits rat intestinal cell growth by regulating cell cycle specific gene expression. Am J Surg 167:14–19; discussion 19–20

Ko TC, Sheng HM, Reisman D, Thompson EA, Beauchamp RD (1995) Transforming growth factor-beta 1 inhibits cyclin D1 expression in intestinal epithelial cells. Oncogene 10:177–184

Koff A, Giordano A, Desai D, Yamashita K, Harper JW, Elledge S, Nishimoto T, Morgan DO, Franza BR, Roberts JM (1992) Formation and activation of a cyclin E-cdk2 complex during the G_1 phase of the human cell cycle. Science 257:1689–1694

Koff A, Ohtsuki M, Polyak K, Roberts JM, Massague J (1993) Negative regulation of G1 in mammalian cells: inhibition of cyclin E-dependent kinase by TGF-beta. Science 260:536–539

Koh J, Enders GH, Dynlacht BD, Harlow E (1995) Tumour-derived p16 alleles encoding proteins defective in cell-cycle inhibition. Nature 375:506–510

Laiho M, De Caprio JA, Ludlow JW, Livingston DM, Massague J (1990) Growth inhibition by TGF-beta linked to suppression of retinoblastoma protein phosphorylation. Cell 62:175–185

Landesman Y, Pagano M, Draetta G, Rotter V, Fusenig NE, Kimchi A (1992) Modifications of cell cycle controlling nuclear proteins by transforming growth factor beta in the HaCaT keratinocyte cell line [published erratum appears in Oncogene 1993 Jan;8(1):229]. Oncogene 7:1661–1665

La Thangue NB (1996) E2F and the molecular mechanisms of early cell-cycle control. Biochem Soc Trans 24:54–59

Li CY, Suardet L, Little JB (1995) Potential role of WAF1/Cip1/p21 as a mediator of TGF-beta cytoinhibitory effect. J Biol Chem 270:4971–4974

Li JM, Nichols MA, Chandrasekharan S, Xiong Y, Wang XF (1995) Transforming growth factor beta activates the promoter of cyclin-dependent kinase inhibitor p15INK4B through an Sp1 consensus site. J Biol Chem 270:26750–26753

Lin HY, Wang XF, Ng-Eaton E, Weinberg RA, Lodish HF (1992) Expression cloning of the TGF-beta type II receptor, a functional transmembrane serine/threonine kinase [published erratum appears in Cell 1992 Sep 18;70(6):following 1068]. Cell 68:775–785

Lopez-Casillas F, Wrana JL, Massague J (1993) Betaglycan presents ligand to the TGF beta signaling receptor. Cell 73:1435–1444

Lukas J, Parry D, Aagaard L, Mann DJ, Bartkova J, Strauss M, Peters G, Bartek J (1995) Retinoblastoma-protein-dependent cell-cycle inhibition by the tumour suppressor p16. Nature 375:503–506

Massague J (1992) Receptors for the TGF-beta family. Cell 69:1067–1070

Massague J, Heino J, Laiho M (1991) Mechanisms in TGF-beta action. Ciba Found Symp 157:51–59; discussion 59–65

Massague J, Cheifetz S, Laiho M, Ralph DA, Weis FM, Zentella A (1992) Transforming growth factor-beta. Cancer Surv 12:81–103

Matsuchime H, Ewen ME, Strom DK, Kato JY, Hanks SK, Roussel MF, Sherr CJ (1992) Identification and properties of an atypical catalytic subunit (p34PSK-J3/cdk4) for mammalian D type G_1 cyclins. Cell 71:323–334

Matsushime H, Quelle DE, Shurtleff SA, Shibuya M, Sherr CJ, Kato JY (1994) D-type cyclin-dependent kinase activity in mammalian cells. Mol Cell Biol 14:2066–2076

Medema RH, Herrera RE, Lam F, Weinberg RA (1995) Growth suppression by p16ink4 requires functional retinoblastoma protein. Proc Natl Acad Sci USA 92:6289–6293

Ohtsubo M, Roberts JM (1993) Cyclin-dependent regulation of G_1 in mammalian fibroblasts. Science 259:1908–1912

Pagano M, Pepperkok R, Verde F, Ansorge W, Draetta G (1992) Cyclin A is required at two points in the human cell cycle. Embo J 11:961–971

Pardee AB (1989) G1 events and regulation of cell proliferation. Science 246:603–608

Park K, Kim SJ, Bang YJ, Park JG, Kim NK, Roberts AB, Sporn MB (1994) Genetic changes in the transforming growth factor beta (TGF-beta) type II receptor gene in human gastric cancer cells: correlation with sensitivity to growth inhibition by TGF-beta. Proc Natl Acad Sci USA 91:8772–8776

Pines J (1994) The cell cycle kinases. Semin Cancer Biol 5:305–313

Pines J (1995) Cyclins, CDKs and cancer. Semin Cancer Biol 6:63–72

Polyak K, Kato JY, Solomon MJ, Sherr CJ, Massague J, Roberts JM, Koff A (1994a) p27Kip1, a cyclin-Cdk inhibitor, links transforming growth factor-beta and contact inhibition to cell cycle arrest. Genes Dev 8:9–22

Polyak K, Lee MH, Erdjument-Bromage H, Koff A, Roberts JM, Tempst P, Massague J (1994b) Cloning of p27Kip1, a cyclin-dependent kinase inhibitor and a potential mediator of extracellular antimitogenic signals. Cell 78:59–66

Quelle DE, Ashmun RA, Shurtleff SA, Kato JY, Bar-Sagi D, Roussel MF, Sherr CJ (1993) Overexpression of mouse D-type cyclins accelerates G_1 phase in rodent fibroblasts. Genes Dev 7:1559–1571

Ravitz MJ, Wenner CE (1997) Cyclin-dependent kinase regulation during G_1 phase and cell cycle regulation by TGF-beta [In Process Citation]. Adv Cancer Res 71:165–207

Resnitzky D, Gossen M, Bujard H, Reed SI (1994) Acceleration of the G1/S phase transition by expression of cyclins D1 and E with an inducible system. Mol Cell Biol 14:1669–1679

Reynisdottir I, Massague J (1997) The subcellular locations of p15(Ink4b) and p27(Kip1) coordinate their inhibitory interactions with cdk4 and cdk2. Genes Dev 11:492–503

Reynisdottir I, Polyak K, Iavarone A, Massague J (1995) Kip/Cip and Ink4 Cdk inhibitors cooperate to induce cell cycle arrest in response to TGF-beta. Genes Dev 9:1831–1845

Riley DJ, Lee EY, Lee WH (1994) The retinoblastoma protein: more than a tumor suppressor. Annu Rev Cell Biol 10:1–29

Rodeck U, Bossler A, Graeven U, Fox FE, Nowell PC, Knabbe C, Kari C (1994) Transforming growth factor beta production and responsiveness in normal human melanocytes and melanoma cells. Cancer Res 54:575–581

Roth SY, Allis CD (1992) Chromatin condensation: does histone H1 dephosphorylation play a role? Trends Biochem Sci 17:93–98

Sanchez I, Dynlacht BD (1996) Transcriptional control of the cell cycle. Curr Opin Cell Biol 8:318–324

Sandhu C, Garbe J, Bhattacharya N, Daksis J, Pan CH, Yaswen P, Koh J, Slingerland JM, Stampfer MR (1997) Transforming growth factor beta stabilizes p15INK4B protein, increases p15INK4B-cdk4 complexes, and inhibits cyclin D1-cdk4 association in human mammary epithelial cells. Mol Cell Biol 17:2458–2467

Schwarz LC, Gingras MC, Goldberg G, Greenberg AH, Wright JA (1988) Loss of growth factor dependence and conversion of transforming growth factor-beta 1 inhibition to stimulation in metastatic H-ras-transformed murine fibroblasts. Cancer Res 48:6999–7003

Serrano M, Hannon GJ, Beach D (1993) A new regulatory motif in cell-cycle control causing specific inhibition of cyclin D/CDK4 [see comments]. Nature 366:704–707

Slansky JE, Farnham PJ (1996) Transcriptional regulation of the dihydrofolate reductase gene. Bioessays 18:55–62

Slingerland JM, Hengst L, Pan CH, Alexander D, Stampfer MR, Reed SI (1994) A novel inhibitor of cyclin-Cdk activity detected in transforming growth factor beta-arrested epithelial cells. Mol Cell Biol 14:3683–3694

Sorrentino V, Bandyopadhyay S (1989) TGF beta inhibits Go/S-phase transition in primary fibroblasts. Loss of response to the antigrowth effect of TGF beta is observed after immortalization. Oncogene 4:569–574

Steiner MS, Barrack ER (1992) Transforming growth factor-beta 1 overproduction in prostate cancer: effects on growth in vivo and in vitro. Mol Endocrinol 6:15–25

Suardet L, Gaide AC, Calmes JM, Sordat B, Givel JC, Eliason JF, Odartchenko N (1992) Responsiveness of three newly established human colorectal cancer cell lines to transforming growth factors beta 1 and beta 2. Cancer Res 52:3705–3712

ten Dijke P, Yamashita H, Ichijo H, Franzen P, Laiho M, Miyazono K, Heldin CH (1994) Characterization of type I receptors for transforming growth factor-beta and activin. Science 264:101–104

Toyoshima H, Hunter T (1994) p27, a novel inhibitor of G1 cyclin-Cdk protein kinase activity, is related to p21. Cell 78:67–74

Tsai LH, Lees E, Faha B, Harlow E, Riabowol K (1993) The cdk2 kinase is required for the G1-to-S transition in mammalian cells. Oncogene 8:1593–1602

Ueki N, Nakazato M, Ohkawa T, Ikeda T, Amuro Y, Hada T, Higashino K (1992) Excessive production of transforming growth-factor beta 1 can play an important role in the development of tumorigenesis by its action for angiogenesis: validity of neutralizing antibodies to block tumor growth. Biochim Biophys Acta 1137:189–196

Wang XF, Lin HY, Ng-Eaton E, Downward J, Lodish HF, Weinberg RA (1991) Expression cloning and characterization of the TGF-beta type III receptor. Cell 67:797–805

Weinberg RA (1995) The retinoblastoma protein and cell cycle control. Cell 81:323–330

Wrana JL, Attisano L, Carcamo J, Zentella A, Doody J, Laiho M, Wang XF, Massague J (1992) TGF beta signals through a heterometric protein kinase receptor complex. Cell 71:1003–1014

Xiong Y, Hannon GJ, Zhang H, Casso D, Kobayashi R, Beach D (1993) p21 is a universal inhibitor of cyclin kinases (see comments). Nature 366:701–704

Zhang H, Hannon GJ, Beach D (1994) p21-containing cyclin kinases exist in both active and inactive states. Genes Dev 8:1750–1758

Big Brothers Are Watching: the Retinoblastoma Family and Growth Control

Peter Stiegler and Antonio Giordano[1]

1 Introduction

Multicellular organisms consist of numerous types of cells and each of these cells has characteristic properties in its pattern of gene expression, its degree of multiplication rate, its state of differentiation, and its life span. The cellular programs are guided by both inner- and extracellular signals converging onto various molecular pathways. Dependent on this information, the cell decides whether to respond with proliferation, differentiation, quiescence, or apoptosis in gene-directed processes to maintain homeostasis. Loss of genetic integrity affecting either growth-promoting (proto oncogenes) or growth-inhibiting (tumor suppressor genes) sources could be the initial step in the multistep development of neoplasia. While heterozygous mutations of proto oncogenes can be sufficient for cellular transformation, only homozygous mutations of tumor suppressor genes, such as the retinoblastoma protein pRB, have been reported to cause cancer in mammalian cells to date. The tumor suppressor p53 is an exception, as dominant negative mutants have been shown to elicit cellular transformation (Bishop 1991). The majority of cells in a developed mammalian organism are in a non-proliferative, either differentiated or quiescent state, and the commitment to become a particular specialized cell necessitates its withdrawal from the cell cycle. The retinoblastoma protein family members pRB, p107, and RB2/p130 have in the past been shown to play a major role for cells to keep these commitments and prevent cells from improperly reentering or improperly staying in the cell-division cycle. In fact, inactivation of the RB family is an integral event for a large number of diverse growth promoting pathways.

In this chapter of *Growth Inhibitors*, we will deal with one of the most interesting and intensively investigated proteins linked to cell-cycle and growth control, the retinoblastoma tumor suppressor gene family. In addition to reporting general aspects concerning the retinoblastoma protein family, the

[1] Department of Pathology, Anatomy and Cell Biology and Sbarro, Institute for Cancer Research and Molecular Medicine, Thomas Jefferson University, Philadelphia, PA 19107, USA

Progress in Molecular and Subcellular Biology, Vol. 20
A. Macieira-Coelho (Ed.)

emphasis will be on trying to connect recent findings on the role of the retinoblastoma family proteins in growth inhibition with general aspects of cellular transformation and gene control.

2
Interaction of Retinoblastoma Family Proteins with the Transcription Factor E2F

In the past years, the tumor-suppressor gene pRB and its cousins p107 and RB2/p130 were mainly described as a source of negative transcriptional growth control of a series of RNA polymerase II-transcribed G1/S phase-activated genes. Repression is due to binding of the RB family proteins in a hypophosphorylated form to, and thereby inhibiting, transcription factors such as E2F (Beijersbergen and Bernards 1996; Cobrinik 1996). The E2F transcription factor family consists of at least five members, E2F 1–5, which form heteromers with DP1 or DP2 at E2F binding sites that are present in many promoters. Among these are several G0-G1 and G1-S phase-activated genes such as pRB, PCNA, c-myc, cyclin E, p107, b-myb and E2F-1, DHFR, TK, TS and RRM2 (Bejersbergen and Bernards 1996; Slanski and Farnham 1996). The different RB family members were shown to have diverse affinity for different E2F members. The present data indicate that pRB complexes preferentially with E2F 1–3 but also E2F4 to a lesser extent (Lees et al. 1993; Moberg et al. 1996), whereas only p107/E2F-4 complexes were described in vivo (Beijersbergen et al. 1994). RB2/p130 is found associated with E2F-4 and E2F-5 and switches from E2F-5 in G0 to E2F-4 complexes as the cell reenters the G1 phase (Ginsberg et al. 1994; Hijmans et al. 1995; Moberg et al. 1996; Slansky and Farnham 1996).

Two general modes of control for E2F-regulated genes have been described so far:

1. Release of RB family members from E2Fs causes activation of transcription through the bound E2F itself; mutations in these E2F sites, for example, result in a transcriptional silent DHFR promoter throughout the cycle (Slansky et al. 1993).
2. At other promoters, such as b-myb, cyclin A, or E2F-1, the E2F transcription factor recruits the RB family members to a promoter and, after release, transcription occurs even without the bound E2F and, therefore, mutations of such E2F sites lead to a derepressed, active promoter (Lam and Watson 1993; Huet et al. 1996; Li et al. 1997).

An important feature of the pRB/E2F interaction in E2F-dependent gene repression is the recent finding that pRB protects E2F from proteolytic degradation via ubiquitinylation (Hateboer et al. 1996; Hofman et al. 1996; Campanero and Flemington 1997). Furthermore, besides overlapping activities, the three members are active in complexes at different times of the cell-division cycle. RB2/p130 is primarily active in arrested G0 or differentiated cells (Smith et al. 1996), active pRB is found in quiescent and differentiated cells as well as in mid to late G1, and p107 complexes are most abundant in cycling cells, in G1/S and S phase complexes (Beijersbergen and Bernards 1996; Slansky and Farnham 1996).

3 Structure and Expression of Retinoblastoma Family Members

The retinoblastoma protein pRB encodes a 105-KDa nuclear phosphoprotein and was the first tumor-suppressor gene to be cloned (Lee et al. 1987). The RB gene is located at chromosome 13q14 and homozygous genetic inactivation of pRB is found in tumors derived from various tissues and occurs by many different types of mutation including deletions in its promoter (Bookstein et al. 1988; Paggi et al. 1996; DeLuca et al. 1996). The gene encoding p107 is located at a genetically stable position on chromosome 20q11.2 (Ewen et al. 1991), while the gene for RB2/p130 is located at 16q12.2, a position susceptible to genetic aberration (Li et al. 1993; Yeung et al. 1993).

The proteins of all three family members are nuclear phosphoproteins and share a significant degree of sequence similarity in several regions. These regions are biochemically and functionally defined as the N-terminal domain, the pocket domain subdivided into subdomains A and B, and the C-terminal domain (Paggi et al. 1996 for review). Besides high structural homology in the A/B pocket domains of the three family members, phylogenetic studies have identified pRB as being specific to vertebrates, whereas p107 and RB/p130, bearing a homologous spacer region between subdomains A and B, form a subfamily which is common to vertebrates, plants, and insects (Cobrinik 1996; Wang 1997). Due to these structural differences, there are also various biochemical and functional differences reported in the recent past, such as the ability of the RB2/p130 and p107 proteins to form stable complexes with E2F and cyclin A-E/CDK2 complexes by two independent binding domains (Zhu et al. 1995b; Woo et al. 1997).

4 Growth Inhibition by Retinoblastoma Family Members

Transient activation of either one of the three family members in diverse tumor-derived cell lines, such as SaOs-2, suppresses proliferation and causes G1 arrest (Goodrich et al. 1991; Zhu et al. 1993; Claudio et al. 1994), but only pRB has been found to be mutated in a high percentage of human tumor specimens. This raises the question whether p107 and RB2/p130 are actually tumor-suppressor genes, despite their structural homology to pRB. Recently, and in contrast to p107, RB2/p130's function has been described as being lost in several different tumors, underlining its tumor-suppressive function (Baldi et al. 1996a; P.P. Claudio et al. 1997, pers. comm.; Helin et al. 1997). Furthermore, only overexpression of RB2/p130 in the glioblastoma cell line T98G was successful in conferring growth arrest, while pRB and p107 had no such effect (Claudio et al. 1994). Similarly, in contrast to pRB, p107 was able to arrest the cervical carcinoma cell line C33, supporting the notion of functional differences between the family members (Zhu et al. 1993). This specific growth-suppressive function of p107 was due to its cyclin-binding domain, as these cells were insensitive to the E2F-binding domain of p107 as well as to pRB (Zhu et al. 1995c).

5 Effects of Deletions of RB Family Members on Mice Development

Genetic analysis showed that mice homozygously deleted for pRB died between 13.5 and 15.5 days of gestation (Clarke et al. 1992; Jacks et al. 1992; Lee et al. 1992), whereas mice deleted for either p107 or RB2/p130 had no overt abnormalities. In the latter cases, the mice still express pRB and either RB2/p130 or p107, respectively. It is likely that either RB2/p130 or p107 can compensate for each other in maintaining a normal phenotype. In contrast, cells deleted for both p107 and RB2/p130 die shortly after birth, showing developmental abnormalities different to the pRB-deleted mice (Cobrinik et al. 1996). Furthermore, mice homozygously deleted for pRB and p107 die at day 10 of gestation and mice heterozygous for pRB and homozygously deleted for p107 show severe growth retardation (Lee et al. 1996). Taking these observations together: each of the three retinoblastoma protein family members has different, but partly overlapping functions in the development and proper maintenance of vertebrate organisms. The tissue-specific damage caused by the lack of different sets of pocket proteins was verified by defining the expression patterns of the RB family members during early development of wild-type mice (Jiang et al. 1997).

Further support for this notion comes from a recent report by Hurford and coworkers, showing that the deletion of different members of the RB family causes different expression patterns of E2F-regulated genes in mouse embryonic fibroblasts (MEF), prepared from embryos of knockout mice at day 13.5 (Hurford et al. 1997). The cells were arrested in G0 and restimulated with serum to progress through the cell division cycle. In agreement with the developmental effects caused by different retinoblastoma family member mutations mentioned above, severe deregulation of E2F-dependent transcription was reported only for RB−/− and p107−/−; RB2/p130−/− double mutants, whereas p107−/− and RB2/p130−/− deletions alone had no effect. However, both the absence of pRB alone and the combined absence of p107 and RB2/p130 caused improper expression of S phase-promoting genes such as cyclin E in RB−/− or b-myb, cyclin A, and E2F1 in p107−/−; RB2/p130−/− MEFs. Importantly, and in agreement with the structural homologies, different sets of E2F-dependent genes were affected in RB−/− and p107−/−; RB2/p130−/− deficient MEF cells. This clearly demonstrates that pRB and the subfamily RB2/p130 and p107 have different functions in regulating the expression of different genes. Unfortunately, the presented data gave no hint in solving the question of whether different E2F complexes, and if so, which, are regulating specific promoters by recruiting different RB family members for proper regulation. This may be explained by the fact that E2F-dependent transcription is not solely dependent on the presence of an E2F site, but that the surrounding promoter environment is important for proper regulation (Karlseder et al. 1996). However, the different effects of diverse retinoblastoma family member mutations on E2F-dependent gene expression may help to explain the diverse developmental defects occurring in the different knock out mice (Hurford et al. 1997).

6 Regulation of RB Family Members by Cyclin/CDK Complexes – the Restriction Point

At the time of the restriction point (R point) in late G1, a cell committed to proliferation is thought to measure its size, nutrition supply, level of DNA precursor pools, and state of DNA before it becomes restricted to entering S phase and finishing the cell division cycle with minor reliance on outside signals and protein synthesis. Several G1-specific cyclin/CDK complexes were shown to transfer pRB family members from its active, inhibiting, hypophosphorylated state to a hyperphosphorylated, inactive state, at a time similar to that of the R point. Therefore, the hypothesis of the R point proposed by Pardee in 1974 may finally be verified at its molecular level in a modified version by including pRB-mediated control mechanisms late in G1 into the model (Pardee 1974, 1989).

The Retinoblastoma gene is constitutively expressed in most vertebrate cells from its promoter, with characteristics of a housekeeping gene, containing binding sites for ATF, SP1, and E2F (Chen et al. 1995 for review). The E2F-1 transcription factor was shown to be part of a negative autoregulatory loop integrating the inhibitory capacity of pRB over E2F-induced transcription (Shan et al. 1994). p107 accumulates at the G1/S transition and this may be due to two E2F sites in its promoter (Beijersbergen et al. 1995; Zhu et al. 1995a). RB2/p130 accumulates if cells are serum-arrested or differentiated, and protein levels drop when cells reenter the cell cycle in late S phase (Mayol et al. 1996). RB2/p130 transcription may be regulated by myo D and SP1 sites in its promoter (Baldi et al. 1996b). The main level of pRB regulation occurs at the posttranslational level through subsequent phosphorylation and dephosphorylation of the protein in G1 and M/G1 phases, respectively (Chen et al. 1989). Up to five different migration forms of pRB between 105 and 115kDa have been detected, whereby the fastest, de- or hypophosphorylated form is converted to more slowly migrating hyperphosphorylated forms by phosphorylation of several serine and threonine residues occurring in late G1 at the G1/S phase transition and during S and S-G2 transition of the cell cycle (Ludlow et al. 1993; Chen et al. 1995). p107 and RB2/p130 are also phosphorylated as the cell passes through G1, and, notably, p107 reappears in a dephosphorylated form in S phase (Baldi et al. 1995; Beijersbergen et al. 1995; Mayol et al. 1995). A major step in the understanding of the control mechanisms of the cell-division cycle was made by the linkage of its major driving force, the cyclin/CDK complexes and its inhibitors to the phosphorylative inactivation of RB family members.

In vitro and in vivo studies have revealed a puzzling picture of G1-activated cyclin/CDK complexes involved in the phosphorylative inactivation of pRB in the late G1 phase (Weinberg 1995). Among these, the cyclin D(1–3)/CDK(4–6) complexes are the best candidates for the early phosphorylation steps of pRB in G1, whereas further phosphorylation in later G1 may be performed by cyclin E/CDK2 and cyclin A/CDK2 in S phase and G2/M (Weinberg 1995; Bartek et al. 1996). However, the actual relevance of cyclin E/CDK2 and cyclin A/CDK2 in phosphorylation of pRB remains unclear (Bartek et al. 1996).

The D-type cyclins, complexed to CDK4/6, bind pRB via their LxCxE motifs at its A/B pocket and phosphorylate pRB preferentially to histone H1 (Ewen et al. 1993; Kato et al. 1993). Another line of pRB regulation comes from studies of CKIs (Cylin-Dependent Kinase Inhibitors), which are divided into two classes due to their structure and preference for different CDKs. p16 and its related proteins are specific for CDK4/6, whereas p21, p27, and p57 are found to complex preferentially with CDK2/4 (MacLachlan et al. 1995). Direct evidence for the specificity of CKI p16 for inhibiting pRB phosphorylation by the cyclin D/CDK4/6 complexes came from experiments showing that p16 inhibition of S phase progression strictly relies on the presence of functional pRB (Serrano et al. 1993). p107 and RB2/p130 are also targets for cyclin CDK complexes in mid G1 (Beijersbergen and Bernards 1996). Besides the ability of p107 and RB2/p130 to form stable complexes with cyclin A-E/CDK2, it is very likely that the main phosphorylation is performed by cyclin D/CDK4 complexes. This is shown in several sets of experiments. First, in vivo phosphorylation occurs at a time before activation of CDK2 by cyclin E or cyclin A. Second, overexpression experiments with cyclin D/CDK4, but not cyclin A-E/CDK2, led to physiologically correct p107 phosphorylation in mid G1 and overcame a p107-mediated cell cycle arrest (Beijersbergen et al. 1995; Zhu et al. 1993; Beijersbergen and Bernards 1996). However, in another study it was demonstrated that RB2/p130-mediated growth arrest could be rescued by cyclin A, cyclin E, as well as cyclin D1-3 overexpression in SaOs-2 cells (Claudio et al. 1996). This indicates that there are differences in the way in which p107 and RB2/p130 are regulated by phosphorylation throughout the cell cycle.

7
Role of pRB in Growth-Promoting Pathways

The abundance of cyclin D1, with a short half-life of 15 min, is strictly related to the availability of etracellular growth stimulating signals (Sherr 1993). Therefore, pRB phosphorylation and further cell cycle progression only occurs in an appropriate environment. Striking evidence for an integrative in vivo involvement of cyclin D/CDK complexes and pRB in mitogen-dependent cell-cycle progression towards S phase comes from several recent reports.

Lukas and coworkers demonstrated in different cellular models that activation of S phase progression by either growth factor-dependent tyrosine kinase receptors, ligand-bound estradiol receptors or cAMP-dependent G-protein coupled thyrotropin receptors strictly relies on functional cyclin D/CDK complexes. S phase entry promoted by the above pathways is inhibited by the injection of the CDK4/6-specific p16 CKI, as well as cyclin D1 neutralizing antibodies (Lukas et al. 1996).

Peeper and coworkers linked the ras proto oncogene-dependent mitogenic signal transduction pathway, essential for G0 exit and G1/S transition, to the cyclin D/CD4/6 dependent inactivation of pRB. A dominant negative ras mutant, RAS^{asn17}, anti ras antisera, and p16 conferred G1arrest in a pRB-dependent manner, whereas a dominant CDK3 mutant and p27 arrested pRB negative cells as well (Peeper et al. 1997).

Another recent report by Leone and coworkers demonstrated that the serum-induced G0/G1/S phase progression of serum-starved fibroblasts is inhibited by ectopic RAS^{asn17} expression. This pRB-dependent block could be bypassed by either the coexpression of cyclin E/CDK2 or E2F, the most prominent downstream effector of pRB inhibited transcription (Leone et al. 1997).

These data, together with another recent report, demonstrate the existence of an pRB-independent, either downstream or parallel occurring cyclinE-dependent S phase-promoting pathway. Using a pRB mutant that is unable to be phosphorylated by CDK-dependent complexes or a dominant negative mutant of DP1, an essential heterodimerizing protein responsible for active E2F, the authors report S phase entry and completion of at least one cell-division cycle even without activation of E2F-dependent transcription of DNA precursor enzymes (Lukas et al. 1997). However, the biological relevance of these findings remains to be clarified, even concerning the fact that the control of E2F-regulated S phase-specific genes incorporates multiprotein complexes other than E2F, which themselves undergo posttranscriptional regulation.

Cyclin E/CDK-induced, cyclin D(1–3)/CDK4/6/pRB/E2F-independent S phase progression is blocked by the cyclin E/CDK2 inhibitor CKI p27, but p27 levels decrease as the cell moves towards S phase. Interestingly, p27 itself has recently been shown to be negatively regulated and inactivated by its target cyclin E/CDK2 (Sheaff et al. 1997).

These data strikingly demonstrate the central position of pRB in growth control, as for such a diversity of growth-promoting pathways, pRB resembles the key to commit a cell passing the R point and finish the cell division cycle. Future studies will reveal the involvement of p107 and RB2/p130 in these pathways.

8 Role of pRB in Growth-Inhibiting Pathways; Apoptosis and Differentiation

Further evidence for the central role of pRB in the regulation of the R point transition comes from the effects of negative growth signals of diverse origins and their reliance on active RB family members.

8.1 pRB as a Downstream Effector of p53

As mentioned above, at the R point cells check the status of DNA and precursor pools and if DNA damage has occurred or the precursor pools are inappropriate, several measuring mechanisms will lead to the stabilization of the tumor-suppressor protein p53, causing a G1 arrest that gives the cell the chance for DNA repair and precursor pool management. p53 itself is a well-described transcription factor which, in response to DNA damage, transactivates the CKI p21 (El Deiri et al. 1993). In a recent report, LaBaer and coworkers reveal a part of the puzzling role of p21 in p53-activated G1 arrest. Dependent on its stoichiometric abundance, p21, in equimolar amounts, is necessary for the formation of active cyclin D/CDK4 complexes, whereas elevated levels of p21, as induced by p53, result in inactivation of cyclin D/CDK4 complexes (La Baer et al. 1997). Therefore, one important pathway to arrest cells is to prevent G1-S phase progression by inhibiting pRB phosphorylation at the R point, characterizing pRB-mediated control as a downstream event of p53. Conclusive with this statement is the fact that p53 induces apoptosis in cells that do not react properly to an ordered arrest. This was reported using in vitro cell culture systems as well as mice models. PALA (phosphonoacetyl-L-aspartate), a drug inhibiting de novo pyrimidine synthesis by inhibiting the CAD (carbamoyl phosphate synthetase, transcarbamylase, dihydroorothase) enzyme complex causes a p53–p21-dependent G1 arrest. p53-negative Li Fraumenti cells fail to arrest and inappropriately enter S phase.

Most of these cells die, but some suffer genome destabilization and give rise to PALA-resistant clones with amplified CAD genes (Livingstone et al. 1992). Of considerable importance, ectopic expression of small DNA tumor virus (TV) proteins specific for binding and inhibiting pRB, such as papillomavirus HPV16/E7 or polyomavirus LT, in p53-positive cells also force cells through the G1 arrest, leading to genome instability and leading to PALA-resistant cells (White et al. 1994; Stiegler et al. 1997). In essence, bypassing p53-mediated growth inhibition by inactivating a downstream mediator like pRB (pRB mutation or small DNA TV) is sufficient to destabilize genomic integrity, which is an essential step in neogenesis. Supportive evidence for such a scenario comes from studies of transgenic mice. Photoreceptor-specific expression of E7 in transgenic mice causes p53-dependent apoptosis but retinal tumors emerge in the absence of p53 (Howes et al. 1994).

Most of the small DNA tumor virus family members have evolved strategies to inactivate p53 (E1B, E6, SV40 LT) as well as pRB (E1A, E7, SV40 LT)-mediated control for successful infection. In support of the idea of a p53-p21-pRB-E2F chain of events are several reports demonstrating that excess pRB can overcome p53-mediated apoptosis and that p53-mediated growth arrest is converted to apoptosis by overexpression of E2F-1 (Wu and Levine 1994; Haupt et al. 1995). Recent reports show that E2F-1-induced apoptosis requires its DNA-binding domain but not its transactivation domain. The latter domain also contains the RB-binding domain, which, if present, prevents apoptosis in the presence of the retinoblastoma protein (Hsieh et al. 1997; Phillips et al. 1997). Therefore, pRB activation seems to be an essential step in the cascade of p53-mediated growth arrest, and as a consequence of inadequate response, conflicting signals force the cell into apoptosis.

8.2 RB Family Members in Differentiation

As mentioned above, mice nullizygous for pRB die in utero and they do this as a consequence of massive apoptosis, whereas mice nullizygous for both pRB and p53 show reduced apoptotic levels but develop tumors (Condorelli and Giordano 1997 for review). pRB also plays a major role in inducing and maintaining myocyte terminal differentiation. In myogenic differentiation, active pRB is thought to bind to a differentiation-specific transcription factor, myoD, and activates its transcription, distinct from the established repressing activities. Differentiated cells contain high levels of hypophosphorylated pRB, RB−/− cells fail to differentiate, and inactivation of pRB by SV40LT induces DNA synthesis in differentiated myotubes (Cardoso et al. 1993; Gu et al. 1993). In this system, CKI p21 is highly activated by the myogenic HLH transcription factor myoD as an early response to induced differentiation (Halevy et al. 1995). Apoptosis is a naturally occurring phenomenon in differentiating tissues; however, it has been shown that the surviving myocytes strictly depend on high levels of p21 (Wang and Walsh 1996). Furthermore, overexpression of cyclin D1, a target of p21, inhibited myogenic differentiation (Skapek et al. 1995; LaBaer et al. 1997). As mentioned above, RB2/p130 levels are induced in arrested and in differentiated cells, and RB2/p130 is found to be specifically bound to the B-*myb* promoter in differentiated neuroblastoma cells (Raschella et al. 1997; P. Stiegler et al., unpubl. results).

pRB also plays an important role in hematopoetic differentiation and adipogenesis differentiation (Chen et al. 1995; Condorelli and Giordano 1997).

8.3 RB Family Members in TGF β-Mediated Growth Inhibition

Additional proof for the central control mediated by pRB, but also of RB2/p130 and p107, comes from studies with the transforming growth factor beta (TGF β). TGF β-induced cell-cycle arrest causes induction of p15 and reduction of CDK4 levels, which results in accumulation of hypophosphorylated RB family members and the subsequent repression of E2F-dependent transcription (Herzinger et al. 1995; Weinberg 1995; Mayol et al. 1996; Li et al. 1997). This repression strictly depends on the presence of a functional E2F site in these promoters, again supporting E2F function as an active repressor by recruiting pocket proteins to promoters (Li et al. 1997).

9 RB2/130 and p107 Confer a Growth-Suppressive Ability, Other than pRB

As mentioned above, the RB2/p130 and p107 subfamily carry one important additional feature in their structure in contrast to pRB, and that is the ability to stably bind cyclin A/CDK 2 or cyclin E/CDK2 complexes. However, the functional consequence of such complexes, which can also include E2F moities, had remained unclear. Recent reports provide the first biochemical insights. It was demonstrated by two groups that RB2/p130 and p107 use p21 similar sequences to bind and inhibit the activity of cyclin/CDK complexes. In vitro and in vivo purified pocket protein/cyclin/CDK complexes have little associated kinase activity; however, dissociation of the pocket protein from the cyclin/CDK complexes with DOC (deoxycholate) restores activity. The N-terminal regions, highly conserved in RB2/p130 and p107, were recently identified to contain growth-suppressive activities in addition to mediating the E2F-dependent suppression (Zhu et al. 1995c). The observed kinase-inhibitory activity was shown to be located in the same region, suggesting that RB2/p130 and p107 confer a part of their growth-inhibitory function by inhibiting important cell-cycle kinases (Woo et al. 1997). Another report provides similar yet distinct results in identifying the RB2/130 spacer region as inhibiting CDK2-dependent kinase activity much more strongly than the RB2/p130 Noterminus (DeLuca et al. 1997).

10 pRB Inhibits Biosynthesis Via Inhibiting RNA pol I- and pol III-Dependent Transcription

Considering the above data that deregulation of S phase-specific RNA pol II-dependent genes by loss of pocket protein control induces cellular proliferation and that inadequate S phase entry in special cases leads to apoptosis, the question remains what other effects may enable tumor growth. One important clue came from reports demonstrating that pRB is also involved in the transcriptional regulation of ribosomal and transfer RNA transcription by RNA pol I and RNA pol III. This repression is most probably due to pRB sequences in pocket A and B which are similar to those on TBP and TFIID, respectively (Hagemeier et al. 1993). Derepression of these genes gives a cell the ability to provide the elevated levels of biosynthesis needed in proliferating cells (Cavenaugh et al. 1995; White et al. 1996; White 1997). Supporting evidence for the importance of pRB-regulated pol I transcription comes from recent reports showing that SV40 LT activates pol I transcription by binding the TBP-TAF1 complex SL1, but it may also affect pRB control (Zhai et al. 1997). Furthermore, papillomavirus type 16 E7 has been shown to be nucleolar-associated in cervical transformed CaSKi cells (Zatsepina et al. 1997). This demonstrates that small DNA tumor viruses have evolved strategies to actively interfere not only with cell cycle-regulating pathways, but also to activate biosynthesis for successful infection.

11 RB Family Members Interact with Several Other Growth-Regulating Genes

In addition to the well-established interactions of pRB, RB2/p130, and p107 with E2Fs and cyclin/CDK complexes, there have been numerous other interactions between the pocket proteins and cellular proteins reported (Wang et al. 1994; Taya 1997; Wang 1997). Among these are MDM-2, a known downregulator of p53, which binds pRB at its C-terminal region and inhibits stable pRB/E2F interaction (Xio et al. 1995). Interaction of the tyrosine kinase c-abl with the C-terminal region of pRB inhibits its kinase activity (Welch and Wang 1995). pRB and p107 were also shown to interact and interfere with SP-1-regulated gene expression (Chen et al. 1994; Datta et al. 1995). A new role in transcriptional activation by pRB binding to a transcription complex was identified by pRB binding to hBrm and BRG1 in glucocorticoid-mediated transcription, leading to cellular differentiation. These genes are yeast SWI2 homologs, which are part of the multiprotein SWI/SNF complex involved in nucleosome positioning (Taya 1997). The position of a nucleosome within a promoter determines which recognition sites are available to be bound by sequence-specific transcription factors (Lewin 1994). Therefore, pRB is able to inhibit proliferating genes and activate differentiation-specific genes, like myoD-mediated transcription (Gu et al. 1993). Another, yet to be clarified, role for pRB in transcriptional regulation comes from data of one of its binding protein family, RbAP48 (retinoblastoma-associated). These proteins were shown to be involved in deacetylation of histones (Roth and Allis 1996; Hassig et al. 1997).

12 Concluding Remarks

Numerous papers in the field of RB family members sometimes draw a puzzling view of their role in growth control and their regulating mechanisms.

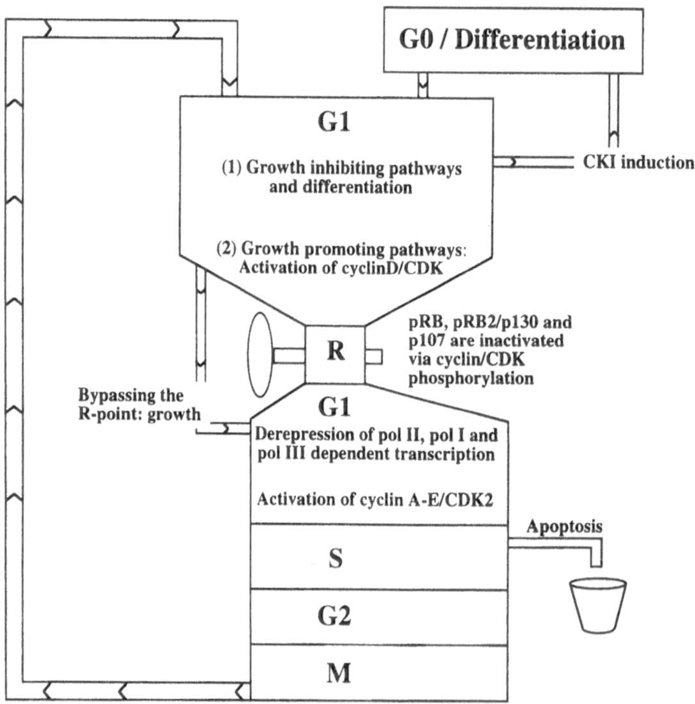

Fig. 1. Summary of the main features in this chapter. During the early stages of the cell cycle, the cell decides whether to stay in or leave the cell-division cycle. The R point is essential in regulating these events. Bypassing it results in cells unable to differentiate or to react to growth arrest. This results in inapropriate S phase entry and leads to apoptosis or growth, and in the long run, to neogenesis

However, recent papers dealing thoroughly with physiological processes reveal a much clearer and better understanding of the importance of pocket proteins in guarding homeostasis of vertebrate organisms. RB family members repress transcription of growth-promoting genes and at the same time induce transcription of growth-inhibiting, differentiation-specific genes. Furthermore, numerous growth-inducing pathways were shown to necessitate RB phosphorylation at the R point in order to commit a cell for proliferation. On the other hand, RB family members are essential for tissue-specific differentiation, and their absence causes severe developmental damage through apoptosis and the absence of terminal differentiation. RB-mediated control was also shown to be a downstream event of p53 control pathways in order to inhibit S phase entry of damaged cells. It is also clear that every family member seems to exhibit a unique, partially overlapping role in upholding cellular growth control, as they are active at different times in a cellular life, and their absence results in damage in different tissues. Figure 1 summarizes essential concepts in the regulatory capacities of the retinoblastoma gene family. Important decisions about which pathway a cell should take are made in early G1, before the R point. Bypassing the R point control by diverse mechanisms causes deregulated growth which promotes neogenesis. However, future studies will contribute to a much better understanding of the central role of the RB family in growth inhibition.

Note added in proof. During the preparation of this manuscript another important aspect of transcriptional regulation by the retinoblastoma protein was revealed. The acetylation status of histones can generally be linked to the activity of chromatin; hypoacetylated histones are found in silent promoters and hyperacetylated histones mark a transcriptionally active region. Three independent laboratories recently reported that the pRB protein can directly interact with a histone deacetylase, HDAC1, through the pocket domain (Brehm et al. 1998; Luo et al. 1998; Magnaghi-Jaulin et al. 1998). This recruitment enables the retinoblastoma protein to silence transcription. Inhibition of HDAC1 using trichostatin A abolishes the repressive action of the retinoblastoma protein (Luo et al. 1998).

Acknowledgment. P. Stiegler is funded by the Sbarro Institute for Cancer research and NIH grants. Many thanks to M. Kasten, C.M. Howard, and P.P. Claudio for thoroughly reading the manuscript.

References

Baldi A, De Luca A, Claudio PP, Baldi F, Giordano GG, Tommasino M, Paggi MG, Giordano A (1995) The RB2/p130 gene product is a nuclear protein whose phosphorylation is cell cycle-regulated. J Cell Biochem 59:402–408

Baldi A, Esposito V, De Luca A, Howard MH, Mazzarella G, Baldi F, Caputi M, Giordano (1996a) Differential expression of the retinoblastoma gene family members pRB/105, p107 and RB2/p130 in lung cancer. Clin Can Res 2:1239–1245

Baldi A, Esposito V, Claudio PP, De Luca A, Giordano A (1996b) Genomic structure of the human retinoblastoma-related RB2/p130 gene. Proc Natl Acad Sci 93:4629–4632

Bartek J, Bartkova J, Lukas J (1996) The retinoblastoma protein pathway and the restriction point. Curr Opin Cell Biol 8:805–814

Beijersbergen RL, Bernards R (1996) Cell cycle regulation by the retinoblastoma family of growth inhibitory proteins. Biochim Biophys Acta 1287:103–120

Beijersbergen RL, Kerkhoven RM, Zhu L, Carlee L, Voorhoeve PM, Bernards R (1994) E2F-4, a new member of the E2F gene family, has oncogenic activity and associates with p107 in vivo. Genes Dev 8:2680–2690

Beijersbergen RL, Carlee L, Kerkhoven RM, Bernards R (1995) Regulation of the retinoblastoma protein related p107 by G1 cyclin complexe. Genes Dev 9:1340–1353

Bishop M (1991) Molecular themes in oncogenesis. Cell 64:235–248

Bookstein R, Lee EY, To H, Young LJ, Sery TW, Hayes RC, Friedmann T, Lee WH (1988) Human retinoblastoma susceptibility gene: genomic orgenization and analysis of heterozygous intragenic deletion mutants. Proc Natl Acad Sci 85:2210–2214

Brehm A, Miska EA, McCane DJ, Reid JL, Bannister AJ, Kouzarides T (1998) Retinoblastoma protein recruits histone deacetylase to repress transcription. Nature 391:597–601

Campanero MR, Flemington EK (1997) Regulation of E2F through ubiquitin-dependent degradation. Proc Natl Acad Sci 94:2221–2226

Cardoso MC, Leonhardt H, Nadal-Ginard B (1993) Reversal of terminal differentiation and control of DNA replication. Cell 74:979–992

Cavenaugh AH (1995) Activity of the polymerase I transcription factor UBF blocked by RB gene product. Nature 374:177–180

Chen LI, Nishinaka T, Kwan K, Kitabayashi I, Yokoyama K, Fu YHF, Gruenwald S, Chiu R (1994) The retinoblastoma gene product RB stimulates SP-1-mediated transcription by liberating SP-1 from a negative regulator. Mol Cell Biol 14:4380–4389

Chen PL, Scully P, Shew JY, Wang JY, Lee WH (1989) Phosphorylation of the retinoblastoma gene product is modulated during the cell cycle and cellular differentiation. Cell 58:1193–1198

Chen PL, Riley DJ, Lee WH (1995) The retinoblastoma protein as a fundamental mediator of growth and differentiation signals. Crit Rev Eucar Gene Expr 5:79–95

Clarke AR, Maandag ER, van Roon M, van der Lugt NM, van der Valk M, Hooper ML, Berns A, te Riele H (1992) Requirement for a functional RB-1 gene in murine development. Nature 359:328–330

Claudio PP, Howard CM, Baldi A, De Luca A, Fu Y, Condorelli G, Sun Y, Colburn N, Calabretta B, Giordano A (1994) p130/pRB2 has growth-suppressive properties similar to yet distinctive from those of the retinoblastoma family members pRB and p107. Cancer Res 54:5556–5560

Claudio PP, De Luca A, Howard CM, Baldi A, Firpo EJ, Paggi MG, Giordano A (1996) Functional analysis of pRB2/p130 interaction with cyclins. Cancer Res 56:2003–2008

Cobrinik D (1996) Regulatory interactions among E2Fs and cell cycle control proteins. Curr Top Microbiol Immunol 208:31–60

Cobrinik D, Lee MH, Hannon G, Mulligan G, Bronson RT, Dyson N, Harlow E, Beach D, Weinberg RA, Jacks T (1996) Shared role of the pRB-related p130 and p107 proteins in limb development. Genes Dev 10:1633–1644

Condorelli G, Giordano A (1997) The synergistic role of E1A-binding proteins and tissue-specific transcription factor in differentiation. J Cell Biol 67:423–431

Datta PK, Raychaudhuri P, Bagchi S (1995) Association of p107 with E2F: genetically separable regions of p107 are involved in regulation of E2F- and SP1-dependent transcription. Mol Cell Biol 15:5444–5452

De Luca A, Esposito V, Baldi A, Giordano A (1996) The retinoblastoma gene family and its role in proliferation, differentiation and development. Histol Histopathol 11:1029–1034

De Luca A, MacLachlan T, Bagella L, Giordano A (1997) A unique domain of pRB2/p130 as an inhibitor of cdk2 kinase activity. J Biol Chem 272:20971–20974

DunaijefJ, Strober BE, Guha S, Khavari PA, Alin K, Luban J, Begeman M, Crabtree GR, Goff SP (1994) The retinoblastoma protein and BRG1 form a complex and coorperate to induce cell cycle arrest. Cell 79:119–130

Dynlacht B, Moberg K, Lees J, Harlow E, Zhu L (1997) Speicific regulation of E2F family members by cyclin-dependent kinases. Mol Cell Biol 17:3867–3875

El Deiri W, Tokino T, Velculescu VE, Levy DB, Parsons R, Trent JM, Lin D, Mercer EW, Kinzler KW, Vogelstein B (1993) WAF 1, a potential mediator of p53 tumor suppression. Cell 57:817–825

Ewen ME, Xing YG, Lawrence JB, Livingston DM (1991) Molecular cloning, chromosomal mapping and expression of the cDNA for p107, a retinoblastoma gene product-related protein. Cell 66:1155–1164

Ewen ME, Sluss HK, Sherr CJ, Matsushime H, Kato JY, Livingston DM (1993) Functional interactions of the retinoblastomaprotein with mammalian D-type cyclins. Cell 73:487–497

Ginsberg D, Vairo G, Chittenden T, Xiao ZX, Xu G, Wyndner GL, DeCaprio JA, Lawrence JB, Livingsto DM (1994) E2F-4, a new member of the E2F transcription factor family, interacts with p107. Genes Dev 8:2665–2679

Goodrich DW, Wang NP, Qian YW, Lee EYH, Lee WH (1991) The retinoblastoma protein regulates progression through the G1 phase of the cell cycle. Cell 67:297–302

Gu W, Schneider JW, Condorelli G, Kaushal S, Mahdavi V, Nadal-Ginard B (1993) Interaction of myogenic factors and the retinoblastoma protein mediates muscle cell commitment and differentiation. Cell 72:309–324

Hagemeier C, Bannister A, Cook A, Kouzarides T (1993) The activation domain of transcription factor PU.1 binds the retinoblastoma (RB) protein and the transcription factor TFID in vitro: RB shows sequence similarity to TFIID and TFIIB. Proc Natl Acad Sci 90:1580–1584

Halevy O, Novitch BG, Spicer DB, Skapek SX, Rhee J, Hannon GJ, Beach D, Lassar AB (1995) Correlation of terminal cell cycle arrest of skeletal muscle with induction of p21 by myoD. Science 267:1018–1020

Hassig CA, Fleischer TC, Bilin AN, Schreiber SL, Ayer DE (1997) Histone deacetylase activity is required for full transcription repression by mSin3A. Cell 89:341–347

Hateboer G, Kerkhoven RM, Shvarts A, Bernards R, Beijersbergen RL (1996) Degradation of E2F by the ubiquitin-proteasome pathway. Genes Dev 10:2960–2970

Haupt Y, Rowan S, Oren M (1995) p53 mediated apoptosis in HeLa cells can be overcome by excess pRB. Oncogene 10:1563–1571

Helin K, Holm K, Niebuhr A, Eiberg H, Tommerup N, Hougaard S, Skovgaard-Poulson H, Spang-Thomsen M, Norgaard P (1997) Loss of the retinoblastoma protein-related p130 in small cell lung carcinoma. Proc Natl Acad Sci 94:6933–6938

Herzinger T, Wolf DA, Eick D, Kind P (1995) The pRB related protein p130 is a possible effector of transforming growth factor β 1 induced cell cycle arrest in keratinocytes. Oncogene 10:2079–2084

Hijmans EM, Voorhoeve PM, Beijersbergen RL, Van't Veer LJ, Bernards R (1995) E2F-5, a new E2F family member that interacts with p130 in vivo. Mol Cell Biol 6:3082–3089

Hofman F, Martelli F, Livingston D, Wang Z (1996) The retinoblastoma gene product protects E2F-1 from degradation by the ubiquiti-proteasome pathway. Genes Dev 10:2949–2959

Howes KA, Ransom N, Papermaster DS, Lasudry JGH, Albert DM, Windle JL (1994) Apoptosis or retinoblastoma: alternative fates of photoreceptor expressing the HPV16 E7 gene in the presence or absence of p53. Genes Dev 8:1300–1310

Hsieh JK, Fredersdorf S, Kouzarides T, Martin K, Lu X (1997) E2F1-induced apoptosis requires DNA binding but not transactivation and is inhibited by the retinoblastoma protein through direct interaction. Genes Dev 11:1840–1852

Huet X, Rech J, Plet A, Vie A, Blanchard JM (1996) Cyclin A expression is under negative transcriptional control during the cell cycle. Mol Cell Biol 16:3789–3798

Hurford RK, Cobrinik D, Lee MH, Dyson N (1997) pRB and p107/p130 are required for the regulated expression of different sets of E2F responsive genes. Genes Dev 11:1447–1463

Jacks T, Fazeli A, Schmitt EM, Bronson RT, Goodel MA, Weinberg RA (1992) Effects of an RB mutation in the mouse. Nature 359:295–300

Jiang Z, Zacksenhausen E, Gallie BL, Phillips RA (1997) The retinoblastoma gene family is differentially expressed during embryogenesis. Oncogene 14:1789–1797

Karlseder J, Rotheneder H, Wintersberger E (1996) Interaction of SP1 with the growth and cell cycle regulated transcription factor E2F. Mol Cell Biol 16:1659–1667

Kato J, Matsushime H, Hiebert SW, Ewen ME, Sherr CJ (1993) Direct binding of cyclin D to the retinoblastoma gene product (pRB) and pRB phosphorylation by the cyclin D-dependent Kinase CDK4. Genes Dev 7:331–342

LaBaer J, Garret MD, Stevenson LF, Slingerland JM, Sandhu C, Chou HS, Fattey A, Harlow E (1997) New functional activities for the p21 family of CDK inhibitors. Genes Dev 11:847–862

Lam EW, Watson RJ (1993) An E2F-binding site mediates cell-cycle regulated repression of mouse B-myb transcription. EMBO 12:2705–2713

Lee EY, Chang CY, Hu N, Wang YC, Lai CC, Herrup K, Lee WH, Bradley A (1992) Mice difficient for RB are nonviable and show defects in neurogenesis and haematopoesis. Nature 359:288–294

Lee MH, Williams BO, Mulligan G, Mukai S, Bronson R, Dyson N, Harlow E, Jacks T (1996) Targeted disruption of p107: functional overlap between p107 and p130. Genes Dev 10:1621–1632

Lee WH, Bookstein R, Hong F, Young LJ, Shew JY, Lee EY (1987) Human retinoblastoma susceptibility gene: cloning, identification, and sequence. Science 235:1394–1399

Lees JA, Saito M, Vidal M, Valentine M, Look T, Harlow E, Dyson N, Helin K (1993) The retinoblastoma protein binds to a family of E2F transkription factors. Mol Cell Biol 13:7813–7825

Leone G, DeGeorgi J, Sears R, Jakoi L, Nevins JR (1997) Myc and Ras collaborate in inducing accumulation of active cyclinE/Cdk2 and E2F. Nature 387:422–426

Lewin B (1994) Chromatin and gene expression: constant questions, but changing answers. Cell 79:397–406

Li JM, Hu-PPH, Shen X, Yu Y, Wang XF (1997) E2F-4RB and E2F-4 p107 complexes suppress gene expression by transforming growth factor β through E2F binding sites. Proc Natl Acad Sci 94:4948–4953

Li Y, Graham C, Lacy S, Duncan AMV, Whyte P (1993) The adenovirus E1A-associated 130 kDa protein is encoded by a member of the retinoblastoma gene family and physically interacts with cyclin A and E. Genes Dev 7:2366–2377

Livingstone LR, White A, Sprouse J, Livianos E, Tylor J, Tlsty TD (1992) Altered cell cycle arrest and gene amplification: potential accompany loss of wild-type p53. Cell 70:923–935

Ludlow JW, Glendening CL, Livingston DM, DeCaprio JA (1993) Specific enzymatic dephosphorylation of the retinoblastoma protein. Mol Cell Biol 13:367–372

Lukas J, Bartkova J, Bartek J (1996) Convergence of Mitotic signalling cascades from diverse classes of receptors at the cyclin D-cyclin-dependent-kinase-pRB-controlled G1 checkpoint. Mol Cell Biol 16:6917–6925

Lukas J, Herzinger T, Hansen K, Moroni MC, Resnitzki D, Helin K, Reed SI, Bartek J (1997) Cyclin E-induced S phase without activation of the pRB/E2F pathway. Genes Dev 11:1479–1492

Luo RX, Postigo AA, Dean DC (1998) Rb interacts with histone deacetylase to repress transcription. Cell 92:463–473

MacLachlan TK, Sang N, Giordano A (1995) Cyclins, cyclin-dependent kinases and CDK inhibitors: implications in cell cycle control and cancer. Crit Rev Eukar Gen Expr 5:127–156

Magnaghi-Jaulin L, Groisman R, Naguibneva I, Robin P, LeVillain JP, Troalen F, Trouche D, Bellan Harel A (1998) Retinoblastoma protein represses transcription by recruiting a histone deacetylase. Nature 391:601–604

Mayol X, Grana X, Baldi A, Sang N, Hu Q, Giordano A (1993) Cloning of a new member of the retinoblastoma gene family (pRB2) which binds to the E1A transforming domain. Oncogene 8:2561–2566

Mayol X, Garriga J, Grana X (1995) G1 cyclin/CDK-independent phosphorylation and accumulation of p130 during the transition from G1 to G0 lead to its association with E2F-4. Oncogene 13:237–246

Mayol X, Garriga J, Grana X (1996) Cell cycle-dependent phosphorylation of the retinoblastoma-related protein p130. Oncogene 11:801–808

Moberg K, Starz MA, Lees JA (1996) E2F-4 switches from p130 to p107 and pRB in response to cell cycle reentry. Mol Cell Biol 16:1436–1449

Paggi MG, Baldi A, Bonetto F, Giordano A (1996) Retinoblastoma family in cell cycle and cancer. J Cell Biochem 62:418–430

Pardee AB (1974) A restriction point for control of normal cell proliferation. Proc Natl Acad Sci 71:1286–1290

Pardee AB (1989) G1 events and regulation of proliferation. Science 246:603–608

Peeper DS, Upton TM, Ladha MH, Neuman E, Zalvide J, Bernards R, DeCaprio JA, Ewen ME (1997) Ras signalling linked to the cell-cycle machinery by the retinoblastoma protein. Nature 386:177–181

Phillips AC, Bates S, Helin K, Vousden KH (1997) Induction of DNA synthesis and apoptosis are separable functions of E2F-1. Genes Dev 11:1853–1863

Raschella G, Tanno B, Bonetto F, Amendola R, Battista T, De Luca A, Giordano A, Paggi MG (1997) Retinoblastoma-related protein pRB2/p130 and its binding to the B-myb promoter increase during human neuroblastoma differentiation. J Cell Biochem 67:297–303

Roth SY, Allis CD (1996) The subunit-exchange model of histone acetylation. Trends Cell Biol 6:371–375

Serrano M, Hannon GJ, Beach D (1993) A new regulatory motive in cell-cycle control causing specific inhibition of cyclin D/CDK4. Nature 366:704–707

Shan B, Chang CY, Jones D, Lee WH (1994) The transcription factor E2F-1 mediates autoregulation of RB gene expression. Mol Cell Biol 14:299–309

Sheaff RJ, Groudine M, Gordon M, Roberts JM, Clurman BE (1997) Cyclin D-CDK2 is a regulator of $p27^{Kip1}$. Genes Dev 11:1464–1478

Sherr CJ (1993) Mammalian G1 cyclins. Cell 73:1059–1065

Singh P, Coe J, Hong W (1995) A role for retinoblastoma protein in potenitating transcriptional activation by the glucocorticoig receptor. Nature 374:562–565

Skapek SX, Rhee J, Spicer DB, Lassar AB (1995) Inhibition of myogenic differentiation in proliferating myoblasts by cyclin D1-dependent kinase. Science 267:1022–1024

Slansky JE, Farnham PJ (1996) Introduction of the E2F family: protein structure and gene regulation. Curr Top Microbiol Immunol 208:1–30

Slansky JE, Li Y, Kaelin GW, Farnham PJ (1993) A protein synthesis-dependent increase in E2F-1 mRNA correlates with growth regulation of the dihydrofolate reductase promoter. Mol Cell Biol 12:5620–5631

Smith EJ, Leone G, DeGeorgi J, Jakol L, Nevins JR (1996) The accumulation of an E2F-p130 complex transcriptional repressor distinguishes a G0 State from a G1 cell state. Mol Cell Biol 16:6965–6976

Stiegler P, Schuechner S, Lestou V, Wintersberger E (1997) Polyomavirus large T antigen-dependent DNA amplification. Oncogene 14:987–995

Taya Y (1997) RB kinase and RB-binding proteins: new points of view. Trends Biol Sci 22:14–17

Wang JYJ (1997) Retinoblastoma protein in growth suppression and death protection. Curr Opin Genet Dev 7:39–45

Wang JYJ, Walsh K (1996) Resistance to apoptosis conferred by CDK inhibitors during myocyte differentiation. Science 273:359–361

Wang JYJ, Knudsen ES, Welch PJ (1994) The retinoblastoma tumor suppressor protein. Adv Cancer Res 64:25–85

Weinberg RA (1995) The retinoblastoma protein and cell cycle control. Cell 81:323–330

Welch P, Wang J (1993) A C-terminal binding domain in the retinoblastoma protein regulates nuclear c-abl tyrosine kinase in the cell cycle. Cell 75:779–790

White AE, Livianos EM, Tlsty TD (1994) Differential disruption of genomic integrity and cell cycle regulation in normal human fibroblasts by the HPV oncoproteins. Genes Dev 8:666–677

White RJ (1997) Regulation of RNA polymerase I and III by the retinoblastoma protein: a mechanism for growth control. Trend Biol Sci 22:77–80

White RJ, Trouche D, Martin K, Jackson SP, Kouzarides T (1996) Repression of RNA polymerase III transcription by the retinoblastoma protein. Nature 382:88–90

Woo MSA, Sanchez I, Dynlacht BD (1997) p130 and p107 use a conserved domain to inhibit cellular cyclin-dependent kinase activity. Mol Cell Biol 17:3566–3579

Wu X, Levine A (1994) p53 and E2F-1 cooperate to mediate apoprtosis. Proc Natal Acad Sci 91:3602–3606

Xio Z, Chen J, Levine A, Modjtahedi N, Xing J, Sellers W (1995) Interaction between the retinoblastome protein and the onco protein MDM2. Nature 375:694–698

Yeung RS, Bell DW, Testa JR, Mayol X, Baldi A, Klinga-Levan K, Knudson AG, Giordano A (1993) The retinoblastoma-related gene, RB2, maps to human chromosome 16q12 and rat chromosome 19. Oncogene 8:3465–3468

Zatsepina O, Braspenning J, Robberson D, Hajibagheri NMA, Blight KJ, Ely S, Hibma M, Spitkovsky D, Trendelenburg M, Crawford L, Tommasino M (1997) The human papillomavirus type 16 E7 protein is associated with the nucleolus in mammalian cells. Oncogene 14:1137–1145

Zhai W, Tuan JA, Comai L (1997) SV40 large T antigen binds to the TBP-TAF1 complex SL1 and coactivates ribosomal RNA transcription. Genes Dev 11:1605–1617

Zhu L, Van Den Heuvel S, Helin K, Fattey A, Ewen M, Livingston DM, Dyson N, Harlow E (1993) Inhibition of cell proliferation by p107, a relative of the retinoblastoma protein. Genes Dev 7:1111–1125

Zhu L, Zhu L, Xie E, Chang LS (1995a) Differential roles of two tandem E2F sites in repression of the human p107 promoter by retinoblastoma and p107 proteins. Mol Cell Biol 15:3552–3562

Zhu L, Enders G, Lees JA Beijersbergen RL, Bernards R, Harlow E (1995b) The pRB-related protein p107 contains two growth suppression domains: independent interactions with E2F and cyclin/cdk complexes. EMBO J 14:1904–1913

Zhu L, Harlow E, Dynlacht D (1995c) p107 uses a p21CIP1-related domain to bind cyclin/cdk2 and regulate interactions with E2F. Genes Dev 9:1740–1752

The Growth-Regulatory Role of p21 (WAF1/CIP1)

Andrei L. Gartel and Angela L. Tyner[1]

1 Introduction

Cyclin kinase inhibitors (CKIs) are proteins that bind to and inhibit the activity of cyclin-dependent kinases (Cdks). One of the first of these to be identified was p21, which binds and inhibits G_1 cyclin/Cdk complexes (Harper et al. 1993; Xiong et al. 1993). The p21 cDNA was cloned independently by several groups using a number of different screening strategies. It was identified as the product of a gene activated by wild-type p53, and it was named WAF1 (wild-type p53-activated factor (E1-Deiry et al. 1993). Microsequencing of a protein that interacted with Cdks led to its cloning using PCR (Xiong et al. 1993). Using a yeast two-hybrid screen, it was also identified as a Cdk-binding protein, and was subsequently named CIP1, for Cdk-interacting protein 1 (Harper et al. 1993). p21 was cloned using an expression screen designed to identify inhibitors of DNA synthesis from senescent fibroblasts, and it was named SDI1 (senescent cell-derived inhibitor (Noda et al. 1994). Using subtractive hybridization, the p21 cDNA was also isolated based on its increased expression in human melanoma cells that were induced to differentiate, and it was termed MDA-6, for melanoma differentiation associated protein (Jiang et al. 1994).

Thus far, seven CKIs have been identified, and they can be divided into two families. p21 is similar to p27 (Polyak et al. 1994; Toyoshima and Hunter 1994) and p57 (Lee et al. 1995; Matsuoka et al. 1995) at the amino terminus, and these CKIs have a broad specificity for Cdks. p16 (Serrano et al. 1993; Kamb et al. 1994), p15 (Hannon and Beach 1994), p18, and p19 (Chan et al. 1995; Hirai et al. 1995) form a second family with a more restricted Cdk specificity. While most of these CKIs act by inhibiting Cdks, p21 has been demonstrated to negatively regulate cell-cycle progression by additional mechanisms. In addition to inhibiting Cdk activity, it has been shown to bind and inhibit activity of proliferating cell nuclear antigen (PCNA) (reviewed in Kelman 1997) and the

[1] Department of Genetics, M/C 669, Molecular Biology Research Building, University of Illinois at Chicago, Chicago, IL 60607, USA

Progress in Molecular and Subcellular Biology, Vol. 20
A. Macieira-Coelho (Ed.)

stress-activated protein kinases or c-Jun amino-terminal kinases (SAPKS/JNKs) (Shim et al. 1996). p21 expression is induced by DNA damage, cellular damage leading to necrosis, a variety of cytokines and growth factors, and is coincident with cell differentiation in many cases. Induction of p21 expression is controlled by both 53-dependent and -independent pathways. In this chapter we summarize recent findings about p21, its regulation, and its role in the cell cycle.

2 p21-Induced Growth Arrest Is Mediated Through Protein-Protein Interactions

2.1 p21 Promotes Cell-Cycle Arrest via Interactions with Cdks and PCNA

Cell-cycle progression is controlled by the Cdks, which are the catalytic partners of the cyclins (reviewed in Morgan 1995). Regulators of G1 progression in mammalian cells include D-type cyclins (D1, D2, D3) and cyclin E (Sherr and Roberts 1995). D-type cyclins are associated with either Cdk4 or Cdk6, while cyclin E is associated with Cdk2 (Bates et al. 1994). These cyclins play important, but probably distinct, roles in the regulation of the G1-S phase transition (Dulic et al. 1992). p21 effectively inhibits Cdk2, Cdk3, Cdk4, and Cdk6, which have a direct role in the G1/S transition, but it is a poor inhibitor of other known Cdks (Harper et al. 1995).

Substrates of the G1 Cdks include the retinoblastoma protein (Rb) and other Rb-related proteins (Reznitzky and Reed 1995). Rb hyperphosphorylation results in the release of E2F, enabling it to activate genes important for S-phase entry. Cdk-2 also plays a key role in regulating early events in DNA replication, including phosphorylation of the single-strand DNA-binding protein RP-A, which can be inhibited by p21 (Cardoso et al. 1993; Fang and Newport 1993; Adachi and Laemmli 1994; Jackson et al. 1995; Yan and Newport 1995).

In normal human diploid fibroblasts, p21 has been shown to be part of quaternary complexes that contain cyclin, Cdk, and proliferating cell nuclear antigen (PCNA), a processivity factor for DNA polymerase δ (Polδ) (Zhang et al. 1993). In these complexes, cyclin kinases may exist in both active and inactive forms (Zhang et al. 1994). The transition between active and inactive states occurs through changes in p21 stoichiometry, particularly when multiple p21 molecules versus a single p21 molecule bind to these complexes (Zhang et al. 1994; Harper et al. 1995). Inhibitory binding by p21 may prevent phosphorylation of Cdks by Cdk-activating kinase (CAK, also called Cdk7) by making the target inaccessible (Aprelikova et al. 1995). The affinity of p21 for Cdks is notably increased when the Cdk is associated with a cyclin, probably because of conformational changes that occur after formation of cyclin/Cdk complexes (Harper et al. 1995). p21 also interacts directly with cyclins, and this interaction is important for its growth-inhibitory activities (I.T. Chen et al. 1996; Lin et al. 1996).

Recently, it has been shown that p21 may act as an assembly factor for Cdk/cyclin complexes. p21 promotes the assembly of Cdk4 with D-type cyclins and in some cells the association of p21 with active cyclinD/Cdk4 complexes is a normal stage of cell-cycle progression (Hiyama et al. 1997). p21 increases the affinity of Cdk4 and cyclin-D1 by 35-fold, and targets Cdk4/cyclin complexes to the nucleus. The role of p21 as an assembly activator or inhibitor depends on its expression level; at low and intermediate concentrations it is an assembly factor, while at high concentrations it is an inhibitor.

PCNA plays important roles in nucleic acid metabolism, DNA replication, and DNA excision repair (reviewed in Kelman 1997). It is assembled into a trimeric ring around the DNA and it forms a sliding clamp that tethers the polymerase to the DNA during DNA synthesis. p21 can bind to PCNA, and when present in excess stoichiometry, can block its ability to activate Polδ, affecting DNA replication through this interaction (Flores-Rozas et al. 1994; Li et al. 1994; Waga et al. 1994). The crystal structure of a 22-residue carboxy terminal p21 peptide and PCNA indicates that p21 binds to PCNA in a 1:1 stoichiometry (Gulbis et al. 1996). Podust et al. found that the Polδ core and p21 compete directly for binding to PCNA (Podust et al. 1995). The structural studies also suggest that p21 may prevent interaction between PCNA and Polδ, as p21 appears to block sites on PCNA that are important for interaction with the polymerase subunit (Gulbis et al. 1996).

In vitro data regarding the role of p21 in regulating DNA repair have been contradictory. In extracts prepared from the RKO human colon carcinoma cell line and healthy human lymphoblasts, addition of recombinant p21 did not influence repair of a UV-irradiated plasmid, and in a reconstituted system dependent upon purified PCNA, p21 inhibition appeared to preferentially inhibit long transcripts (Li et al. 1994). PCNA-dependent DNA repair also proceeded in *Xenopus* egg extracts in the presence of excess p21 (Shivji et al. 1994). Since p21 appears to inhibit the synthesis of long DNA templates, but not of short DNA fragments as required for repair (Flores-Rozas et al. 1994; Li et al. 1994; Podust et al. 1995), p21-bound PCNA may still be competent for repair. In contrast, repair of DNA damaged by UV irradiation or alkylating agents was inhibited when p21 was added to HeLa cell extracts in a two to four fold stoichiometry relative to PCNA; PCNA addition reversed this inhibition (Pan et al. 1995). p21 was also shown to inhibit mismatch repair through its binding to PCNA (Umar et al. 1996). p21 appears to compete with the protein Fen1, a 5′ nuclease that removes RNA primers during DNA replication and is important for DNA repair in yeast, for binding to PCNA (U. Chen et al. 1996; Warbrick et al. 1997). In vivo experiments have also provided evidence for a repair defect in p21-deficient cells; lower levels of reporter gene expression were detected following transfection of damaged constructs into p21−/− cells when compared with p21+/+ cells (McDonald et al. 1996). At this point, further study is still required to clarify the role of p21 in repair.

2.2 Proteins Other than Cdks and PCNA that Bind p21

p21 has been shown to bind and inhibit the stress-activated protein kinases (SAPKs), also known as c-jun amino terminal kinases (JNKs) (Shim et al. 1996). JNK phosphorylation of c-jun is important for the activation of many immediate early gene promoters containing AP-1 sites. p21 interacts with and inhibits JNK in vitro and in vivo (Shim et al. 1996), and an additional function for p21 may be the ability to attenuate the SAPK group of mitogen-activated protein kinases (MAPKs).

p21 also interacts with Gadd45 (growth arrest and DNA damage) (Kearsey et al. 1995), which has several features in common with p21. Its expression is induced in response to UV irradiation in a p53-dependent manner (Canman et al. 1995), and it interacts with PCNA and can compete with p21 for binding to PCNA (I.T. Chen et al. 1995). In an ELISA-based pepscan assay, peptides spanning amino acids 1 to 20, 141 to 160, and 144 to 164 of human p21 interact with Gadd45 (Kearsey et al. 1995). The carboxy terminal region of p21 that binds Gadd45 overlaps with, but is probably distinct from, the region of p21 that interacts with PCNA (Warbrick et al. 1995). In a yeast two-hybrid system, Gadd45 also interacted with the amino terminus of p21 (Kearsey et al. 1995). It is possible that Gadd45 may regulate the availability of p21 and modulate its ability to interact with PCNA.

Protein kinase CK2 (also known as casein kinase II) is a ubiquitous serine/threonine protein kinase present in the nucleus and cytoplasm. In vitro, p21 was shown to bind to the regulatory β-subunit of CK2, but not to the catalytic α-subunit (Gotz et al. 1996). Activity of the CK2 holoenzyme was also found to be reduced in the presence of p21 (Guerra et al. 1997).

2.3 Protein Interaction Domains of p21

The p21 cDNA encodes a 164-amino acid (aa) protein, and the Cdk and PCNA inhibitory activities have been mapped to different domains of the protein using a variety of cDNA expression constructs. The amino terminus of p21 binds cyclins (aa 21–26) and Cdks (aa 49–71), while the carboxy terminus binds PCNA (aa 124–164) (Y. Chen et al. 1995; Luo et al. 1995; Nakanishi et al. 1995b; Warbrick et al. 1995; Zakut and Givol 1995; Y. Chen et al. 1996; Lin et al. 1996). The PCNA-binding activity was localized to 8 residues which, when contained within a 20-amino acid peptide, retained the ability to bind PCNA and inhibit SV40 DNA replication in vitro (Warbrick et al. 1995). The amino terminus (aa 1–84) was also able to inhibit SAPK activity, but less well than the full-length p21 protein (Shim et al. 1996).

While transfection of either the amino or carboxy terminal regions of p21 into R-1B/L17 cells resulted in inhibition of DNA synthesis, the amino terminus of p21 was as effective as the entire protein, but the carboxy terminus had only 25 to 40% of the inhibitory effect (Luo et al. 1995). Suppression of tumor cell growth appears to be regulated by the amino terminal half of p21 (Chen J et al. 1995; Hall et al. 1995; Luo et al. 1995; Nakanishi et al. 1995b; Zakut and Givol 1995). Amino acids 49–71 were found to bind Cdk2, but sequences from amino acids 22 to 71 were found to be required for inhibition of DNA synthesis (Nakanishi et al. 1995). A cyclin-binding site was mapped to amino acids 21 through 24 in a stretch of sequence conserved among p21, p27, and p57 (Chen IT et al. 1996; Lin et al. 1996). Mutant p21 proteins with defects in residues 21 to 24 failed to suppress tumor cell growth in culture (Lin et al. 1996). Deletion of residues 17 through 24 were also shown to produce a p21 protein that failed to suppress cell growth (Chen J et al. 1996).

The presence of sequences related to the p21 cyclin-binding domain in other cell-cycle regulators suggests that p21 and related CKIs may act by blocking interactions of the cyclin-Cdk complexes with their substrates. An eight-residue peptide derived from E2F1 was able to block both E2F and p21 binding to cyclin A and E-Cdk complexes (Adams et al. 1996). Interactions of p21 and the Rb-related protein p107 with cyclin A/Cdk2 or cyclin E/Cdk2 are mutually exclusive, and sequences required for p107 interaction were mapped to a short sequence motif with homology to p21 (Zhu et al. 1995). The cyclin-binding domain of p21 also shares similarity with a domain in Cdc25A, a phosphatase which removes inhibitory phosphate groups from Cdc2 (Saha et al. 1997). A p21 peptide consisting of the cyclin-binding domain can disrupt interaction between Cdc25A and cyclin E- or cyclin A-Cdk2 (Saha et al. 1997). p21 has also been shown to disrupt interaction between Cdk2 and the E2F-p130 complex (Shiyanov et al. 1996).

3
p21 Expression During Development and Differentiation

p21 expression in the developing mouse embryo is correlated with cell-cycle arrest that precedes terminal differentiation in a variety of tissues (Parker et al. 1995). p21 transcripts were first detected in the day 8.5 postcoitum (pc) embryo in the midline of the neural tube and hindgut. p21 transcripts were later detected in postmitotic muscle cells in the myotome by day 10 pc, and in a variety of terminally differentiating epithelia between days 13 to 15 pc. Although expression of p53 was detected in all cells of the 8.5 to 10.5-pc embryo

(Schmid et al. 1991), p21 expression was more restricted and detected only in differentiating cells, demonstrating that p53 alone is not sufficient for induction of p21. Expression of p21 appeared normal in embryonic tissues of mice lacking a functional p53 gene (p53−/−), indicating that induction of p21 transcription during development of most tissues does not require p53 (Parker et al. 1995).

Expression of p21 in tissues of the adult mouse was also localized to terminally differentiating cells. In most tissues examined except the spleen, expression of p21 is p53-independent (Macleod et al. 1995; Parker et al. 1995). Highest levels of p21 expression were found in tissues with high turnover rates, such as the skin and linings of the gastrointestinal tract (Huppi et al. 1994; Macleod et al. 1995; Gartel et al. 1996). The intestinal epithelium continuously regenerates, requiring rapid proliferation of large numbers of cells in the crypts. Differentiation of the four principal intestinal lineages is accompanied by cell-cycle arrest and induction of high levels of p21 transcription. p21 transcripts and protein have been localized to differentiated cells of the small intestinal villus epithelium and upper crypt epithelium in the colon (El-Deiry et al. 1995; Parker et al. 1995; Gartel et al. 1996). This striking compartmentalization of p21 to the nonproliferating cells becomes disorganized early during the development of benign adenomas (El-Deiry et al. 1995).

The intestinal epithelium was used as a model for examining the role of p21 in regulating differentiation and cell proliferation in vivo. Chimeric mice composed of $p21^{-/-}$ and $p21^{+/+}$ cells were generated and intestinal development and differentiation were studied (Brugarolas et al. 1995). No changes in morphogenesis, differentiation, or cell proliferation could be detected in $p21^{-/-}$ epithelial cells. Mice deficient in p21 in all tissues were also found to develop normally up to 7 months of age (Deng et al. 1995). Thus, although p21 appears to play an important role in regulating cell-cycle progression, its activity may be functionally redundant in the animal.

The correlation of p21 induction with terminal differentiation and the cessation of proliferation has been also been demonstrated in a variety of in vitro systems (Michieli et al. 1994; Steinman et al. 1994; Halevy et al. 1995; Missero et al. 1995; Gartel et al. 1996). In a few in vitro systems, p21 actually appears to induce or be required for differentiation to take place. The hormonal form of vitamin D activates the p21 promoter and leads to the expression of terminal differentiation markers in the myelomonocytic U937 cell line. Transient overexpression of p21 and/or p27 in these cells in the absence of vitamin D also resulted in induction of terminal differentiation (Liu et al. 1996(b)). In another study, ectopic exprssion of p21 or its N-terminal domain (aa 1 to 75) led to differentiation of the CMK megakaryoblastic leukemia cell line (Matsumura et al. 1997). When primary keratinocytes derived from p21-deficient mice were induced to differentiate in vitro, expression of a subset of differentiation markers was significantly reduced (Missero et al. 1996). These in vitro data suggest a possible role for p21 during differentiation, which was not evident from analyses of the p21 knockout mice.

4 Growth Suppression by p21

p21 has been shown to inhibit proliferation both in vitro and in vivo. Introduction of p21 expression constructs into normal (Harper et al. 1993) and tumor cell lines (El-Deiry et al. 1993) results in cell-cycle arrest in G1 (Harper et al. 1995). Transfection of p21 into the SW480 colon carcinoma cell line in an expression vector conferring hygromycin resistance resulted in strong growth suppression, as demonstrated by a 10 to 20-fold decrease in the number of hygromycin-resistant colonies (El-Deiry et al. 1993). Overexpression of p21 in a malignant melanoma cell line resulted in accumulation of cells in G0/G1 phase of cell cycle and cell differentiation (Yang et al. 1995). Infection of lung and breast normal and tumor cell lines with recombinant adenovirus expressing p21 led to a decline in the percentage of cells in S phase, and an accumulation of cells in G1 phase (Katayose et al. 1995). Overexpression of p21 in colon and prostate carcinoma cell lines led to suppression of tumor cell proliferation and anchorage-independent cell growth (Chen YQ et al. 1995). When p53 or p21 genes were introduced into p53-deficient mouse prostate cancer cells, p21 was much more effective for growth inhibition in vivo and in vitro than wild-type p53 (Eastham et al. 1995). Apparently, these two inhibitors do not necessarily behave identically, and p21 may be a stronger proliferation inhibitor in some cell types. Expression of p21 antisense RNA in growth-arrested cells resulted in entry into S phase, further demonstrating the importance of p21 for maintaining of cells in G0 and G1 (Nakanishi et al. 1995a).

Several cases illustrate the ability of p21 to inhibit tumor growth in vivo (Y. Chen et al. 1995; Eastham et al. 1995; Yang et al. 1995). Following introduction of a viral p21 expression construct into murine renal carcinoma cells, growth of tumors was suppressed when these cells were inoculated into mice (Yang et al. 1995). Injection of p21 expressing adenovirus into tumor nodules in mice also resulted in striking tumor regression (Yang et al. 1995). Inducible expression of p21 in colon and prostate carcinoma cell lines resulted in a significant decrease in tumor formation when these cells were injected into nude mice (Y. Chen et al. 1995).

In normal cells, expression of oncogenic ras does not lead to transformation, but to G1 arrest accompanied by accumulation of p53, p21, and p16 (Lloyd et al. 1997; Serrano et al. 1997). Inactivation of p53 as well as introduction of E1A or SV40 large T-antigen prevented activation of p21 and cell-growth arrest and led to cell proliferation (Lloyd et al. 1997; Serrano et al. 1997). A similar effect was observed after inactivation of p16 (Serrano et al.

1997). Activation of p21 by ras through raf and p53 could be a protective mechanism of the cell designed to allow it to escape tumorigenesis. Experiments with p21-deficient keratinocytes support this idea. When primary p21 −/− keratinocytes were transformed with a ras oncogene and introduced into nude mice, aggressive tumors formed; this was not observed with control wild-type keratinocytes (Missero et al. 1996).

Further evidence that p21 is important for growth arrest following stimulation of the MAPK pathway came from studies with nerve growth factor. Nerve growth factor (NGF) inhibited growth of NIH 3T3 cells expressing TrkA, and this correlated with the induction of p21 protein expression and a 10–20-fold increase in levels of p21 complexed with Cdk2 and Cdk4 (Decker 1995). Treatment of these cells with the MAPK kinase (MEK)/MAPK pathway inhibitor PD98059 reversed growth inhibition by NGF (Pumiglia and Decker 1997). Activated raf also induced p21 expression and growth arrest in NIH 3T3 cells, and this effect again was reversed by PD98059. These data suggest that the MEK/MAPK pathway negatively regulates Cdk activity and mediates cell-cycle arrest through activation of p21 (Pumiglia and Decker 1997).

Transforming growth factor-β (TGF-β) inhibits proliferation of many cell types, including epithelial cells (Derynck 1994). This is, in part, due to its ability to regulate the Cdk inhibitors p21 and p27 (Kip1) and p15 (Reynisdottir et al. 1995). p21 expression is induced by TGF-β by p53-independent pathways in a human keratinocyte cell line (Datto et al. 1995a) and ovarian cancer cell lines (Elbendary 1994). Basic fibroblast growth factor (bFGF) is a classical mitogen, but it inhibits proliferation of MCF-7 and other breast tumor cells. The stimulation of MCF-7 cells by bFGF led to induction of p21 mRNA and protein and results in inactivation of cyclin-D1-Cdk4, cyclin E-Cdk2, and cyclin A-Cdk2 complexes (H. Wang et al. 1997).

Expression of human p21 posterior to the morphogenetic furrow in the *Drosophila* eye imaginal disk inhibited mitosis that gives rise to precursors of photoreceptor cells R1, R6, and R7, the cone cells, the pigment cells, and precursors of the sensory bristles (de Nooij and Hariharan 1995). The decrease in the number of precursor cells available as a result of p21 inhibition of mitosis resulted in the absence of some cell types, although cell fate determination itself did not appear to be influenced by ectopic expression of p21.

Inhibition of hepatocyte proliferation in vivo by p21 has been demonstrated in transgenic mice, by ectopic expression of p21 in the liver using transthyretin regulatory elements (Wu et al. 1996). Hepatocytes expressing p21 consistently failed to divide, and did not respond to partial hepatectomy; instead, oval cells from the stem-cell compartment, which did not express the transgene, and isolated nodules of non-p21-expressing hepatocytes, which perhaps had deleted or inactivated the transgene, were responsible for most of the growth and regeneration observed. Many of the remaining nuclei labeled with bromodeoxyuridine after partial hepatectomy were abnormally large, as if arrested at some intermediate point in the cell cycle.

p21 expression has been shown to increase during liver regeneration in response to both partial hepatectomy (Albrecht et al. 1997) and carbon tetrachloride toxicity (Serfas et al. 1997). Carbon tetrachloride-induced liver damage leads to two waves of p53-independent p21 expression. The first appears rapidly, between 4 and 8h after carbon tetrachloride administration in pericentral hepatocytes that will die by necrosis. The second wave occurs at 1 to 2 days, and is consistent with postmitotic exprssion. Both waves of p21 expression probably represent barriers to cell division, as the signals that regulate liver growth during regeneration must include factors that stop cell proliferation, once the appropriate size has been attained (Serfas et al. 1997).

Normal cells have a finite life-span in culture. After a fixed number of generations, they will stop dividing and eventually die, by a process known as cellular senescence. Increased expression of p21 in senescent cells led to its cloning as a possible regulator of growth arrest associated with senescence (Noda et al. 1994). In senescent cells, p21 was found complexed with the E2F transcription factor and it was found that p21 can inhibit E2F driven transcription (Afshari et al. 1996). Disruption of the p21 gene by homologous recombination led to a significant increase in the life-span of cultured cells, providing evidence that p21 is a regulator of senescence (Brown et al. 1997).

5
p21 Expression and Apoptosis

Conflicting results obtained by a number of investigators regarding the role of p21 in apoptosis suggest that its role may depend on multiple factors, including the expression of other genes, the cell type, and growth conditions. Using isolated thymocytes from p21 null mice, p21 was found not to be required for p53-dependent apoptosis of γ-irradiated thymocytes (Dent et al. 1995). Several groups have data that suggest that p21-induced growth arrest actually protects cells from apoptosis. Irradiation of Baf-3 murine hematopoietic cells in the presence of interleukin-3 (IL-3) induces G1 arrest, while radiation in the absence of IL-3 results in apoptotic cell death (Canman et al. 1995). No significant changes in the levels of Bcl-2, Bcl-x_L, or Bax proteins were associated with IL-3 withdrawal, but a decrease in the absolute levels of p21 was noted, although p21 was induced by p53 after irradiation. These data led to the conclusion that p21 inhibits apoptosis in this system and also provides a possible explanation for why p21 is mutated so rarely in human tumors. Mutation and inactivation of p21 in cells with wild-type p53 would increase the tendency of a cell to undergo apoptosis and, therefore, this cell would not have a growth advantage during tumorigenesis (Canman et al. 1995).

Growth factor withdrawal from murine C2C12 myoblasts induces differentiation and extensive cell death (Wang and Walsh 1996). During differentiation, the appearance of an apoptosis-resistant phenotype correlated with expression of p21. Transient overexpression of p21 inhibited apoptosis during myocyte differentiation in wild-type (Wang and Walsh 1996), but not in Rb−/−myocyte cultures (J. Wang et al. 1997).

RKO human colorectal carcinoma cells normally undergo apoptosis in response to cyclopentenone prostaglandin A2 (PGA_2) and exhibit low levels of p21. In contrast, NIH 3T3 cells (Hitomi et al. 1996) and MCF-7 cells express high levels of p21 and undergo G1 arrest in response to PGA_2. Similar p53-independent upregulation of p21 and G1 arrest were observed after treatment of MCF-7 cells with phenylacetate (Gorospe et al. 1996a). In both cases, reduction of endogenous p21 expression in MCF-7 cells by antisense p21 RNA overexpression attenuates the growth arrest and promotes cell death. Similarly, the treatment of differentiating neuroblastoma cells with p21 antisense oligonucleotides decreased both expression of p21 protein and cell survival (Poluha et al. 1996). Interestingly, overexpression of p21 in an adenoviral vector in RKO cells rescued these cells from PGA_2-induced apoptotic cell death (Gorospe et al. 1996b).

Polyak and colleagues identified two types of colorectal cancer cell lines with different responses to overexpression of exogenous p53. In response to p53, type A cell lines were found to undergo growth arrest, while type D cell lines underwent apoptosis (Polyak et al. 1996). Inactivation of p21 by homologous recombination in A-type cells converts them to D-type cells, suggesting that p21 protects A-type cells from apoptosis. p53 overexpression was also highly toxic for p21-deficient mouse embryonal fibroblasts, but ectopic expression of exogenous p21 protected cells from p53-induced apoptosis (Gorospe et al. 1997).

In contrast to the examples cited above, several groups have found apoptosis accompanied by expression of p21, and in some cases p21 even appeared to induce apoptosis. In MCF-7 cells, okadaic acid (OA) induced apoptosis that was coupled with overexpression of endogenous p53 and p21, with no evidence of G1 arrest. p53 did not appear necessary, because OA also induced apoptosis and induction of p21 in variants of MCF-7 cells where p53 function was disrupted (Saeed Sheikh et al. 1996). In contrast to these data, nitric oxide (NO) induced apoptosis and p21 only in human cancer cells that contain wild-type p53 (Ho et al. 1996). Similarly, extensive apoptosis accompanied by increased levels of p53 and p21 occurs in central nervous system of mouse embryos with targeted deletion of the Rb gene (Macleod et al. 1996).

Addition of the genotoxic drug adriamycin to HTLV-1-transformed lymphocytes induced both p53-dependent and p53-independent apoptosis. In both cases, apoptosis was accompanied by induction of p21 mRNA and protein (Gartenhaus et al. 1996). In malignant glioma cells, p21 was induced during p53-dependent apoptosis by cisplatin, and overexpression of p21 in these cells led to apoptosis without cisplatin treatment. However, induction of apoptosis by cisplatin in another glioma cell line with inactive p53 did not lead to p21 upregulation (Kondo et al. 1996).

Expression of the Wilms tumor suppressor gene WT1 in osteosarcoma cells led to apoptosis, which was preceded by p53-independent induction of p21 and cell-cycle arrest (Englert et al. 1997). A novel retinoid AHPN inhibits proliferation of the retinoic acid (RA)-resistant, estrogen receptor (ER)-negative leukemic cell line HL-60R, and induces p53-independent apoptosis in these cell lines. Significantly, before apoptosis in these cells, very rapid activation of p21 and GADD45 mRNAs was observed (Hsu et al. 1997). In serum-deprived quiescent mouse 3T3 fibroblasts, apoptosis is accompanied by increased levels of p21 mRNA and protein, and apoptosis was delayed after stable introduction of a p21 antisense construct (Duttaroy et al. 1997).

In the N417 small cell lung carcinoma cell line that contains inactive p53 and Rb, overexpression of exogenous p53 led to induction of p21 and apoptosis, but not to G1 arrest (Adachi et al. 1996). A similar effect was observed in Saos-2 and HeLa cells which also have inactive p53 and Rb after overexpression of wild-type p53 (Yonish-Rouach 1996). In contrast, introduction of exogenous p21 expression in the p53 null cell lines Saos2 and H1299 led only to growth arrest, while expression of exogenous p53 in one of the sublines led to a comparable level of p21 expression and growth arrest and subsequent apoptosis. Here p21-mediated growth arrest was not sufficient to protect cells from apoptosis, because eventually all cells from this subline underwent cell death (X. Chen et al. 1996).

Removal of serum from cultured mammary epithelial cells which express high levels of c-myc and cyclin D1 when grown in the absence of a basement membrane resulted in the induction of p21 expression and apoptosis. Apoptosis also occurred after overexpression of p21 in cycling cells, or following overexpression of c-myc in quiescent cells (Boudreau et al. 1996). To explain this, it was proposed that a combination of opposing signals which promote cell proliferation and growth arrest may lead to apoptosis (Boudreau et al. 1996).

6 Regulation of p21 Expression

6.1 p53-Dependent Regulation of p21 Transcription

The p53 tumor-suppressor protein is a transcription factor required for the transactivation of a number of genes involved in growth control (Clarke et al.

1993; Lowe et al. 1993). Sequence comparison of 4.5 kb of upstream sequence of the rat, mouse, and human genes revealed conservation of two p53 recognition sequences. At least one of the p53 sites was required for p53 responsiveness (El-Deiry et al. 1995). Inactivation of wild-type p53 function can lead to a growth advantage and tumor progression. p21 was identified in a screen for genes activated by p53 (El-Deiry et al. 1993). It has been demonstrated that DNA damage (irradiation) activates p21 transcription in a p53-dependent manner in human diploid fibroblasts (Dulic et al. 1994) and human thyroid epithelial cells (Namba et al. 1995), leading to cell-cycle arrest. Following exposure of mice to ionizing radiation, induction of p21 was also found to be p53-dependent in most tissues, with the exception of the intestine, where high levels of p21 are constitutively expressed (Macleod et al. 1995).

The redundancy of pathways leading to cell-cycle arrest following DNA damage was demonstrated following disruption of the p21 gene in mice (Brugarolas et al. 1995; Deng et al. 1995). The same number of cells died in intestinal crypts derived from both p21$^{-/-}$ and p21$^{+/+}$ cells, following irradiation of chimeric mice (Brugarolas et al. 1995). There was also a similar decrease in proliferation in the intestinal crypts regardless of the p21 status. These data indicated that p21 is not required for DNA damage-induced cell death and cell-cycle arrest in the mouse intestine, although p53 appears to play a role in normal response to radiation in this tissue (Clarke et al. 1994; Merritt et al. 1994).

In contrast, cultured p21-deficient mouse embryonic fibroblasts were compromised in their ability to undergo G1 arrest in response to DNA damage (Brugarolas et al. 1995; Deng et al. 1995). p21$^{-/-}$ cells had a phenotype intermediate between p53$^{-/-}$ and wild-type cells, suggesting that p53 may induce an additional gene that participates in cell-growth arrest. Homozygous deletion of p21 in human colon carcinoma cell line HCT116 completely abrogated the G1 checkpoint following γ-irradiation (Waldman et al. 1995), suggesting that p21 is solely responsible G1 arrest in these cells. The absolute requirement of p21 for G1 exit as a result of γ-irradiation in human cells, versus the partial requirement for it in murine embryonic fibroblasts, may be explained by differences in species and cell types.

p53-dependent induction of p21 has also been observed without DNA damage. The correlation between p21 RNA and protein levels and expression of wild-type p53 was examined in breast epithelial cells transfected with a human papilloma virus E6 protein expression construct (Gudas et al. 1995). The E6 protein binds p53 and induces its degradation through an ubiquitin-dependent pathway (Scheffner et al. 1993). Decreased levels of both p53 and p21 protein were detected in E6 protein-expressing clones, although in some cases p21 RNA levels remained high (Gudas et al. 1995).

Zta (also known as BZLF1, EB1, Zebra), the immediate-early lytic switch transactivator of Epstein-Barr virus (EBV), plays a critical role in the initiation of the lytic cascade of EBV. Zta induces cell-cycle arrest in several epithelial cell lines through posttranscriptional upregulation of p53 and subsequent transcriptional activation of p21 (Cayrol and Flemington 1996).

Neu differentiation factor (NDF)/heregulin (HRG), a ligand of erbB-3 and erbB-4, stimulates tyrosine phosphorylation of its receptors, inhibits proliferation, and induces differentiation in breast cancer cell lines. It upregulates p53 expression by stabilizing the protein and induces p21 expression by a p53-dependent mechanism, because p21 was observed only in cell lines with wild-type p53 (Bacus et al. 1996). These data suggest that breast cancer cell growth inhibition via erbB receptor activation includes stabilization of p53 protein and transcriptional activation of p21.

Ribonucleotide biosynthesis inhibitors, such as PALA, pyrazofurin, cyclopentenylcytosine, and others, induce p53-dependent p21 expression in the absence of DNA damage, resulting in hypophosphorylation of Rb in normal human fibroblasts and cell growth arrest (Linke et al. 1996). Thus, p53 may serve as a metabolite sensor activated by depletion of ribonucleotides, but the exact mechanism of p53 activation is unknown.

6.2 p53-Independent Activation of p21 Promoter

p21 expression is often activated during differentiation in a p53-independent manner. In differentiating muscle cells, the basic helix-loop-helix transcription factor MyoD regulates p21 expression during muscle cell differentiation. Expression of both p21 RNA and protein was induced during differentiation of C2 muscle cells, but not in 10T1/2 fibroblasts (Halevy et al. 1995; Missero et al. 1995). Forced expression of cyclin D1 inhibits MyoD-mediated transactivation of muscle-specific genes, but this inhibition can be overcome by overexpression of p21 (Skapek et al. 1995). Stable introduction of MyoD into 10T1/2 cells converts these cells to myoblasts and leads to p21 induction (Halevy et al. 1995; Parker et al. 1995). Although high levels of p21 expression could be detected in wild-type embryonic fibroblasts when they were grown in both high and low serum media, expression of p21 in p53-deficient fibroblasts could be detected only when cells were infected with a retroviral MyoD expression construct and grown in low-serum differentiation-promoting conditions. In this case, p21 induction appeared MyoD-dependent and did not require p53 (Halevy et al. 1995).

Signal transduction initiated by ligand binding to cellular receptors often involves the JAK-STAT pathway (reviewed in Ihle 1996). Three STAT transcription factor binding sites are present in the p21 promoter. Proteins in nuclear extracts from EGF treated A431 cells expressing high levels of STAT1 and STAT3 bandshift and supershift probes for these sits in a manner consistent with STAT1 homodimer binding at the 5′ sites, and binding of STAT1/STAT3 heterodimer or either homodimer at the 3′ site (Chin et al. 1996). Likewise, γ-interferon stimulation of three cell types results in STAT1 induction and STAT1 complex formation at all three sites. In each of these cases, p21 is induced and replication is inhibited.

Abnormal activation and nuclear translocation of STAT1, as a result of the constitutive tyrosine kinase activity of a mutated fibroblast growth-factor receptor (FGFR), results in induction of p21 expression and cell-cycle arrest. In this case, aberrant expression of p21 may be responsible for growth retardation in bone development and may induce thanatophoric dysplasia, a common form of dwarfism in humans that is caused by different mutations in FGFR (Su et al. 1997). A novel hematopoietic growth factor, thrombopoietin (TPO), also induces megakaryocytic differentiation and transcriptional activation of p21 by STAT5 through the two distal STAT-binding sites identified in p21 promoter (Matsumura et al. 1997).

Activation of p21 transcription by the hormonal form of vitamin D occurs in a p53-independent manner in U937 cells (Liu et al. 1996b). A functional vitamin D3-responsive element has been identified, centered at −771 bp of the p21 promoter. p21 was also transcriptionally activated by retinoic acid (RA) in U937 cells, and a functional RA-responsive element was localized between −1212 and −1194 bp in the p21 promoter (Liu et al. 1996a).

Phorbol ester, 12-0-tetradecanoyl-phorbol-13-acetate (TPA) activates p21 expression and induces cell-cycle arrest in hematopoietic cell lines, and phosphatase inhibitor okadaic acid (OA) induces p21 and growth arrest in different cell types. The treatment of p53-null human leukemia cells by OA caused cell-growth arrest and monocyte/macrophage differentiation and concomitant induction of p21 (Zhang et al. 1995). TPA induces transcription initiation of p21 and OA induces transcription initiation as well as posttranscriptional stabilization (Zeng and E1-Deiry 1996). Protein kinase C (PKC) inhibitor, staurosporine, interferes with cell-cycle arrest and p21 induction by TPA, suggesting that TPA needs PKC for p21 activation (Zeng and E1-Deiry 1996). It was shown that TPA and OA both activate p21 expression through an AP-2 consensus binding site located between −121 and −95 bp from transcription initiation site and AP-2 binds to this site (Zeng et al. 1997). Overexpression of AP2 in HepG2 human hepatoblastoma cells and SW480 human colon adenocarcinima cells inhibited cell division and nearly 75% of stable colony formation (Zeng et al. 1997).

Sp1 is a zinc-finger DNA-binding protein that has been shown to play an important role in the regulation of p21 transcription. It mediates induction of the p21 gene in response to phorbol ester and okadaic acid (Biggs et al. 1996). Both PMA and OA induce differentiation of human leukemic cells U937 toward macrophages, and stimulate p21 transcription via the region between

−154 and +16 bp of the p21 promoter. This region contains five consensus Sp1 sites. Deletion or mutation of two sites between −117 and −100 bp reduces the response of the p21 promoter to PMA 3–5-fold and to OA 5–10-fold. The same sites were crucial for activation of p21 promoter by exogenous Sp1 in Sp1-deficient *Drosophila* SL2 cells (Biggs et al. 1996).

Addition of nerve growth factor (NGF) to PC12 cells induces neuronal differentiation and p21 expression. In these cells, transient expression of the adenovirus E1A protein inhibits induction of p21, and overexpression of the p300 transcriptional coactivator reversed its effect. Stable expression of E1A inhibited p21 expression, cell-cycle arrest and neuronal differentiation (Billon et al. 1996). Similar results were obtained when examining p53-independent activation of p21 transcription in cultured keratinocytes which differentiate in response to calcium (Missero et al. 1995). Deletion-mapping studies showed that both Sp1 and Sp3 sites located near the TATA box are responsible for basal activity of p21 promoter in keratinocytes. Only the transcription factor Sp3 was involved in the induction of the p21 promoter during keratinocyte differentiation (Prowse et al. 1997). In these cells, induction of p21 during differentiation was abolished by expression of the adenovirus E1A oncoprotein. Overexpression of p300 suppressed the E1A effect, independent of its binding to E1A. In the absence of E1A, exogenous p300 could not significantly activate p21 expression (Missero et al. 1995).

The sequences between −83 and −54 bp of the p21 promoter are sufficient for induction of transcription by TGF-β (Datto et al. 1995b). This region contains three Sp1-binding sites, and mutation of each of these sites diminished activation by TGF-β. Gel shift assays with the probe from −84 to −74 bp demonstrated that several transcription factors, including Sp1 and Sp3, bind this element in vitro. This sequence was also sufficient to drive TGF-β-mediated transcription from previously nonresponsive promoter (Datto et al. 1995b). Activins are members of the TGF-β family, and activin A induces G1 cell-cycle arrest in mouse B cell hybridoma cells. Cell-cycle arrest was caused by suppression of cyclin D2 and activation of p21 on transcriptional and posttranscriptional levels (Yamato et al. 1997).

Induction of CCAAT/enhancer binding protein-α (C/EBPα) expression in HT1 cells results in the inhibition of cell proliferation and a 12- to 20-fold increase in levels of p21 protein (Timchenko et al. 1996). The dramatic induction of p21 protein levels by C/EBPα was shown to be due to a transient threefold increase in p21 mRNA levels, followed by stabilization of p21 protein.

Deregulated homeobox gene expression may lead either to cell proliferation or to inhibition of cell growth. The gax homeobox gene is expressed in adult cardiovascular tissues, including vascular smooth muscle cells (VSMC). Overexpression of gax with a replication-deficient adenovirus vector led to cell-cycle arrest of VSMCs and fibroblasts, and p53-independent activation of p21. p21 was associated with Cdk2, and Cdk2 activity decreased. Gax overexpression in p21$^{-/-}$ fibroblasts did not result in cell-growth arrest or a decrease in Cdk2 activity (Duttaroy et al. 1997).

Bromodeoxyuridine (BrdUrd) inhibits growth and activity of cyclin A-Cdk2 complexes in human melanoma cells. Inhibition of growth was due to p53-independent activation of p21, and a decrease in cyclin A and PCNA. BrdUrd did not affect Cdk2 or cyclin D1 in these cells (Strasberg Rieber et al. 1996). Cyclin D1 may activate p21 transcription through an E2F1-dependent mechanism (Hiyama et al. 1997). We have found that E2F1 and E2F3, but not E2F4 activate the p21 promoter via the short 117-bp region close to start of transcription that contains multiple Sp1-binding sites (A.L. Gartel and A.L. Tyner, unpubl. data).

It was shown earlier that interferons (IFNs) have cell growth-inhibitory activity. Recently, this effect was attributed to p21 induction in different cell types. In two hematopoietic cell lines, Daudi and U-266, IFN-α activated p53-independent transcription of p21 and cell-growth arrest (Sangfelt et al. 1997). In these cell lines, induction of p21 RNA was accompanied by an increase in p21 protein and a decrease in Cdk2 activity. In a third cell line, H9, p21 RNA was also upregulated after IFN-α addition, but p21 protein was not expressed and there was no growth arrest (Sangfelt et al. 1997). IFN-α treatment of the prostate cancer cell line DU145, which expresses mutant p53, inhibited growth and colony formation and cell-cycle progression from G1 to S phase. IFN-α induced p21 expression, increased the amounts of p21 complexed with Cdk2, and reduced Cdk2 activity (Hobeika et al. 1997). Addition of IFN-β and the antileukemic compound mezerin to HO-1 human melanoma cells induces growth arrest and differentiation. This is accompanied by induction of p21 at the mRNA and protein levels, while a decrease in the amount of p53 protein was observed (Jiang et al. 1994).

In p53-negative hematopoietic and hepatoma cells, p21 was induced during differentiation after these cell lines were exposed to butyrate or other differentiation agents (Steinman et al. 1994). In quiescent embryonic p53-deficient fibroblasts, p21 is induced as an immediate early gene by serum or a variety of purified growth factors including PDGF, FGF, and EGF (Michieli et al. 1994). Nanomolar concentrations of epidermal growth factor (EGF) induce p21 and cell growth arrest in A431 squamous carcinoma cells, where growth arrest was induced via inhibition of Cdk2 and Cdk6 activities by p21 and binding of p21 to PCNA (Jakus and Yeudall 1996).

Treatment of TE85 osteosarcoma and MDAH041 Li-Fraumeni patient-derived cell lines, containing no functional p53, with hydroxyurea or hydrogen peroxide leads to transcriptional activation of p21 and growth arrest (Johnson et al. 1994). Administration of a novel synthetic retinoid, AHPN, to the human breast cell line MDA-MB-231 that contains mutant p53, also induces G0/G1 arrest and simultaneous upregulation of p21 (Shao et al. 1995). In p53 null cell line HL-60 monocytic differentiation is induced by TNF-α, which also induces p53-independent transcriptional activation of p21 (Yoshida et al. 1996). TNF-α treatment causes a six fold increase in p21 mRNA stability and a moderate increase in transcription rate in KG-1 myeloid leukemic cells, an effect which does not require protein synthesis (Shiohara et al. 1996).

In the MCF-7 mammary epithelial cell line, the protein kinase inhibitors and antimitogenic agents H7 and staurosporine induced p21 expression via a p53-independent mechanism (Jeoung et al. 1995). The protein kinase C inhibitor and anticancer drug UCN-01 (7-hydroxyl-staurosporine) induced G1 arrest in the A431 squamous carcinoma cell line, which was linked to p53-independent activation of p21 and dephosphorylation of Cdk2 and Rb proteins (Akiyama et al. 1997).

6.3 Repression of p21 Promoter Activity

In both NIH 3T3 cells and immortalized human keratinocytes, the E5 proteins of human papilloma viruses type 11 and type 16 repressed activity of p21 promoter. The repressor activities of E5 mutants correlated with their transforming activities. The E5 proteins probably repress p21 promoter activity through c-jun, since c-jun is activated by E5 and c-jun represses the p21 promoter (Tsao et al. 1996).

Transient expression of adenovirus E1A protein inhibits p53-independent induction of p21 in NGF-treated PC12 cells (Billon et al. 1996) and in calcium-treated keratinocytes (Missero et al. 1995), and overexpression of p300 reversed its effect. The viral oncoprotein E1A blocks also TGF β-mediated (Datto et al. 1997) and p53-mediated (Zeng et al. 1997) induction of p21 through its p300/CBP-interacting region. These data show that p300 is necessary for transcriptional activity of p53 (Lill et al. 1997) and that E1A efficiently represses p53-dependent and -independent activation of p21.

Recently, a novel Rb and p130-binding protein termed HBP1 (HMG-box protein 1) was cloned using a yeast two-hybrid screen (Tevosian et al. 1997). HBP1 is a sequence-specific HMG box containing transcriptional repressor that can down-regulate transcription from the N-MYC promoter. We have found that this HMG box protein can also repress p21 transcription by acting through the multiple Sp1-binding sites adjacent to the TATA box (A.L. Gartel and A.L. Tyner, unpubl. data).

The bcl-2 gene, which usually prevents apoptosis, suppressed p53-dependent activation of p21 in the MCF 10A breast epithelial cell line, but not mitogen-mediated induction (Upadhyay et al. 1995). Since bcl-2 can sequester p53 in the cytoplasm during G1, this finding is not unexpected (Ryan et al. 1994).

7
p21 and Transcriptional Activation of Downstream Genes

Cellular damage, serum factors, phorbol esters, and okadaic acid activate both p21 and the nuclear transcription factor NF-kB, and induction of both of these proteins is associated with growth arrest and cellular differentiation. Interestingly, transfection of p21 alone or p21 and p300 strongly activated NF-kB-driven transcription in Jurkat T-leukemia cells. Simultaneous expression of IkB α inhibited stimulation by p21, and simultaneous expression of RelA enhanced stimulation of NF-kB-dependent gene activation by p21 (Perkins et al. 1997).

Accumulation of wild-type p53 protein in murine or human cells induced cyclin D1 synthesis. A similar effect was observed after overexpression of p21 in the same cells. The mechanism of transcriptional activation of cyclin D1 by p21 is currently unknown (X. Chen et al. 1995). p21 also suppressed the activity of E2F responsive promoters, including the E2F1 promoter in cells with or without functional Rb (Dimri et al. 1996).

8
p21 and Cancer

Since p21 has properties of a tumor-suppressor protein, it was hypothesized that it may be mutated in human cancers. Analysis of the status of p21 in a number of tumor cell lines and tumors of different origin revealed polymorphisms in the p21 gene (Chedid et al. 1994; Shiohara et al. 1994; Law and Deka 1995; Li et al. 1995; Mousses et al. 1995). A frequent polymorphism is a single-base change of C to A at codon 31 that converts a serine to an arginine (Lukas et al. 1997). This mutation was not associated with the loss of the tumor-suppressor activity of p21 (Chedid et al. 1994) and was not associated with a predisposition to developing colorectal cancer (Li et al. 1995).

Shiohara et al. examined the status of p21 in 351 DNA samples from 14 different kinds of human malignancies and 36 human transformed cell lines using single-strand conformation polymorphism analysis of PCR amplified regions of these DNAs, and no somatic mutations were found (Shiohara et al. 1994). These data were confirmed later for lung and pancreatic cancers (Shimizu et al. 1996), malignant melanomas (Vidal et al. 1995), brain tumors (Koopman et al. 1995; Tsumanuma et al. 1997), colorectal cancers (Li et al. 1995), renal cancers (Papandreou et al. 1997), and hepatocellular carcinomas (Furutani et al. 1997).

While they appear infrequent, some mutations or deletions of the p21 gene in tumors have been reported. Five out of 40 thyroid papillary carcinoma specimens had deletions of the p21 gene (Shi et al. 1996). Another example was found in prostate tumors, where four mutations were found in the p21 gene in 3 patients, when the status of p21 was examined in 18 prostate cancer patients with wild-type p53. Two of these mutations were single-base insertions which would produce a frame shift and protein truncation at codon 35 (Gao et al. 1995). This truncated p21 protein would not be able to inhibit either PCNA or Cdks. The other two mutations were a silent mutation in codon 39 and a substitution of glutamine for proline in codon 4 (Gao et al. 1995). Only few additional mutations or deletions of p21 in different types of cancers that could affect the ability of p21 to suppress cell growth have been described. A phenylalanine to leucine mutation at codon 63 was identified in the Burkitt's lymphoma cell line DH978. This mutation was shown to affect the growth-inhibitory properties of p21 in transfection experiments (Bae et al. 1995). A mutation of Arg at position 94 to Trp in the p21 protein that was identified in breast carcinoma impaired the ability of p21 to inhibit Cdks (Balbin et al. 1996). A deletion of 37 amino acids from codon 65 to 101 of the p21 protein was found in a human adrenocortical adenoma (Iida et al. 1997).

Although the p21 gene is activated by p53, it does not appear that p21 a primary role in the tumor-suppressor function of p53. The p53 gene is mutated in a wide variety of human cancers (Hollenstein et al. 1991), and mice lacking a functional p53 gene develop spontaneous tumors at an early age (Donehower et al. 1992; Harvey et al. 1993; Jacks et al. 1994). In contrast, mutations in the p21 gene are rare, and mice lacking a functional p21 gene do not have a propensity for developing tumors (Brugarolas et al. 1995; Deng et al. 1995). The data suggest that other targets of the p53 gene may be responsible for its tumor-suppression properties.

While mutation of the p21 gene appears relatively rare in human carcinomas, changes in the expression pattern of p21 in tumors were more widespread phenomena. In dysplastic aberrant crypt foci and in small colorectal adenoma, decreased p21 immunostaining and loss of compartmentalization of p21 and Ki67 expression was observed (Polyak et al. 1996). It was also found that reduced expression of p21 colorectal cancers coincided with later stages, while cancers with higher p21 expression represented early stages of tumor progression (A. Wang et al. 1997). In breast tumors (Bukholm et al. 1997) and hepatocellular carcinomas (Hui et al. 1997), a strong correlation was found between abnormalities in p53 and lack of p21 expression.

Somewhat paradoxical results were reported by Jung and coauthors (Jung et al. 1995). They found that p21 expression was low in normal and reactive non-neoplastic brain tissue. In contrast, the majority of gliomas of all malignancy grades with either wild-type or mutant p53 expressed increased levels of p21 (Jung et al. 1995). It was not determined whether the p21 gene is mutated in gliomas. Although the Cdk inhibitor p16 was found frequently deleted in gliomas, no deletion or rearrangement of p21 was seen in 13 primary brain tumors. The authors proposed that expression of high levels of p21 in gliomas may be the result of a feedback mechanism designed to halt proliferation, and this may account for the slow proliferative rate of gliomas (Jung et al. 1995). Overexpression of p21 was also observed in human cutaneous malignant melanomas (Trotter et al. 1997) and nonsmall cell lung carcinomas (Marchetti et al. 1996).

Acknowledgments. Research performed by the authors was supported by NIH Grant DK48836 (A. L. T.) and an award from the American Cancer Society, Illinois Division (A. L. G.). We thank Michael S. Serfas for his helpful comments.

References

Adachi J, Ookawa K, Shiseki M, Okazaki T, Tsuchida S, Morishita K, Yokota J (1996) Induction of apoptosis but not G1 arrest by expression of the wild-type p53 gene in small cell lung carcinoma. Cell Growth Differ 7:879–886

Adachi Y, Laemmli UK (1994) Study of the cell cycle-dependent assembly of the DNA pre-replication centres in *Xenopus* egg extracts. EMBO J 13:4153–4164

Adams PD, Sellers WR, Sharma SK, Wu AD, Nalin CM, Kaelin W Jr (1996) Identification of a cyclin-cdk2 recognition motif present in substrates and p21-like cyclin-dependent kinase inhibitors. Mol Cell Biol 16:6623–6633

Afshari CA, Nichols MA, Xiong Y, Mudryl M (1996) A role for a p21-E2F interaction during senescence arrest of normal human fibroblasts. Cell Growth Differ 7:979–988

Akiyama T, Yoshida T, Tsujita T, Shimizu M, Mizukami T, Okabe M, Akinaga S (1997) G1 phase accumulation induced by UCN-01 is associated with dephosphorylation of Rb and CDK2 proteins as well as induction of CDK inhibitor p21/Cip1/WAF1/Sdi1 in p53-mutated human epidermoid carcinoma A431 cells. Cancer Res 57:1495–1501

Albrecht JH, Meyer AH, Hu MY (1997) Regulation of cyclin-dependent kinase inhibitor p21 (WAF1/Cip1/Sdi1) gene expression in hepatic regeneration. Hepatology 25:557–563

Aprelikova O, Xiong Y, Liu ET (1995) Both p16 and p21 families of cyclin dependent kinase inhibitors block the phosphorylation of cyclin-dependent kinases by the CDK-activating kinase. J Biol Chem 270:18195–18197

Bacus SS, Yarden Y, Oren M, Chin DM, Lyass L, Zelnick CR, Kazarov A, Toyofuku W, Gray-Bablin J, Beerli RR, Hynes NE, Nikiforov M, Haffner R, Gudkov A, Keyomarsi K (1996) Neu differentiation factor (Heregulin) activates a p53-dependent pathway in cancer cells. Oncogene 12:2535–2547

Bae I, Fan S, Bhatia K, Kohn KW, Fornace AJ, O'Connor PM (1995) Relationships between G1 arrest and stability of the p53 and p21Cip1/Waf1 proteins following gamma-irradiation of human lymphoma cells. Cancer Res 5585:2387–2393

Balbin M, Hannon GJ, Pendas AM, Ferrando AA, Vizoso F, Fueyo A, Lopez-Otin C (1996) Functional analysis of a p21 WAF1, CIP1, SDI1 mutant (Arg94 → Trp) identified in a human breast carcinoma. Evidence that the mutation impairs the ability of p21 to inhibit cyclin-dependent kinases. J Biol Chem 271:15782–15786

Bates S, Bonetta L, Macallan D, Parry D, Holder A, Dickson C, Peters G (1994) CDK6 and CDK4 are distinct subset of the cyclin-dependent kinases that associate with cyclin D1. Oncogene 9:71–79

Biggs JR, Kudlow JE, Kraft AS (1996) The role of the transcription factor Sp1 in regulating the expression of the WAF1/CIP1 gene in U937 leukemic cells. J Biol Chem 271:901–906

Billon N, van Grunsven LA, Rudkin BB (1996) The CDK inhibitor p21WAF1/Cip1 is induced through a p300-dependent mechanism during NGF-mediated neuronal differentiation of PC12 cells. Oncogene 13:2047–2054

Boudreau N, Werb Z, Bissell MJ (1996) Suppression of apoptosis by basement membrane requires three-dimensional tissue organization and withdrawal from the cell cycle. Proc Natl Acad Sci USA 93:3509–3513

Brown JP, Wei W, Sedivy JM (1997) Bypass of senescence after disrupton of $p21^{CIP1/WAF1}$ gene in normal diploid human fibroblasts. Science 277:831–834

Brugarolas J, Chadrasekaran C, Gordon J, Beach D, Jacks T, Hannon G (1995) Radiation-inuced cell cycle arrest compromised by p21 deficiency. Nature 377:552–557

Bukholm IK, Nesland JM, Karesen R, Jacobsen U, Borresen AL (1997) Relationship between abnormal p53 protein and failure to express p21 protein in human breast carcinomas. J Pathol 181:140–145

Canman CE, Gilmer TM, Coutts SB, Kastan MB (1995) Growth factor modulation of p53-mediated growth arrest versus apoptosis. Genes Dev 9:600–611

Cardoso MC, Leonhardt H, Nadal-Ginard B (1993) Reversal of terminal differentiation and control of DNA replication: cyclin A and Cdk2 specifically localize at subnuclear sites of DNA replication. Cell 74:979–992

Cayrol C, Flemington EK (1996) The Epstin-Barr virus bZIP transcription factor Zta causes G0/G1 cell cycle arrest through induction of cyclin-dependent kinase inhibitors. EMBO J 15:2748–2759

Chan FK, Zhang J, Cheng L, Shapiro DN, Winoto A (1995) Identification of human and mouse p19, a novel CDK4 and CDK6 inhibitor with homology to p16ink4. Mol Cell Biol 15:2682–2688

Chedid M, Michieli P, Lengel C, Huppi K, Givol D (1994) A single nucleotide substitution at codon 31 (Ser/Arg) defines a polymorphism in a highly conserved region of the p53-inducible gene WAF1/CIP1. Oncogene 9:3021–3024

Chen IT, Smith ML, O'Connor PM, Fornance AJJ (1995) Direct interaction of Gadd45 with PCNA and evidence for competitive interaction of Gadd45 with $p21^{Waf1/Cip1}$ with PCNA. Oncogene 11:1931–1937

Chen IT, Akamatsu M, Smith ML, Lung FD, Duba D, Roller PP, Fornace A Jr, O'Connor PM (1996) Characterization of p21Cip1/Waf1 peptide domains required for cyclin E/Cdk2 and PCNA interaction. Oncogene 12:595–607

Chen J, Jackson PK, Kirschner MW, Dutta A (1995) Separate domains of p21 involved in the inhibition of Cdk kinase and PCNA. Nature 374:386–388

Chen J, Saha P, Kornbluth S, Dynlacht BD, Dutta A (1996) Cyclin-binding motifs are essential for the function of p21CIP1. Mol Cell Biol 16:4673–4682

Chen W, Ko LJ, Jayaraman L, Prives C (1996) p53 levels, functional domains, and DNA damage determine the extent of the apopotic response of tumor cells. Genes Dev 10:2438–2451

Chen X, Bargonetti J, Prives C (1995) p53, through p21(WAF1/CIP1), induces cyclin D1 synthesis. Cancer Res 55:4257–4263

Chen U, Chen S, Saha P, Dutta A (1996) p21 Cip1/Waf1 disrupts the recruitment of human Fenl by proliferating-cell nuclear antigen into the DNA replication complex. Proc Natl Acad Sci USA 93(21):11597–11602

Chen YQ, Cipriano SC, Arenkiel JM, Miller FR (1995) Tumor suppression by p21 (WAF1) Cancer Res 55:4536–4539

Chin YE, Kitagawa M, Su W-CS, You Z-H, Iwamoto Y, Fu X-Y (1996) Cell growth arrest and induction of cyclin-dependent kinase inhibitor p21 WAF1/CIP1 mediated by STAT1. Science 272:719–722

Clarke AR, Purdie CA, Harrison DJ, Morris RG, Bird CC, Hooper ML, Wyllie AH (1993) Thymocyte apoptosis induced by p53-dependent and -independent pathways. Nature 362:849–852

Clarke AR, Gledhill S, Hoper ML, Bird CC, Wyllie AH (1994) p53 dependence of early apoptotic and proliferative responses within the mouse intestinal epithelium following gamma irradiation. Oncogene 9:1767–1773

Datto M, Li Y, Panus J, Howe D, Xiong Y, Wang X (1995a) Transforming growth factor β induces the cyclin-dependent kinase inhibitor p21 through a p53-independent mechanism. Proc Natl Acad Sci USA 92:5545–5549

Datto MB, Yu Y, Wang X-F (1995b) Functional analysis of the transforming growth factor beta-responsive elements in the WAF1/Cip1/p21 promoter. J Biol Chem 270:28623–28628

Datto MB, Hu PP, Kowalik TF, Yingling J, Wang XF (1997) The viral oncoprotein E1A blocks transforming growth factor beta-mediated induction of p21/WAF1/Cip1 and p15/INK4B. Mol Cell Biol 17:2030–2037

Decker SJ (1995) Nerve growth factor-induced growth arrest and induction of p21 Cip1/Waf1 in NIH-3T3 Cells expressing TrkA. J Biol Chem 270:30841–30844

Deng C, Zhang P, Harper J, Elledge S, Leder P (1995) Mice lacking p21 undergo normal development, but are defective in G1 checkpoint control. Cell 82:675–684

de Nooij JC, Hariharan IK (1995) Uncoupling cell fate determination from patterned cell division in the *Drosophila* eye. Science 270:983–985

Derynck R (1994) TGF-β-receptor-mediated signaling. TIBS 19:548–553

Dimri GP, Nakanishi M, Desprez PY, Smith JR, Campisi J (1996) Inhibition of E2F activity by the cyclin-dependent protein kinase inhibitor p21 in cells expressing or lacking a functional retinoblastoma protein. Mol Cell Biol 16:2987–2997

Donehower LA, Harvey M, Slagle BL, McArthur MJ, Montgomery CA, Butel JS, Bradley A (1992) Mice deficient for p53 are developmentally normal but susceptible to spontaneous tumors. Nature 356:215–221

Dulic V, Lees E, Reed S (1992) Association of human cyclin E with a periodic G1-S phase protein kinase. Science 257:1958–1961

Dulic V, Kaufman WK, Wilson SJ, Tlsty TD, Lees E, Harper JW, Elledge SJ, Reed SI (1994) p53-dependent inhibition of cyclin-dependent kinase activities in human fibroblasts during radiation-induced G1 arrest. Cell 76:1013–1023

Duttaroy A, Qian JF, Smith JS, Wang E (1997) Up-regulated P21CIP1 expression is part of the regulation quantitatively controlling serum deprivation-induced apoptosis. J Cell Biochem 64:434–446

Eastham JA, Hall SJ, Sehgal I, Wang J, Timme TL, Yang G, Connell-Crowley L, Elledge SJ, Zhang WW, Harper JW, Thompson TC (1995) In vivo gene therapy with p53 or p21 adenovirus for prostate cancer. Cancer Res 55:5151–5155

Elbendary E (1994) Transforming growth factor β can induce CIP1/WAF1 expression independent of the p53 pathway in ovarian cancer cells. Cell Growth Differ 5:1301–1307

El-Deiry W, Tokino T, Velculescu V, Levy D, Parsons R, Trent J, Lin D, Mercer W, Kinzler K, Vogelstein B (1993) WAF1, a potential mediator of p53 tumor suppression. Cell 75:817–825

El-Deiry WS, Tokino T, Waldman T, Oliner JD, Velculscu VE, Burrel M, Hill DE, Healy E, Rees JL, Hamilton SR, Kinzler KW, Vogelstein B (1995) Topological control of p21 (WAF1/CIP1) expression in normal and neoplastic tissues. Cancer Res 55:2910–2919

Englert C, Maheswaran S, Garvin AJ, Kreidberg J, Haber DA (1997) Induction of p21 by the Wilms' tumor suppressor gene WT1. Cancer Res 57:1429–1434

Fang F, Newport JW (1993) Distinct roles of cdk2 and cdc2 in RP-A phosphorylation during the cell cycle. J Cell Sci 106:983–994

Flores-Rozas H, Kelman Z, Dean FB, Pan ZQ, Harper JW, Elledge SJ, O'Donnell MO, Hurwitz J (1994) Cdk-interacting protein 1 directly binds with proliferating cell nuclear antigen and inhibits DNA replication catalyzed by the DNA polymeras delta holoenzyme. Proc Natl Acad Sci USA 91:8655–8659

Furutani M, Arii S, Tanaka H, Mise M, Niwano M, Harada T, Higashitsuji H, Imamura M, Fujita J (1997) Decreased expression and rare somatic mutation of the CIP1/WAF1 gene in human hepatocellular carcinoma. Cancer Lett 111:191–197

Gao X, Chen YQ, Wu N, Grignon DJ, Sakr W, Porter AT, Honn KV (1995) Somatic mutations of the WAF1/CIP1 gene in primary prostate canner. Oncogene 11:1395–1398

Gartel AL, Serfas MS, Gartel M, Goufman E, Wu GS, El-Deiry WS, Tyner AL (1996) p21 (WAF1/CIP1) expression is induced in newly nondividing cells in diverse epithelial and during differentiation of the Caco-2 intestinal cell line. Exp Cell Res 227:171–181

Gartenhaus RB, Wang P, Hoffmann P (1996) Induction of the WAF1/CIP1 protein and apoptosis in human T-cell leukemia virus type I-transformed lymphocytes after treatment with adriamycin by using a p53-independent pathway. Proc Natl Acad Sci USA 93:265–268

Gibbs E, Kelman Z, Gulbis JM, O'Donnell M, Kuriyan J, Burgers P, Hurwitz J (1997) The influence of the proliferating cell nuclear antigen-interacting domain of p21 (CIP1) on DNA synthesis catalyzed by the human and *Saccharomyces cerevisiae* polymerase delta holoenzymes. J Biol Chem 272:2373–2381

Gorospe M, Shack S, Guyton K, Samid D, Holbrook N (1996a) Up-regulation and functional role of p21 (WAF1/CIP1) during growth arrest of human breast carcinima MCF-7 cells by phenylacetate. Cell Growth Differ 7:1609–1615

Gorospe M, Wang X, Guyton K, Holbrook N (1996b) Protective role of p21 (WAF1/CIP1) against prostaglandin A2-mediated apoptosis of human colorectal carcinima cells. Mol Cell Biol 16:6654–6660

Gorospe M, Cirielli C, Wang X, Seth P, Capogrossi M, Holbrook N (1997) P21 (WAF1/CIP1) protects against p53-mediated apoptosis of human melanoma cells. Oncogene 14:929–935

Gotz C, Wagner P, Issinger OG, Montenarh M (1996) p21 WAF1/CIP1 interacts with protein kinase CK2. Oncogene 13:391–398

Gudas J, Nguyen H, Li T, Hill D, Cowan K (1995) Effects of cell cycle, wild-type p53 and DNA damage on p21 CIP1/WAF1 expression in human breast epithelial cells. Oncogene 11:253–261

Guerra B, Gotz C, Wagner P, Montenarh M, Issinger OG (1997) The carboxy terminus of p53 mimics the polylysine effect of protein kinase CK2-catalyzed MDM2 phosphorylation. Oncogene 14:2683–2688

Gulbis JM, Kelman Z, Hurwitz J, O'Donnell M, Kuriyan J (1996) Structure of the C-terminal region of p21 (WAF1/CIP1) complexed with human PCNA. Cell 87:297–306

Halevy O, Novitch BG, Spicer DB, Skapek SX, Rhee J, Hannon GJ, Beach D, Lassar AB (1995) Correlation of terminal cell-cycle arrest of skeletal muscle with induction of p21 by MyoD. Science 267:1018–1021

Hall M, Bates S, Peteres G (1995) Evidence for different modes of action of cyclin-dependent kinase inhibitors: p15 and p16 bind to kinases, p21 and p27 bind to cyclins. Oncogene 11:1581–1588

Hannon GJ, Beach D (1994) $p15^{ink4b}$ is a potential effector of cell cycle arrest mediated by TGF-β. Nature 371:257–261

Harper JW, Adami G, Wei N, Keyomarsi K, Elledge S (1993) The p21 cdk-interacting protein Cip1 is a potent inhibitor of G1 cyclin-dependent kinases. Cell 75:805–816

Harper JW, Elledge SJ, Keyomarsi K, Dynlacht B, Tsai L, Zhang P, Dobtowolski S, Bai C, Connell-Crowley L, Swindell E, Fox MP, Wei N (1995) Inhibition of cyclin-dependent kinases by p21. Mol Biol Cell 6:387–400

Harvey M, McArthur MJ, Montgomery CA Jr, Butel JS, Bradley A, Donehower LA (1993) Spontaneous and carcinogen-induced tumorigenesis in p53-deficient mice. Nat Genet 5:225–229

Hirai H, Roussel MF, Kato JY, Ashmun RA, Sherr CJ (1995) Novel INK4 proteins, p19 and p18, are specific inhibitors of the cyclin D-dependent kinases CDK4 and CDK6. Mol Cell Biol 15:2672–2681

Hitomi M, Shu J, Strom D, Hiebert SW, Harter ML, Stacey DW (1996) Prostaglandin A2 blocks the activation of G1 phase cyclin-dependent kinase without altering mitogen-activated protein kinase stimulation. J Biol Chem 271:9376–9383

Hiyama H, Iavarone A, LaBaer J, Reeves SA (1997) Regulated ectopic expression of cyclin D1 induces transcriptional activation of the cdk inhibitor p21 gene without altering cell-cycle progression. Oncogene 14:2533–2542

Ho YS, Wang YJ, Lin JK (1996) Induction of p53 and p21/WAF1/CIP1 expression by nitric oxide and their association with apoptosis in human cancer cells. Mol Carcinogen 16:20–31

Hobeika AC, Subramaniam PS, Johnson HM (1997) IFNalpha induces the expression of the cyclin-dependent kinase inhibitor p21 in human prostate cancer cells. Oncogene 14:1165–1170

Hollenstein M, Sidransky D, Vogelstein B, Harris CC (1991) p53 mutation in human cancer. Science 253:49–53

Hsu CA, Rishi AK, Su-Li X, Gerald TM, Dawson MI, Schiffer C, Reichert U, Shroot B, Poirer GC, Fontana JA (1997) Retinoid-induced apoptosis in leukemia cells through a retinoic acid nuclear receptor-independent pathway. Blood 89:4470–4479

Hui AM, Kanai Y, Sakamoto M, Tsuda H, Hirohashi S (1997) Reduced p21 (WAF1/CIP1) expression and p53 mutation in hepatocellular carcinomas. Hepatology 25:575–579

Huppi K, Siwarski D, Dosik J, Michieli P, Chedid M, Reed S, Mock B, Givol D, Mushinski JF (1994) Molecular cloning, sequencing, chromosomal location and expression of mouse p21 (Wafl). Oncogene 9:3017–3020

Ihle JN (1996) STATs and MAPKs: obligate or opportunistic partners in signaling. Bioessays 18:95–98

Iida S, Fujii H, Moriwaki K (1997) A somatic mutation of the p21 (Wafl/Cipl) gene in a human adrenocortical adenoma. Anticancer Res 17:633–636

Jacks T, Remington L, Williams BO, Schmitt EM, Halachmi S, Bronson RT, Weinberg RA (1994) Tumor spectrum analysis in p53-mutant mice. Curr Biol 4:1–7

Jackson PK, Chevalier S, Philippe M, Kirschner MW (1995) Early events in DNA replication require cycin E and are blocked by $p21^{CIP1}$. J Cell Biol 130:755–769

Jakus J, Yeudall WA (1996) Growth inhibitory concentrations of EGF induce p21 (WAF1/Cip1) and alter cell cycle control in squamous carcinoma cells. Oncogene 12:2369–2376

Jeoung DI, Tang B, Sonenberg M (1995) Induction of tumor suppressor p21 protein by kinase inhibitors in MCF-7 cells. Biochem Biophys Res Commun 214:361–366

Jiang H, Liu J, Su Z, Collart F, Huberman E, Fisher P (1994) Induction of differentiation in human promyelocytic HL-60 leukemia cells activates p21, WAF1/CIP1, expression in the absence of p53. Oncogene 9:3397–3406

Jiang H, Lin J, Su Z, Herlyn M, Kerbel R, Weissman B, Welch D, Fisher P (1995) The melanoma differentiation-associated gene mda-6, which encodes the cyclin-dependent kinase inhibitor p21, is differentially expressed during growth, differentiation and progression in human melanoma cells. Oncogene 10:1855–1864

Johnson M, Dimitrov D, Vojta P, Barrett J, Noda A, Pereira-Smith O, Smith J (1994) Evidence for a p53-independent pathway for upregulation of SDI1/CIP1/WAF!/p21 RNA in human cells. Mol Carcinogen 11:59–64

Jung J-M, Bruner JM, Ruan S, Langford LA, Kyritis AP, Kobayashi T, Levin VA, Zhang W (1995) Increased levels of $p21^{WAF1/Cip1}$ in human brain tumors. Oncogene 11:2021–2028

Kamb A, Gruis N, Weaver-Feldhaus J, Liu Q, Harshman K, Tavtigian S, Stockert E, Day III R, Johnson B, Skolnick M (1994) A cell-cycle regulator involved in genesis of many tumor types. Science 264:436–440

Katayose D, Wersto R, Cowan K, Seth P (1995) Consequences of p53 gene Expression by adenovirus vector on cell-cycle arrest and apoptosis in human aortic vascular smooth muscle cells. Biochem Biophys Res Commun 215:446–451

Kearsey JM, Coates PJ, Prescott AR, Warbrick E, Hall PA (1995) Gadd45 is a nuclear cell-cycle-regulated protein which interacts with $p21^{Cip1}$. Oncogene 11:1675–1681

Kelman Z (1997) PCNA: structure, functions and interactions. Oncogene 14:629–640

Kondo S, Barna BP, Kondo Y, Tanaka Y, Casey G, Liu J, Morimura T, Kaakaji R, Peterson JW, Werbel B, Barnett GH (1996) WAF1/CIP1 increases the susceptibility of p53 nonfunctional malignant glioma cells to cisplatin-induced apoptosis. Oncogene 13:1279–1285

Koopman J, Maintz D, Schild S, Schramm J, Lousi DN, Wiestler OD, von Deimling A (1995) Multiple polymorphisms, but no mutations, in the WAF1/CIP1 gene in human brain tumors. Br J Cancer 72:1230–1233

Law JC, Deka A (1995) Identification of a PstI polymorphism in the p21 Cip1/Waf1 cyclin-dependent kinase inhibitor gene. Hum Genet 95:459–460

Lee M, Reynisdottir I, Massague J (1995) Cloning of p57, a cyclin-dependent kinase inhibitor with unique domain structue and tissue distribution. Genes Dev 9:639–649

Li R, Waga S, Hannon G, Beach D, Stillman B (1994) Differential effects by the p21 CDK inhibitor on PCNA dependent DNA replication and repair. Nature 371:534–537

Li YJ, Laurent-Puig P, Salmon RJ, Thomas G, Hamelin R (1995) Polymorphisms and probable lack of mutation in the WAF1-CIP1 gene in colorectal cancer. Oncogene 10:599–601

Lill NL, Grossman SR, Ginsberg D, DeCaprio J, Livingston DM (1997) Binding and modulation of p53 by p300/CBP coactivators. Nature 387:823–827

Lin J, Reichner C, Wu X, Levine AJ (1996) Analysis of wild-type and mutant $p21^{WAF\text{-}1}$ gene activities. Mol Cell Biol 16:1786–1793

Linke SP, Clarkin KC, Di Leonardo A, Tsou A, Wahl GM (1996) A reversible, p53-dependent G0/G1 cell-cycle arrest induced by ribonucleotide depletion in the absence of detectable DNA damage. Genes Dev 10:934–947

Liu M, Iavarone A, Freedman LP (1996a) Transcriptional activation of the human p21 (WAF1/CIP1) gene by retinoic acid receptor. Correlation with retinoid induction of U937 cell differentiation. J Biol Chem 271:31723–31728

Liu M, Lee M-H, Cohen M, Bommakanti M, Freedman LP (1996b) Transcriptional activation of the Cdk inhibitor p21 by vitamin D_3 leads to the induced differentiation of the myelomonocytic cell line U937. Genes Dev 10:142–153

Lloyd AC, Obermuller F, Staddon S, Barth CF, McMahon M, Land H (1997) Cooperating oncogenes converge to regulate cyclin/cdk complexes. Genes Dev 11:663–677

Lowe SW, Schmidt EM, Smith SW, Osborne BA, Jacks T (1993) p53 is required for radiation induced apoptosis in mouse thymocytes. Nature 362:847–849

Lukas J, Groshen S, Saffari B, Niu N, Reles A, Wen WH, Felix J, Jones LA, Hall FL, Press MF (1997) WAF1/Cip1 gene polymorphism and expression in carcinomas of the breast, ovary, and endometrium. Am J Pathol 150:167–175

Luo Y, Hurwitz J, Massague J (1995) Cell-cycle inhibition by independent CDK and PCNA binding domains in p21. Nature 375:159–161

Macleod K, Sherry N, Hannon G, Beach D, Tokino T, Kinzler K, Vogelstein B, Jacks T (1995) p53-dependent and -independent expression of p21 during cell growth, differentiation, and DNA damage. Genes Dev 9:935–944

Macleod KF, Hu Y, Jacks T (1996) Loss of Rb activates both p53-dependent and -independent cell death pathways in the developing mouse nervous system. EMBO J 15:6178–6188

Marchetti A, Doglioni C, Barbareschi M, Buttitta F, Pellegrini S, Bertacca G, Chella A, Merlo G, Angeletti CA, Dalla Palma P, Bevilacqua G (1996) p21 RNA and protein expression in non-small cell lung carcinomas: evidence of p53-independent expression and association with tumoral differentiation. Oncogene 12:1319–1324

Matsumura I, Ishikawa J, Nakajima K, Oritani K, Tomiyama Y, Miyagawa J, Kato T, Miyazaki H, Matsuzawa Y, Kanakura Y (1997) Thrombopoietin-induced differentiation of a human megakaryoblastic leukemia cell line, CMK, involves transcriptional activation of p21 (WAF1/Cip1) by STAT5. Mol Cell Biol 17:2933–2943

Matsuoka S, Edwards M, Bai C, Parker S, Zhang P, Baldini A, Harper J, Elledge S (1995) p57, a structurally distinct member of the p21 cdk inhibitor family, is a candidate tumor suppressor gene. Genes Dev 9:650–662

McDonald ER 3rd, Wu GS, Waldman T, El-Deiry WS (1996) Repair defect in p21 WAF1/CIP1 −/− human cancer cells. Cancer Res 56:2250–2255

Merritt AJ, Potten CS, Kemp CJ, Hickman JA, Balmain A, Lane DP, Hall PA (1994) The role of p53 in spontaneous and radiation-induced apapotosis in the gastrointestinal tract of normal and p53-deficient mice. Cancer Res 5420:614–617

Michieli P, Chedid D, Lin D, Pierce J, Mercer W, Givol D (1994) Inhibition of WAF1/CIP1 by a p53-independent pathway. Cancer Res 54:3391–3395

Missero C, Calautti E, Eckner R, Chin J, Tsai L, Livingston D, Dotto GP (1995) Involvement of the cell-cycle inhibitor Cip1/WAF1 and the E1A-associated p300 protein in terminal differentiation. Proc Natl Acad Sci USA 92:5451–5455

Missero C, Di Cunto F, Kiyokawa H, Koff A, Dotto GP (1996) The absence of p21Cip1/WAF1 alters keratinocyte growth and differentiation and promotes ras-tumor progression. Genes Dev 10:3065–3075

Morgan DO (1995) Principles of CDK regulation. Nature 374:131–134

Mousses S, Ozcelik H, Lee PD, Malkin D, Bull SB, Andrulis IL (1995) Two variants of the CIP1/WAF1 gene occur together and are associated with human cancer. Hum Mol Genet 4:1089–1092

Nakanishi M, Adami G, Robetorye R, Noda A, Venable S, Dimitrov D, Pereira-Smith O, Smith J (1995a) Exit from Go and entry into the cell cycle of cells expressing p21 antisense RNA. Proc Natl Acad Sci USA 92:4352–4356

Nakanishi M, Robetorye RS, Adami R, Pereira-Smith OM, Smith JR (1995b) Identification of the active region of the DNA synthesis inhibitory gene p21Sdi1/CIP1/WAF1. EMBO J 14:555–563

Namba H, Hara T, Tukazaki T, Migita K, Ishikawa N, Ito K, Nagataki S, Yamashita S (1995) Radiation-induced G1 arrest is selectively mediated by the p53-WAF1/Cip1 pathway in human thyroid cells. Cancer Res 55:2075–2080

Noda A, Ning Y, Venable SF, Pereira-Smith OM, Smith JR (1994) Cloning of senescent cell-derived inhibitors of DNA synthesis using an expression screen. Exp Cell Res 211:90–98

Pan ZQ, Reardon JT, L. L, Flores-Rozas H, Legerski R, Sancar A, Hurwitz J (1995) Inhibition of nucleotide excision repair by the cyclin-dependent kinase inhibitor p21. J Biol Chem 270: 22008–22016

Papandreou CN, Bogenrieder T, Loganzo F, Albino AP, Nanus DM (1997) Expression and sequence analysis of the p21 (WAF1/CIP1) gene in renal cancers. Urology 49:481–486

Parker S, Eichele G, Zhang P, Rawls A, Sands A, Bradley A, Olson E, Harper J, Elledge S (1995) p53-independent expression of p21 in muscle and other terminally differentiating cells. Science 267:1024–1027

Perkins ND, Felzien LK, Betts JC, Leung K, Beach DH, Nabel GJ (1997) Regulation of NF-kappaB by cyclin-dependent kinases associated with the p300 coactivator. Science 275:523–527

Podust VN, Podust LM, Goubin F, Ducommun B, Hubscher U (1995) Mechanism of inhibition of proliferating cell nuclear antigen-dependent DNA synthesis by the cyclin-dependent kinase inhibitor p21. Biochemistry 34:8869–8875

Poluha W, Poluha DK, Chang B, Crosbie NE, Schonhoff CM, Kilpatrick DL, Ross AH (1996) The cyclin-dependent kinase inhibitor p21 (WAF1) is required for survival of differentiating neuroblastoma cells. Mol Cell Biol 16:1335–1341

Polyak K, Lee M, Erdjument-Bromage H, Koff A, Roberts J, Tempst P, Massague J (1994) Cloning of p27, a cyclin-dependent kinase inhibitor and a potential mediator of extracellular antimitogenic signals. Cell 78:59–66

Polyak K, Waldman T, He TC, Kinzler KW, Vogelstein B (1996) Genetic determinants of p53-induced apoptosis and growth arrest. Genes Dev 10:1945–1952

Prowse DM, Bolgan L, Molnar A, Dotto GP (1997) Involvement of the Sp3 transcription factor in induction of p21Cip1/WAF1 in keratinocyte differentiation. J Biol Chem 272:1308–1314

Pumiglia KM, Decker SJ (1997) Cell cycle arrest mediated by the MEK/mitogen-activated protein kinase pathway. Proc Natl Acad Sci USA 94:448–452

Reynisdottir I, Polyak K, Iavorne A, Massague J (1995) Kip/Cip and Ink4 Cdk inhibitors cooperate to induce cell cycle arrest in response to TGF-beta. Genes Dev 9:1831–1845

Reznitzky D, Reed S (1995) Different roles for cyclins D1 and E in regulation of the G1-to-S transition. Mol Cell Biol 15:3463–3469

Ryan JJ, Prochownik E, Gottlieb CA, Apel IJ, Merino R, Nunez G, Clarke MF (1994) c-myc and bcl-2 modulate p53 function by altering p53 subcellular trafficking during the cell cycle. Proc Natl Acad Sci 91:5878–5882

Saeed Sheikh M, Garcia M, Zhan Q, Liu Y, Fornace AJJ (1996) Cell cycle-independent regulation of p21(WAF1/CIP1) and retinoblastoma protein during okadaic acid-induced apoptosis is coupled with induction of Bax protein in human breast carcinoma cells. Cell Growth Differ 7:1599–1607

Saha P, Eichbaum Q, Siberman ED, Mayer BJ, Dutta A (1997) p21 (Cip1) and Cdc25A: competition between an inhibitor and an activator of cyclin-dependent kinases. Mol Cell Biol 17:4338–4345

Sangfelt O, Erickson S, Einhorn S, Grander D (1997) Induction of Cip/Kip and Ink4 cyclin dependent kinase inhibitors by interferon-alpha in hematopoietic cell lines. Oncogene 14:415–423

Scheffner M, Huibregste JM, Vierstra RD, Howley PM (1993) The HPV-16 E6 and E6-AP complex functions as a ubiquitin-protein ligase in the ubiquitination of p53. Cell 75:495–505

Schmid P, Lorenz A, Hameister H, Montenarh M (1991) Expression of p53 during mouse embryogenesis. Development 113:857–865

Serfas MS, Goufman E, Feuerman MH, Gartel AL, Tyner AL (1997) p53-independent induction of $p21^{WAF1/CIP1}$ expression in pericentral hepatocytes following carbon tetrachloride intoxication. Cell Growth Differ 8:951–961

Serrano M, Hannon G, Beach D (1993) A new regulatory motif in cell-cycle control causing specific inhibition of cyclinD/cdk4. Nature 366:704–707

Serrano M, Lin AW, McCurrach ME, Beach D, Lowe SW (1997) Oncogenic ras provokes premature cell senescence associated with accumulation of p53 and p16INK4a. Cell 88:593–602

Shao ZM, Dawson MI, Li XS, Rishi AK, Sheikh MS, Han QX, Ordonez JV, Shroot B, Fontana JA (1995) p53-independent G0/G1 arrest and apoptosis induced by a novel retinoid in human breast cancer cells. Oncogene 11:493–504

Sherr C, Roberts J (1995) Inhibitors of mammalian G1 cyclin-dependent kinases. Genes Dev 9:1149–1163

Shi Y, Zou M, Farid NR, al-Sedairy ST (1996) Evidence of gene deletion of p21 (WAF1/CIP1), a cyclin-dependent protein kinase inhibitor, in thyroid carcinomas. Br J Cancer 74:1336–1341

Shim J, Lee H, Park J, Kim H, Choi E (1996) A non-enzymatic p21 protein inhibitor of stress-activated protein kinases. Nature 381:804–807

Shimizu T, Miwa W, Nakamori S, Ishikawa O, Konishi Y, Sekiya T (1996) Absence of a mutation of the p21/WAF1 gene in human lung and pancreatic cancers. Jpn J Cancer Res 87:275–278

Shiohara M, El-Deiry WS, Wada M, Nakamaki T, Takeuchi S, Yang R, Chen DL, Vogelstein B, Koeffler HP (1994) Absence of WAF1 mutations in a variety of human malignancies. Blood 8484:3781–3784

Shiohara M, Akashi M, Gombart A, Yang R, Koeffler H (1996) Tumor necrosis factor alpha: posttanscriptional stabilization of WAF1 mRNA in p53-deficient human leukemic cells. J Cell Physiol 166:568–576

Shivji MK, Grey SJ, Strausfeld UP, Wood RD, Blow JJ (1994) Cip1 inhibits DNA replication but not PCNA-dependent nucleotide excision-repair. Curr Biol 4:1062–1068

Shiyanov P, Bagchi S, Adami G, Kokontis J, Hay N, Arroyo M, Morozov A, Raychaudhuri P (1996) p21 disrupts the interaction between cdk2 and the E2F-p130 complex. Mol Cell Biol 16:737–744

Skapek SX, Rhee J, Spicer DB, Lassar AB (1995) Inhibition of myogenic differentiation in proliferating myoblasts by cyclin D1-dependent kinase. Science 267:1022–1024

Steinman R, Hoffman B, Iro A, Guillouf C, Liebermann D, El-Houseini M (1994) Induction of p21(WAF-1/CIP1) during differentiation. Oncogene 9:3389–3396

Strasberg Rieber M, Welch DR, Miele ME, Rieber M (1996) p53-independent increase in p21WAF1 and reciprocal down-regulation of cyclin A and proliferating cell nuclear antigen in bromodeoxyuridine-mediated growth arrest of human melanoma cells. Cell Growth Differ 7: 197–202

Su WC, Kitagawa M, Xue N, Xie B, Garofalo S, Cho J, Deng C, Horton WA, Fu XY (1997) Activation of Stat 1 by mutant fibroblast growth-factor receptor in thanatophoric dysplasia type II dwarfism. Nature 386:288–292

Tevosian SG, Shih HH, Mendelson KG, Sheppard KA, Paulson KE, Yee AS (1997) HBP1: a HMG box transcriptional repressor that is targeted by the retinoblastoma family. Genes Dev 11:383–396

Timchenko NA, Wilde M, Nakanishi M, Smith JR, Darlington GJ (1996) CCAAT/enhancer-binding protein α (C/EBPα) inhibits cell proliferation through the p21 (WAF-1/CIP-1/SDI-1) protein. Genes Dev 10:804–815

Toyoshima H, Hunter T (1994) p27, a novel inhibitor of G1 cyclin/cdk protein kinase activity, is related to p21. Cell 78:67–74

Trotter MJ, Tang L, Tron VA (1997) Overexpression of the cyclin-dependent kinase inhibitor p21 (WAF1/CIP1) in human cutaneous malignant melanoma. J Cutan Pathol 24:265–271

Tsao YP, Li LY, Tsai TC, Chen SL (1996) Human papillomavirus type 11 and 16 E5 represses p21 (Wafl/Sdil/Cipl) gene expression in fibroblasts and keratinocytes. J Virol 70:7535–7539

Tsumanuma I, Tanaka R, Abe S, Kawasaki T, Washiyama K, Kumanishi T (1997) Infrequent mutation of Waf1/p21 gene, a CDK inhibitor gene, in brain tumors. Neurol Med Chir 37:150–156

Umar A, Buermeyer AB, Simon JA, Thomas DC, Clark AB, Liskay RM, Kunkel TA (1996) Requirement for PCNA in DNA mismatch repair at a step preceding DNA resynthesis. Cell 87:65–73

Upadhyay S, Li G, Liu H, Chen YQ, Sarkar FH, Kim H-RC (1995) bcl-2 suppresses expression of $p21^{WAF1/CIP1}$ in breast epithelial cells. Cancer Res 55:4520–4524

Vidal MJ, Loganzo F Jr, de Oliveira AR, Hayward NK, Albino AP (1995) Mutations and defective expression of the WAF1 p21 tumour-suppressor gene in malignant melanomas. Melanoma Res 5:243–250

Waga S, Hannon G, Beach D, Stillman B (1994) The p21 inhibitor of cyclin-dependent kinases controls DNA replication by interaction with PCNA. Nature 369:574–578

Waldman T, Kinzler KW, Vogelstein B (1995) p21 is necessary for the p53-mediated G1 arrest in human cancer cells. Cancer Res 55:5187–5190.

Wang A, Yoshimi N, Ino N, Tanaka T, Mori H (1997) WAF1 expression and p53 mutations in human colorectal cancers. J Cancer Res Clin Oncol 123:118–123

Wang H, Rubin M, Fenig E, DeBlasio A, Mendelsohn J, Yahalom J, Wieder R (1997) Basic fibroblast growth factor causes growth arrest in MCF-7 human breast cancer cells while inducing both mitogenic and inhibitory G1 events. Cancer Res 57:1750–1757

Wang J, Walsh K (1996) Resistance to apoptosis conferred by cdk inhibitors during myocyte differentiation. Science 273:359–361

Wang J, Guo K, Wills KN, Walsh K (1997) Rb functions to inhibit apoptosis during myocyte differentiation. Cancer Res 57:351–354

Warbrick E, Lane DP, Glover DM, Cox LS (1995) A small peptide inhibitor of DNA replication defines the site of interatction between the cyclin-dependent kinase inhibitor p21WAF1 and proliferating cell nuclear antigen. Curr Biol 5:275–282

Warbrick E, Lane DP, Glover DM, Cox LS (1997) Homologous regions of Fen1 and p21Cip1 compete for binding to the same site on PCNA: a potential mechanism to co-ordinate DNA replication and repair. Oncogene 14:2313–2321

Wu H, Wade M, Krall L, Grisham J, Xiong Y, Van Dyke T (1996) Targeted in vivo expression of the cyclin-dependent kinase inhibitor p21 halts hepatocyte cell-cycle progression, postnatal liver development, and regeneration. Genes Dev 10:245–260

Xiong Y, Hannon G, Zhang H, Casso D, Kobayashi R, Beach D (1993) p21 is a universal inhibitor of cyclin kinases. Nature 366:701–704

Yamato K, Koseki T, Ohguchi M, Kizaki M, Ikeda Y, Nishihara T (1997) Activin A induction of cell-cycle arrest involves modulation of cyclin D2 and p21CIP1/WAF1 in plasmacytic cells. Mol Endocrinol 11:1044–1052

Yan H, Newport J (1995) An analysis of the regulation of DNA synthesis by cdk2, Cip1, and licensing factor. J Cell Biol 129:1–15

Yang ZY, Perkins ND, Ohno T, Nabel EG, Nabel GJ (1995) The p21 cyclin-dependent kinase inhibitor suppresses tumorigenicity in vivo. Nat Med 1:1052–1056

Yonish-Rouach E (1996) The p53 tumour-suppressor gene: a mediator of a G1 growth arrest and of apoptosis. Experientia 52:1001–1007

Yoshida K, Murohashi I, Hirashima K (1996) p53-independent induction of p21 (WAF1/CIP1) during differentiation of HL-60 cells by tumor necrosis factor alpha. Int J Hematol 65:41–48

Zakut R, Givol D (1995) The tumor suppression function of p21 Waf is contained in its N-terminal half ("half-WAF"). Oncogene 11:393–395

Zeng YX, El-Deiry WS (1996) Regulation of p21WAF1/CIP1 expression by p53-independent pathways. Oncogene 12:1557–1564

Zeng YX, Somasundaram K, El-Deiry WS (1997) AP2 inhibits cancer cell growth and activates p21WAF1/CIP1 expression. Nature Genet 15:78–82

Zhang H, Xiong Y, Beach D (1993) Proliferating cell nuclear anitgen and p21 are components of multiple cell cycle kinase complexes. Mol Biol Cell 4:897–906

Zhang H, Hannon G, Beach D (1994) p21-containing cyclin kinases exist in both active and inactive states. Genes Dev 8:1750–1758

Zhang W, Grasso L, McClain C, Gambel A, Cha Y, Travali S, Deisseroth A, Mercer W (1995) p53-independent induction of WAF1/CIP1 in human leukemia cells is correlated with growth arrest accompanying monocyte/macrophage differentiation. Cancer Res 55:668–674

Zhu L, Harlow E, Dynlacht BD (1995) p107 uses a p21CIP1-related domain to bind cyclin/cdk2 and regulate intractions with E2F. Genes Dev 9:1740–1752

Mechanisms of Cell Cycle Blocks at the G2/M Transition and Their Role in Differentiation and Development

Michael R.A. Mowat[1,2,3] and Nancy Stewart[1,2]

1 Introduction

One of the most critical events in the life of a cell is segregation of chromosomes at meiosis and mitosis. Improper segregation of the genetic material during both mitosis and meiosis would be detrimental to the daughter cells. Therefore, cells have evolved many controls and checkpoints to ensure that a critical size is met, DNA is undamaged and correctly replicated, and that the centrosome and spindle apparatus are intact before cell division occurs (Hartwell and Weinert 1989; Murray 1992). A principal point for control of the cell cycle is at the Gap2/mitosis (G2/M) transition. This chapter will concentrate on the mechanisms of negative growth control at the G2/M transition in multicellular eukaryotes. Also, we will review how cells alter these G2/M controls in both cell differentiation and development. Since the genetics of cell-cycle checkpoints in yeast have been recently reviewed, these studies will not be extensively covered; instead, the yeast studies will be referred to where they best illustrate the point (Stewart and Enoch 1996; Elledge 1996).

2 Cdk Kinases and Cell-Cycle Control

It is now firmly established that the transitions through the cell cycle are controlled by the cyclin-dependent kinases (cdks) (for reviews see Forsburg and Nurse 1991; Norbury and Nurse 1992; Sherr 1993). The cdk kinases are universal regulators of the eukaryotic cell cycle from yeast to humans. Cdc2 is the prototypical cyclin-dependent kinase first described in the yeasts *Schizosaccharomyces pombe* and *Saccharomyces cerevisiae* (*Cdc28*). The Cdc2 protein is needed for progression through the cell cycle in G1 and for completion of mitosis (for review see Forsburg and Nurse 1991). The human Cdc2

Manitoba Institute of Cell Biology, Manitoba Cancer Treatment and Research Foundation[1] and the Departments of Human Genetics[2] and Physiology,[3] University of Manitoba, 100 Olivia Street, Winnipeg, Manitoba, Canada, R3E 0V9

Progress in Molecular and Subcellular Biology, Vol. 20
A. Macieira-Coelho (Ed.)

kinase was identified through complementation of a *Cdc2* mutant in yeast and as a protein that cross-reacts with antibodies to yeast Cdc2 (Draetta et al. 1987; Lee and Nurse 1987). Cdc2 and cyclin B are also components of the maturation promoting factor (MPF) that was first described as a cytoplasmic factor that induces meiotic maturation in *Xenopus* oocytes (Dunphy et al. 1988; Gautier et al. 1988). Cyclins were first described as proteins whose level increases during the cell cycle and then are destroyed at mitosis (Evans et al. 1983). Cyclin binding is needed for $p34^{Cdc2}$ kinase activation and specificity (Brizuela et al. 1989).

Cdc2 kinase and related kinases phosphorylate critical proteins involved in S phase and M phase transitions resulting in DNA synthesis, chromosome condensation, cytoskeletal reorganization, nuclear lamin disassembly, and cell shape changes (for a review, see Nigg 1993). Some targets include histone H1, nuclear lamins, vimentin, a microtubule-associated protein 115, p53, and others (Nigg 1993; Stukenberg et al. 1997). In animal cells different members of the cdk family of cyclin-dependent kinases (Cdk2 to Cdk7), each with its own cyclin partners, regulate different parts of the cell cycle (for review, see Pines 1993; Sherr 1993). For example, Cdk4 or Cdk6 are active with partners cyclin D or E at the G1/S cell boundary before DNA replication. Whereas, Cdc2 binding with cyclins A, B1, and B2 controls the transition from S phase to mitosis (for reviews see Sherr 1993; Grana and Reddy 1995). Cdc2 is the major enzyme controlling G2/M transition and the basic controls for its activity will first be reviewed.

3 Basic Controls of Cdc2 Kinase Activity

3.1 Positive Control of Cdc2 Kinase

Activation of Cdc2 requires both cyclin binding and phosphorylation at position threonine 161 (Ducommun et al. 1991; Gould et al. 1991; Norbury et al. 1991). Phosphorylation at Thr 161 may enhance binding of some cyclins to Cdc2 (Ducommun et al. 1991; Desai et al. 1995). Recent structural analysis of the Cdk2 and Cdk2/cyclin A complex shows that the regulatory T loop (residues 146–166) blocks protein substrate binding to the catalytic site (De Bondt et al. 1993; Jeffrey et al. 1995; Russo et al. 1996). Cyclin binding reorientates the ATP-binding site and allows movement of the T loop to allow substrate binding and make it more accessible for phosphorylation by cyclin-activating kinase (Cak) (De Bondt et al. 1993; Russo et al. 1996). Phosphorylation of Thr 160 (equivalent to Thr 161 in Cdc2) in the T loop moves the loop further and stabilizes Cdk2-cyclin A binding by increasing the number of contacts (Russo et al. 1996).

Cak kinase was first found in *Xenopus* egg extracts (Solomon et al. 1992) and is identical to the *Xenopus* pMO15 kinase, with sequence similarity to Cdc2, now called Cdk7 (Fesquet et al. 1993; Poon et al. 1993; Solomon et al. 1993). The cyclin partner of Cdk7 has been identified and called cyclin H (Fisher and Morgan 1994; Makela et al. 1994). A third protein called MAT1 binds to and stabilizes the Cdk7/cylin H complex (Devault et al. 1995). Cdk7/cyclin H is also part of the TFIIH general transcription factor complex involved in both RNA polymerase II transcription initiation and in DNA excision repair (Roy et al. 1994; Serizawa et al. 1995; Shiekhattar et al. 1995; Nigg 1996). The same kinase complex phosphorylates both Cdc2 and the carboxyl terminal domain of RNA polymerase II (RNA polII CTD) in vertebrates and *S. pombe*. However, different kinases may be responsible for Cak and RNA pol II CTD activities in *S. cerevisiae* (for review see Draetta 1997). Phosphorylation of Thr 160/161 on Cdc2/Cdk2 rises and falls during the cell cycle in parallel with cyclin binding (Krek and Nigg 1991a; Gu et al. 1992). This may be due to cyclin binding, since phosphorylation of Cdk2 by Cak can be stimulated by cyclin B binding although cyclin binding is not necessary for phosphorylation (Solomon et al. 1993). Cak levels and activity do not change during the cell cycle or development (Krek and Nigg 1991a; Gu et al. 1992; Brown et al. 1994; Matsuoka et al. 1994). These results suggest that Cak levels or activity are not the rate-limiting step for cdk activation during the cell cycle and that cyclin levels may be one important control for Cak activation.

Other factors may also be important in controlling Cdk activation. The Cdk2/Cdc2-associated phosphatase KAP dephosphorylates monomeric Cdk2 at position Thr160 in the absence of cyclin A binding (Poon and Hunter 1995). This dephosphorylation of Thr160 by KAP may prevent immediate activation of cdks in the next cycle when cyclins are made (ibid.). Also, cdk inhibitors (see below) can prevent the Cak phosphorylation when bound to cdks (Kato et al. 1994; Aprelikova et al. 1995).

3.2 Negative Control of Cdc2

3.2.1 Inhibitory Phosphorylations

Cdc2 is negatively controlled by phosphorylation at positions tyrosine 15 (Tyr 15) and in multicellular eukaryotes also at threonine 14 (Thr 14, for review see Forsburg and Nurse 1991; Draetta 1993). This phosphorylation is also dependent on cyclin binding and does not affect ATP binding to Cdc2 (Solomon et al. 1990; Atherton-Fessler et al. 1993). Phosphorylation of Cdc2 at Tyr 15 and Thr 14 is at its maximum at the G2 phase of the cell cycle and decreases suddenly at mitosis when H1 kinase activity is maximal (Solomon et al. 1990; Krek and

Nigg 1991a). Phosphorylation at Tyr 15 and Thr 14 downregulates Cdc2 kinase activity and Wee1 and the redundant mik1 kinases phosphorylate Cdc2 only at Tyr 15 (Lundgren et al. 1991; Parker and Piwnica-Worms 1992; McGowan and Russell 1993; Watanabe et al. 1995). Both human and *Xenopus* membrane-associated kinase (Myt1), which phosphorylates Cdc2 at both Thr14 and Tyr 15, have recently been cloned (Mueller et al. 1995; Liu et al. 1997). This kinase is found in both the endoplasmic reticulum and Golgi complex (Liu et al. 1997) whereas Wee1 is predominantly in the nucleus (Heald et al. 1993).

Upon completion of DNA synthesis, dephosphorylation of Tyr15 and Thr 14 by the Cdc25 phosphatase activates Cdc2 kinase activity and allows mitosis to progress (Gould and Nurse 1989; Dunphy and Kumagai 1991; Krek and Nigg 1991b; Norbury et al. 1991; Lee et al. 1992). In higher eukaryotes, three different Cdc25 genes have been found (Galaktionov and Beach 1991; Nagata et al. 1991; Kakizuka et al. 1992). Cdc25A appears to function in G1/S progression (Jinno et al. 1994; Hoffmann et al. 1994), whereas Cdc25C is associated with G2 progression showing an increase in phosphorylation at the G2/M boundary (Kumagai and Dunphy 1992; Izumi et al. 1992). Cdc25B expression occurs during prophase and appears responsible for activating cytoplasmic Cdc2/CyclinB resulting in microtubule nucleation before mitosis (Gabrielli et al. 1996).

3.2.2 Control of the Controllers

The regulatory enzymes that control Cdc2 phosphorylation are also subject to control of their activity through phosphorylation. In yeast, the Wee1 kinase is negatively regulated by the Nim1 kinase (Coleman et al. 1993; Parker et al. 1993; Wu and Russell 1993). Recently, a Nim1-binding protein that inhibits its activity, named Nif1, has been identified in the fission yeast *S. pombe* (Wu and Russell 1997). Nif1 deletion mutants divide at a smaller cell size showing that Nif1 acts as a negative regulator of the onset of mitosis (ibid.).

Cdc25C phosphatase is activated by phosphorylation at the G2/M boundary (Izumi et al. 1992; Kumagai and Dunphy 1992; Hoffmann et al. 1993). Cdc2 kinase phosphorylates Cdc25 and may be part of a positive feedback loop (Hoffmann et al. 1993; Izumi and Maller 1993; Strausfeld et al. 1994). However, other kinases may also play a role as the trigger for Cdc25 activation (Izumi and Maller 1995). Recently, a Cdc25 activating kinase, *Plx1*, has been purified and cloned from *Xenopus* egg extracts (Kumagai and Dunphy 1996). The Plx1 kinase is a member of the polo kinase gene family first described in *Drosophila* and being needed for normal mitosis (Llamazares et al. 1991). These phosphorylation events are subject to control by phosphatases such as type-2A that maintains Cdc25 in the inactive state and Wee1 in the active unphosphorylated state (Clarke et al. 1993; Kinoshita et al. 1993).

3.3 Negative Regulators of Cdc2

In recent years, a group of negative regulator proteins of G1-specific cdks has been described, namely the cdk inhibitors (for review see Harper and Elledge 1996). To date, no cdk inhibitor specific for Cdc2 acting at G2/M transition has yet been described. p21WAF1/CIP1 protein is a potent inhibitor of G1 cdks such as cyclinA/Cdk2, cyclinE/Cdk2, cyclinD/Cdk4, but a weaker inhibitor of cyclinB/Cdc2 kinase (Harper et al. 1993, 1995; Xiong et al. 1993). p21WAF1/CIP1 can inhibit mitosis in *Xenopus* egg extracts through inhibition of Cdc25 phosphorylation and Cdc2/cyclin B kinase activation (Guadagno and Newport 1996). Addition of Cdk2/cyclinE can overcome this inhibition, suggesting a positive function for this kinase in regulation of mitosis in early embryo cells. In contrast, cells from p21WAF1/CIP1-deficient mice still showed evidence of a G2 block after radiation or drug treatment (Brugarolas et al. 1995; Deng et al. 1995). Therefore, p21WAF1/CIP1 probably does not play a major role at G2/M transition in somatic cells.

Other regulators of Cdc2 include the cdk-binding protein Suc1/Cks (Patra and Dunphy 1996). Suc1 can inhibit Cdc2 tyrosine dephosphorylation at high concentrations, but is also needed for normal dephosphorylation of Cdc2 and destruction of Cyclin B for completion of anaphase (ibid.). These results suggested that Suc1 may function as a docking molecule for regulators that interact with Cdc2. The fission yeast cdk inhibitor Rum1 functions to inhibit both G1 and G2 progression (for review see Labib and Moreno 1996). Rum1 can inhibit the G1 cyclin Cig2/Cdc2 kinase complex to prevent premature DNA synthesis and the G2 cyclin Cdc13/Cdc2 complex to prevent premature mitosis in G1 (Moreno and Nurse 1994; Correa-Bordes and Nurse 1995). Elevated expression of Rum1 in G2 inhibits Cdc2/Cdc13 kinase activity, resulting in multiple rounds of DNA synthesis without mitosis (Correa-Bordes and Nurse 1995). These results suggest that Cdc2/cyclin B activity plays a role in preventing multiple rounds of DNA synthesis and mitosis. The budding yeast *S. cerevisiae* has an analogous but nonidentical protein called Sic1 that appears to function after the start of the cell cycle to prevent premature S phase (Mendenhall 1993; Donovan et al. 1994; Nugroho and Mendenhall 1994; Schwob et al. 1994). The vertebrate equivalents of either Sic1 or Rum1 proteins have not been identified.

4 Mechanisms of Negative Control at G2/M

Early studies in fission yeast have shown that overexpression of Cdc25 phosphatase or use of a Cdc2 mutant, which is independent of Cdc25 activation, could overcome the normal cell cycle checkpoint induced by a block in DNA synthesis (Enoch and Nurse 1990). The yeast genes Cdc25 and Wee1 are not essential for this DNA replication-dependent checkpoint but still can modulate this checkpoint (Enoch et al. 1992). In *Xenopus* egg cell-free extracts mitosis is dependent on completion of DNA synthesis (Dasso and Newport 1990). This block in mitosis results in tyrosine phosphorylation of Cdc2 kinase that can be overcome by addition of Cdc25 phosphatase or addition of either caffeine or the phosphatase inhibitor okadaic acid. These treatments result in reduced tyrosine phosphorylation of Cdc2 kinase (Kumagai and Dunphy 1991; Smythe and Newport 1992).

Other studies have not always found this DNA replication checkpoint to be dependent on tyrosine 15 phosphorylation of Cdc2. In budding yeast, mutation of tyrosine 19 in Cdc28 (equivalent to Tyr 15 in Cdc2) to phenylalanine does not affect the G2 arrest after inhibition of DNA synthesis or DNA damage (Amon et al. 1992; Sorger and Murray 1992). This may reflect the fact that in budding yeast the controls for initiation of the cell cycle occur mostly at the G1 part of the cell cycle, in contrast to fission yeast, which controls entrance into mitosis at G2 (Forsburg and Nurse 1991). Recently, it has been found that a morphogenesis checkpoint in budding yeast that coordinates mitosis with bud formation needs Tyr 19 phosphorylation of Cdc28 and this checkpoint can be manipulated by Swe1 (Wee1) and Mih1 (Cdc25) expression (Lew and Reed 1995; Sia et al. 1996). In another study using DNA synthesis-blocked *Xenopus* egg extracts, inhibition of Cdc2/cyclin B activity was not dependent on altered phosphorylation at amino acids Thr 14, Tyr 15, or Thr 161 of Cdc2 (Kumagai and Dunphy 1995). This study also described an inhibitor of Cdc2/cyclin B that could be inhibited competitively with catalytically inactive Cdc2/cyclin B. Recently a membrane associated Cdc2 inhibitor complex was biochemically characterized from interphase *Xenopus* extracts but whether this inhibitor plays a role in checkpoint control is unknown at this time (Lee and Kirschner 1996). Consistent with these results, others found that expression of a phosphorylation sites mutant *Cdc2AF* in HeLa cells would not induce premature mitosis in cells arrested in early S phase (Jin et al. 1996). However, in *Cdc2AF*-expressing cells progressing through S phase, approximately 25% displayed signs of premature mitosis. These cells also exhibited a reduced mitotic delay after X-irradiation. Overall, these results suggest that Cdc2 maybe subject to control by an inhibitor protein(s) early in the cell cycle, and negative phosphorylation may play a more important role later in the cell cycle.

Inhibition of Cdc2 kinase in human cells after UV irradiation is mediated through changes in Tyr 15 phosphorylation of Cdc2 (Herzinger et al. 1995; Poon et al. 1996). Recently, Src-like tyrosine kinases have been implicated in phosphorylation of Cdc2 after DNA damage. X-irradiation of hematopoietic cells results in an inhibitory tyrosine 15 phosphorylation of Cdc2 kinase that is dependent on Lyn tyrosine kinase binding and phosphorylation (Kharbanda et al. 1994a,b; Uckun et al. 1996). The Lyn tyrosine kinase also induces a G2/M arrest after irradiation of hematopoietic cells (ibid.). Elevated expression of the p210 BCR-Abl tyrosine kinase fusion protein in hematopoietic cells induces a G2/M arrest and protects cells from radiation and chemotherapeutically induced apoptosis (Bedi et al. 1995). The increased tyrosine phosphorylation of Cdc2 seen in B leukemic cells after X-irradiation is not due to Wee 1 kinase (Tuel-Ahlgren et al. 1996). Interestingly, the *Chk1* checkpoint control gene from fission yeast needs Weel kinase for its activity (O'Connell et al. 1997). Chkl kinase also phosphorylates Weel kinase in vitro but does not affect its activity (ibid.). It is not known at this time whether this phosphorylation affects the subcellular localization of Weel. These results suggest that after DNA damage, tyrosine phosphorylation of Cdc2 is an important checkpoint control and it may be carried out by tyrosine kinases other than Wee 1 in mammalian cells.

Other mechanisms for negatively controlling Cdc2 kinase activity after DNA damage have also been described. In nitrogen mustard treated lymphoma cells hyperphosphorylation and activation of Cdc25C phosphatase was blocked. Also, Cdc25C failed to interact with Cdc2 kinase compared with untreated control cells (O'Connor et al. 1994). Another study has shown that low-dose gamma radiation inhibits phosphorylation of Cdc25C as recognized by the mitosis-specific phosphorylation monoclonal antibody MPM-2 (Barth et al. 1996). It is not known whether this checkpoint is working at the level of the phosphatases or kinases that control Cdc25 phosphorylation or through control of the interaction between Cdc2 kinase with Cdc25. After irradiation of cells, the levels of cyclin B mRNA and protein decrease at G2 (Kao et al. 1997). The decrease of cyclin B levels after irradiation may be controlled by alteration of the mRNA stability (Maity et al. 1995). This radiation G2 delay can be diminished by overexpression of cyclin B (Kao et al. 1997).

In addition to tyrosine phosphorylation of Cdc2, the *BIME* (blocked-in-mitosis E) gene from the fungus *Aspergillus nidulans* is also needed for a fully functional replication checkpoint control (James et al. 1995; Ye et al. 1996). BIME appears to alter the activity and phosphorylation of the NIMA (not-in-mitosis A) kinase (Ye et al. 1996). NIMA is an essential mitotic promoting kinase that appears to act downstream of Cdc2 kinase (for review see Osmani and Ye 1996). The BIME protein in *Xenopus* has recently been found as part of the ubiquitin ligase complex called anaphase-promoting complex (APC, Peters et al. 1996). APC plays an important role in the degradation of cyclin B, NIMA and other proteins such as those involved in sister chromatid adhesion (Deshaies 1995; Chen et al. 1996). This allows anaphase to proceed for the completion of mitosis (ibid). Besides its role in exit from mitosis, BIME as part of APC may also prevent initiation of mitosis by preventing NIMA kinase activation, by altering phosphorylation through degradation of the kinase that may act on NIMA.

5 Oncogenes and Tumor-Suppressor Genes in G2 Arrest

Elevated expression of the *H-Ras* oncogene in fibroblasts can cause a G2/M arrest that is dependent on p53 and $p16^{INK4}$ tumor-suppressor functions (Hirakawa and Ruley 1988; Hicks et al. 1991; Serrano et al. 1997). This growth arrest by H-Ras in fibroblasts is associated with cellular senescense (Serrano et al. 1997). Transfection of Ras can also cause growth arrest at G1/S and G2/M in Schwann cells (Ridley et al. 1988). SV40 large T antigen, which binds p53 and retinoblastoma proteins, can overcome this growth arrest (ibid). In another study it was shown that activated H-Ras but not C-Ras protein could block cell cycle progression into M phase using activated *Xenopus* egg extracts and bacterially expressed H-ras protein (Pan et al. 1994). A reduction of p34Cdc2 kinase activity was also associated with this block. Recent work has shown that activation of an unknown 96-kDa histone H2b kinase correlates with the reduction in $p34^{Cdc2}$ kinase activity (Chen and Pan 1994), and this $p96^{H2b}$ kinase can also be activated by the MEK kinase (Pan et al. 1997). Overexpression of the *Mos* oncogene and other members of the MAP kinase pathway including activated *Ras*, *Raf*, and *MEK* oncogenes induce growth arrest and apoptosis in fibroblasts (Fukasawa and Vande Woude 1997). This lethality is greatly reduced in fibroblasts from $p53^{-/-}$ mice. Therefore, constitutive overexpression of members of the Ras/MAPK pathway can result in inhibition of Cdc2 kinase, possibly mediated through the p53 tumor suppressor. The mechanism for this arrest is uncertain at this time. Interestingly p53 can be phosphorylated by MAP kinase but the consequences of this phosphorylation are unknown (Milne et al. 1994).

p53 was originally found to control the G1/S part of the cell cycle, but it is now shown that it also functions at G2/M transition (Vikhanskaya et al. 1994; Agarwal et al. 1995; Aloni Grinstein et al. 1995; Stewart et al. 1995). p53 also plays a part in spindle checkpoint control (Cross et al. 1995). Damage to the mitotic spindle will result in cell arrest at G2/M but, in cells lacking p53, multiple rounds of DNA synthesis occur without cell division (ibid). p53 also plays a role in controlling centrosome duplication (Fukasawa et al. 1996). Fibroblasts from p53-deficient mice show multiple copies of centrosomes, which results in abnormal segregation of chromosomes (ibid). Alterations in the spindle checkpoint control and centrosome duplication in p53 mutant cells may play an important role in the genetic instability seen in p53-deficient cells (Harvey et al. 1993; Tsukada et al. 1993). Growth arrest at G2/M caused by p53 overexpression may be mimicking the downstream events of the spindle check point control. Whether p53 also functions as part of the G2/M checkpoint control due to unreplicated or damaged DNA is not clear. Cells deficient in p53 still show a G2/M arrest after irradiation (Kuerbitz et al. 1992). It is possible that other checkpoint controls, such as induction of the tyrosine kinases (see above), may still be operational at G2/M in p53-deficient cells obscuring the effects of p53. How spindle damage activates p53 activity and by what mechanism p53 controls the G2/M part of the cell cycle is unknown at present.

Delay or cell cycle block at G2/M is a common feature in many developmental and differentiation pathways of multicellular eukaryotes. The reasons for G2/M delay varies between cell types, and are only partially understood at present. The following is a summary of a few mechanisms involved in G2/M developmental and differentiation pathways.

6 G2/M Blocks in Meiosis

6.1 Oocyte Maturation

Meiosis is a unique form of cell division involving two divisions with only one round of DNA synthesis. Maturation of both oocytes and spermatocytes involves two blocks to cell cycle in or near G2 phase for most animals (Sagata 1996). Control over the two G2/M phase blocks and cell division ultimately occurs at the level of the Cdc2-Cyclin B or MPF (Gebauer and Richter 1997; Sagata 1997). These processes have best been studied in *Xenopus*, thus for the purposes of this chapter, the following will be based on this research. In some invertebrate species, there is an additional metaphase I arrest point before meiosis II which may be related to the formation of chiasmata. This phenomenon is not observed in *Xenopus* and will not be discussed further. The authors direct the reader to the review by Sagata (1996) for further information. For a more thorough review of all the processes involved in meiosis, the reader is directed to the following articles (Gebauer and Richter 1997; Page and Orr Weaver 1997; Sagata 1996, 1997).

During oogenesis, immature oocytes undergo two arrest points. The first is an arrest of immature oocytes before meiosis I in a late G2- or prophase I-like state (Sagata 1996; Page and Orr Weaver 1997; Gebauer and Richter 1997). This first arrest state can last from a few years in *Xenopus*, and up to decades in humans, and this may allow oocyte growth before fertilization and embryonic development (Sagata 1996). Release from this arrest into the first meiotic division occurs after hormonal stimulation that induces ovulation, which in *Xenopus* involves the hormone progesterone (Sagata 1996, 1997). Physiologically, progesterone-induced release is soon followed by germinal vesicle breakdown (GVBD), spindle formation, and completion of meiosis I with extrusion of the first polar body. There is a brief interkinesis without any DNA replication, and for most vertebrates a second arrest induced by the cytostatic factor at metaphase II of meiosis until fertilization.

6.2 Release from Meiosis I

Release from meiosis I arrest is accomplished by the stimulation of MPF activity by Mos kinase. Biochemically, progesterone stimulates a decrease in cyclic AMP (cAMP) levels. This, in turn, inactivates the cAMP-dependent protein kinase A (PKA) and results in de novo synthesis of Mos (Daar et al. 1991; Matten et al. 1994). PKA acts as a negative regulator at several points in oocyte maturation. Injection of the catalytic subunit of PKA into *Xenopus* oocytes can prevent the progesterone-induced synthesis of endogenous Mos (Matten et al. 1994). PKA can also inhibit MPF activity by preventing Cdc25 activation (ibid).

Hormonal stimulation of de novo synthesis of Mos kinase is required for release from G2 arrest and reentry into meiosis (Sagata et al. 1989a; Kanki and Donoghue 1991; Yew et al. 1991). Mos is a serine/threonine protein kinase expressed at high levels in germ cells, rarely expressed in somatic cells, and capable of inducing oncogenic transformation when overexpressed (Yew et al. 1993; Gebauer and Richter 1997; Sagata 1997). Both injection of Mos protein (Yew et al. 1992) and Mos mRNA (Freeman et al. 1989; Sagata et al. 1989a) will initiate oocyte maturation without progesterone stimulation. Further, Mos kinase activation occurs prior to initiation of GVBD and is required for this and all subsequent events (Yew et al. 1991). Injection of antisense oligonucleotides against Mos will prevent GVBD (Sagata et al. 1988; Kanki and Donoghue 1991).

Mos has several functions in regulating *Xenopus* meiosis, acting as an M phase activator in meiosis I and an M phase repressor in meiosis II (see below). Mos action in meiosis I involves activation of MPF by stimulation of the mitogen-activating protein kinase (MAPK) pathway (Nebreda and Hunt 1993; Kosako et al. 1994; Gotoh et al. 1995; Matten et al. 1996; Roy et al. 1996). For example, addition of a Mos fusion protein to *Xenopus* cell free extracts activates MAPK (Nebreda and Hunt 1993). De novo synthesis of Mos in vivo is required for stimulation of MAPK (Matten et al. 1996). Further, antisense oligonucleotides against Mos abolish activation of MAPK by progesterone, while reinjection of GST-Mos restored MAPK activation and GVBD (Roy et al. 1996).

MAPK is a protein kinase that responds to a number of stimuli including cell proliferation and differentiation (for review see Seger and Krebs 1995; Graves et al. 1995). Mos activation of the MAPK pathway functions through its interaction with the Map kinase kinase (MEK1). Activation of MAPK by Mos and oocyte maturation can be blocked by the addition of anti-MEK1 antibodies (Kosako et al. 1994; Gotoh et al. 1995). MEK1 activation is due to direct phosphorylation by Mos (Posada et al. 1993; Nebreda et al. 1995).

MAPK activation of MPF is required for both meiotic divisions, although the precise mechanism MAPK uses is currently unknown (Honigberg et al.

1993; Gebauer and Richter 1997). Regulation of Cdc2 kinase may not involve Cak, as the catalytic subunit is constitutively active during oogenesis, meiotic maturation, and the first stages of embryonic development (Brown et al. 1994). Activation of MPF by Mos/MAPK promotes GVBD and chromosome condensation (Gebauer and Richter 1997). MPF may act in a positive feedback loop on MAPK activation (Ferrell et al. 1991; Gotoh et al. 1991; Nebreda et al. 1995). MAPK, in turn, may act in a positive feedback loop with Mos, to stimulate accumulation and stabilization of Mos for completion of meiosis, as injection of constitutively active MAPK induces synthesis of Mos (Roy et al. 1996). MAPK also phosphorylates Mos at Ser-3 in vitro (Matten et al. 1996). Thus all three components interact to enhance the stability of Mos protein and through its action on MPF propel the oocytes through meiosis. Mos/MAPK activation of the MPF appears in the oocytes of other species such as frogs, clams, and starfish, suggesting it is a uniform mechanism of control (Abrieu et al. 1997). The timing of MAPK activation varies between species, such as in starfish, where it occurs after MPF activation and GVBD, suggesting that hormonal response does not always require MAPK for initiation and uses some alternate pathway (ibid).

After progression through meiosis I, oocytes undergo a brief interstitial stage without DNA synthesis, prior to arresting before the second meiotic division in metaphase II. The oocytes will remain in this arrested state until fertilization, after which MPF activity is lost, and cyclin B and Mos protein are degraded (Watanabe et al. 1989; Weber et al. 1991; Nishizawa et al. 1993). DNA synthesis is prevented after meiosis I through Mos reactivation of MPF, which also drives the cells into meiosis II (Furuno et al. 1994). Suppression of Mos allows *Xenopus* oocytes to enter interphase and replicate their DNA (ibid). This is similar to what is seen in *S. pombe*, where active Cdc2 restrains initiation of another S phase and prevents endoreduplication (Stern and Nurse 1996). This restraint is lost when the cells enter mitosis and Cdc2 kinase activity is lost with the degradation of cyclin B (Hayles et al. 1994; Stern and Nurse 1996).

In summary, control over meiosis I G2 arrest in *Xenopus* occurs at Cdc2/MPF activation. Stimulation of Mos kinase by hormones causes it to induce the MAPK pathway through phosphorylation of MEK1. This allows activation of Cdc2/MPF and release from meiosis I arrest, through the first meiotic division and into meiosis II without a DNA synthesis step. This action of Mos stands in contrast to its function in maintaining the second meiotic arrest point, outlined below.

6.3 Control of Meiosis II

Further division at meiosis II is prevented by the action of the cytostatic factor (CSF) (Sagata et al. 1989b; Daar et al. 1991) and Mos is most likely CSF

(Watanabe et al. 1989; Daar et al. 1991; Roy et al. 1996; Kanki and Donoghue 1991). For example, oocytes depleted of Mos RNA before GVBD but forced into meiosis with the addition of exogenous MPF continued through meiosis I and GVBD, but were unable to arrest at meiosis II (Daar et al. 1991). Mos may, in part, induce meiosis II arrest by mimicking a meiotic spindle checkpoint control via the stimulation of MAPK/ERK2 in *Xenopus* (Minshull et al. 1994). The spindle checkpoint in *Xenopus* prevents cyclin degradation and possibly other proteins associated with sister-chromatid separation until spindle assembly is completed ((Holloway et al. 1993; Murray 1995) for review, see (Deshaies 1995)). *Xenopus* oocytes and early embryos lack the cell-cycle checkpoint that keeps anaphase from occurring before spindle assembly is complete (Gerhart et al. 1984; Minshull et al. 1994), but can still arrest at meiosis II until fertilization (Sagata et al. 1989b). However, the addition of sperm nuclei to a cell-free oocyte extract allows microtubule depolymerization to arrest mitosis (Minshull et al. 1994). In this system, MAPK blocked cyclin B degradation, and inactivation of MAPK by the addition of the MAP kinase-specific phosphatase MKP-1 prevented both initiation and maintenance of the mitotic arrest (ibid). Destruction of cyclin is strongly inhibited during meiosis II arrest, suggesting that Mos as the CSF may act to suppress APC activity and, in turn, degradation of proteins needed for anaphase progression (Murray et al. 1989; Sagata 1997). Thus, Mos may mimic the spindle checkpoint pathway by activation of MAPK to stop cells from entering metaphase and completing meiosis.

Other proteins may be involved in stabilization of MPF, as Mos does not induce arrest at meiosis I, and the appearance of CSF does not occur until after GVBD (Kanki and Donoghue 1991). It has been suggested that Cdk2 kinase might be essential for MPF activation and regulation in *Xenopus* oocyte maturation (Gabrielli et al. 1993; Rempel et al. 1995; Guadagno and Newport 1996). Activity of Cdk2-cyclin E peaks during meiosis II in *Xenopus* oocytes (Rempel et al. 1995). Injection of antisense oligonucleotides against Cdk2 kinase into *Xenopus* oocytes prevented the meiosis II block, while readdition of Cdk2 restored the arrest (Gabrielli et al. 1993). Addition of Cdk2 in cell-free extracts allowed for MPF activation, while depletion or inhibition of Cdk2 prevented MPF activation (Guadagno and Newport 1996). However, recent results using an intact cell system suggest the opposite. Immature *Xenopus* oocytes injected with p21WAF1/CIP1 could complete meiosis I and II normally (Furuno et al. 1997). More than 98% of Cdk2 kinase activity was inhibited, while normal Cdc2 kinase (MPF) activity was not inhibited (ibid.). Moreover, the p21WAF1/CIP1 expressing oocytes were maintained at meiosis II for the correct length of time, unlike those injected with neutralizing anti-Mos antibodies released prematurely from meiosis II (ibid). Thus, whether Cdk2 plays a role in activation of the MPF complex, and whether this action influences meiosis II progression remains to be determined.

6.4
Mouse Oogenesis and Spermatogenesis

Oogenesis

Mos expression in mouse oogenesis differs slightly from *Xenopus*, and may not be required for release from meiosis I arrest (Araki et al. 1996; Verlhac et al. 1996). Instead, activation occurs after GVBD (Zhao et al. 1991; Araki et al. 1996). Mos expression is still required to induce MAPK activity, as Mos$^{-/-}$ oocytes fail to activate MAPK during meiosis (Choi et al. 1996a). In mice, Mos may be more important for maintenance of the meiosis II arrest. Mouse oocytes contain a CSF-like activity required for meiosis II arrest, maintained by stabilization of Cyclin B levels (Kubiak et al. 1993). Female Mos$^{-/-}$ mice fail to arrest their oocytes at meiosis II during meiosis, suggesting that for mammals Mos may function to prevent spontaneous parthenogenetic activation of unfertilized eggs (Colledge et al. 1994; Hashimoto et al. 1994; Choi et al. 1996a).

As previously mentioned, Mos may act through the spindle fiber checkpoint machinery to induce arrest at the metaphase II stage. In mice, an intact metaphase spindle is required for Cyclin B degradation in metaphase and completion of meiosis (Kubiak et al. 1993). Meiotic spindle formation in mouse Mos$^{-/-}$ cells is altered and the chromatin is poorly condensed (Araki et al. 1996; Choi et al. 1996a). The spindles also fail to translocate to the cortex, creating an altered cleavage plane (Choi et al. 1996a). Injection of Mos or a constitutively activated form of MEK1 into oocytes under conditions of inactive MPF induces formation of multiple microtubule arrays (Choi et al. 1996b). Therefore, Mos/MEK1 appears to affect the formation of the meiotic spindle and spindle poles (ibid). Mos$^{-/-}$ oocytes also form abnormally large first polar bodies and can undergo an additional cleavage instead of degenerating, suggesting Mos may have a role in polar body formation (Choi et al. 1996a).

Spermatogenesis

Spermatogenic control is not as well understood as oocyte maturation. However, it seems to employ a different set of controls over G2 and Cdc2 kinase/MPF than oogenesis. For example, during mouse development, Mos is not important for spermatogenesis, as *Mos*$^{-/-}$ male mice were completely fertile (Colledge et al. 1994). In *Drosophila* spermatogenesis, spermatogonii undergo four rounds of mitotic division to generate 16 spermatocytes. These spermatocytes undergo a premeoitic G2 arrest that can last more than 2 days before undergoing two rounds of meiotic division to produce 64 spermatids (Eberhart and Wasserman 1995). Progression from the meiosis I G2 arrest requires the action of the Cdc25C homolog *Twine* (Alphey et al. 1992; Courtot et al. 1992). Loss of *Twine* results in sterility, as meiotic divisions are unable to occur within the spermatocytes, and the cells remain as 16-cell cysts (Alphey et al. 1992; Courtot et al. 1992). *Twine* mutant females contain many eggs that have undergone abnormal divisions, indicating *Twine* may act during the premeiotic cell cycle or during meiosis (Alphey et al. 1992; Courtot et al. 1992).

Progression through meiosis I block in spermatocytes also requires the action of a second gene termed *Pelota* (Eberhart and Wasserman 1995). *Drosophila* males mutant for *Pelota* also go through the normal mitotic divisions during spermatogenesis, but arrest prior to the first meiotic division similar to *Twine* mutants (ibid). Partial chromosomal condensation occurs, but the cells fail to form spindles and begin nuclear envelope breakdown, suggesting a requirement for *Pelota* in these functions (ibid). How *Pelota* influences these functions has yet to be determined.

In conclusion, there exists a large variation in the timing and functions of Mos, MAPK/MEK1, Cdk2, Cdc25, cyclins, Pelota, and other, yet undefined proteins in controlling MPF and G2 blocks in meiosis. This suggests different mechanisms may be in use for G2/M meiotic arrests both in oogenesis/spermatogenesis, and between various species. More research is needed before there is a complete understanding of these processes.

7 G2 Blocks During Development

7.1 *String, Twine,* and *Drosophila* Development

Another area where altered control over G2 is important is during the early stages of *Drosophila* development. A brief description of *Drosophila* development follows, a more thorough review can be found elsewhere (Reed 1995). The first 13 cycles of *Drosophila* development after fertilization do not contain gap phases (G1, G2), but instead alternate between S phase and mitosis (Edgar and O'Farrell 1989). These divisions are driven by maternal products and do not require the production of zygotic proteins (Edgar and O'Farrell 1989; Reed 1995). Upon the 14th cycle, $Cdc25^{string}$, the *Drosophila* Cdc25 isoform becomes limiting, with its mRNA disappearing after mitosis 13 resulting in a G2 delay (Edgar and O'Farrell 1989, 1990; Edgar and Datar 1996; Edgar et al. 1994). $Cdc25^{string}$ is the only requirement for mitosis at this stage. Induction of $Cdc25^{string}$ expression by heat shock during the G2 phase induces premature entry into mitosis (Edgar and O'Farrell 1990). $Cdc25^{string}$ mutant embryos arrest at G2 phase of cycle 14 (ibid). Expression of $Cdc25^{string}$ also correlates with the sites of mitosis in the subsequent divisions within the embryo (Edgar and O'Farrell 1989). Further, the promoter for *String* contains dynamic cis response elements which are tissue-specific (Edgar et al. 1994), and without *String* expression the cells remain in G2 and will not divide. The loss of Cdc25 phosphatase activity allows the inhibitory phosphorylation of Cdc2 kinase by Dweel kinase, the *Drosophila* Wee 1 kinase equivalent, affecting G2 arrest (Edgar and Datar 1996).

Although important for meiosis, mutation of the *Twine* Cdc25C phosphatase also affects syncytial development. *Twine* mutant embryos contain nuclei that appear to undergo DNA synthesis without mitosis, resulting in fragmented nuclei at the syncytial stage (Alphey et al. 1992; Courtot et al. 1992). Deletion of *Twine* will also lead to premature arrest of the embryo at G2 at mitosis 12, while increasing the maternal Twine concentration can increase the number of mitoses before G2 arrest at mitosis 14 (Edgar and Datar 1996). Thus, reduction of String and Twine expression induces a G2 delay that occurs at the time when maternal components are being exhausted. This suggests that the G2 delay mediated by depletion of String and Twine protein is important to allow time for the switch between maternal and embryonic cellular components, and continued growth of the developing embryo (Reed 1995).

Another potential kinase may function in a checkpoint at G2 during syncytial cycles of early *Drosophila* development. *Drosophila grapes (Grp)* gene product is homologous to the *Chk1/Rad27* kinase gene (Fogarty et al. 1997). As mentioned, *Chk1* phosphorylates the Wee 1 kinase to inhibit its activity in G2 arrest (O'Connell et al. 1997). Grp may function to induce a progressively longer interphase during late syncytial divisions (divisions 12–13), as *Grp*$^{-/-}$ embryos fail to develop longer interphase (Fogarty et al. 1997). The embryos also failed to pass through the regulatory Cdc2 kinase step at division 14 (ibid.).

Transition from an early embryonic cell cycle with short or nonexistent G1 and G2 phases to a complete cell cycle in later *Xenopus* development is accompanied by changes in other cell cycle regulators. Both cyclin A and E disappear after the midblastula transition (Hartley et al. 1996). Tyrosine phosphorylation of Cdc2 kinase is not present during cycles 2–12, and its activity correlating with the rise and fall of cyclin B levels. This suggests control over Cdc2 does not occur through phosphorylation during early embryogenesis (ibid).

In summary, G2 delay during embryonic development is a common feature in many species, generally occurring at the switch from maternal to embryonic proteins. This delay probably occurs to allow time for the embryonic genome to replace the maternal proteins which to that point had driven cell division. The timing of this switch is species specific, and is accomplished by control (downregulation) over the activity of Cdc2 kinase. This control is accomplished by at least three mechanisms. First by depletion of the Cdc2 activators Cdc25^{string} and Cdc25^{twine}, second by induction of a potential Cdc2 inhibitor *grapes*, and third by the downregulation of cyclin partners.

8
G2 Blocks in Differentiation

8.1
Polyploidy and Endoreduplication

Polyploidy, or the blocking of the normal progression through mitosis during G2 followed by multiple rounds of DNA synthesis (termed endoreduplication) has been reported in many organisms and cell types (for a review, see Brodsky and Uryvaeva 1977; Zybina and Zybina 1996). In mammals, binucleated cells and cells with varying degrees of polyploidy are common in many tissues, including potentially up to 50–70% of the liver (Carriere 1967; Brodsky and Uryvaeva 1977; Roszell et al. 1978; Papa et al. 1987; Brill et al. 1993; Kudryavtsev et al. 1993; Sigal et al. 1995), trophoblast giant cells (Varmuza et al. 1988; Hoffman and Wooding 1993; Zybina and Zybina 1996), atrial and cardiac myocytes (Oberpriller et al. 1983; Rosenberg and Pfitzer 1983; W.Y. Brodsky et al. 1985; V.Y. Brodsky et al. 1994), and during differentiation of megakaryocytes (Erusalimsky and Martin 1993; Wang et al. 1995; Zhang et al. 1996).

The nature of endoreduplication and its potential function may vary, depending on the tissue in question. In hepatocytes, fetal cells progress from predominantly mononuclear cells with a few binucleated cells to roughly 20% binucleated in the adult (Carriere 1967; Roszell et al. 1978; Papa et al. 1987; Brill et al. 1993; Kudryavtsev et al. 1993; Sigal et al. 1995). Adult livers may also contain a significant number of mononucleated tetraploid and octaploid cells (ibid). Polyploidy in liver cells is associated with the state of differentiation, although what function this provides is currently unknown (ibid). In the heart, cells are mainly mononucleated (80%), the rest being polyploid cells with binuclei (Brodsky et al. 1985; Oberpriller et al. 1983). Development of extreme polyploidy is associated with the development of heart disease (Oberpriller et al. 1983; Rosenberg and Pfitzer 1983). Atrial myocytes have a low level of binucleated cells (15–20%), which increases (40–77%) in adaptive response to injury/stress (infarction) (Oberpriller et al. 1983). Excessive polyploidy (15–20 N) develops during normal myocyte differentiation, associated with congenital or acquired childhood diseases (Brodsky et al. 1994). This suggests that cells blocking G2 and undergoing endoreduplication in heart tissue may be an attempt to improve functional ability of individual cells without the delay or expense of further cell division (ibid).

A number of trophoblast subtypes exist. These cells can contain two diploid nuclei (binucleated cells), single nuclei with amplified DNA (rodent and rabbit trophoblast giant cells), or multiple nuclei (human placental bed giant cells) (Hoffman and Wooding 1993). This may reflect both the individual tissues of origin, the specific functions of each tissue type, and different methods for duplication of DNA content in trophoblast cells. Several theories have been proposed for why trophoblastic cells allow a block at G2 and subsequent DNA duplication. Similar to liver and heart models, duplication may allow the acquisition of a large cell mass and the increased quantity of genes while saving materials, time, and space compared with normal cell division (Brodsky and Uryvaeva 1977; Varmuza et al. 1988; Hoffman and Wooding 1993; Zybina and Zybina 1996). For trophoblast cells involved with the placenta, this might be important to provide the necessary components for the development of the embryo until the development of the embryo's own organ systems (Zybina and Zybina 1996). Finally, as with the liver, endoreduplication accompanies differentiation, suggesting some interrelatedness between the two (Brodsky and Uryvaeva 1977; Zybina and Zybina 1996). It remains to be shown whether one, some, or all of these theories are correct. It is not known how cells become binucleated. A recent study has shown that Mos overexpression in 4N cells, and presumably MAPK overexpression, can result in binucleated cells (Fukasawa and Vande Woude 1995).

Although the processes behind endoreduplication are not well understood in every situation, this retention of genetic material most likely results from different aberrations in the mitotic process. The production of endoreduplication potentially involves modulation of Cdc2 kinase. Moderate levels of Cdc2 kinase activity during G2 phase has been proposed to restrain initiation of another S phase and prevent endoreduplication in *S. pombe* (Stern and Nurse 1996). This restraint is lost when the cells enter mitosis and Cdc2 kinase activity is lost (Hayles et al. 1994; Stern and Nurse 1996). Further, when Cdc2 activity is lost, endoreduplication can occur (Hayles et al. 1994). Elevated expression of the cdk inhibitor Rum1, also results in endoreduplication (Correa-Bordes and Nurse 1995).

Two examples for which at least part of the specific mechanism for the block at G2 and subsequent endoreduplication is known are provided as follows. Many postmitotic cells in *Drosophila* enter an endoreduplication stage where they undergo multiple S phases without subsequent mitosis, the best known being the giant polytene chromosomes of the ovary and salivary glands (Brodsky and Uryvaeva 1977; King et al. 1981). This suggests loss of the control which prevents entry into S phase before mitosis is complete. Loss of this control allows conversion from a mitotic to an endoreduplication cycle, which in *Drosophila* involves the altered expression of cyclin proteins. DmcycE transcript levels undergo a short pulse of activity to induce S phase during endoreduplication, and are then downregulated by a self-feedback mechanism (Sauer et al. 1995). Normal endoreduplication fails to occur in *Drosophila* mutants which have stable DmcycE (*Drosophila* Cyclin E) mRNA (Knoblich et al. 1994). However, these DmcycE mutants still induce S phase-specific genes while failing to undergo the multiple rounds of DNA replication, suggesting that downregulation of cyclinE-Cdc2 activity is needed for endoreduplication

(Knoblich et al. 1994; Sauer et al. 1995). Also, mutation of cyclin A results in a switch from mitotic to endoreduplicative cycles in thoracic cells, suggesting that cyclin A is required for preventing inappropriate endoreduplication (Sauer et al. 1995).

The second system where something is known about the regulation of polyploidy is the production of megakaryocytes. Megakaryocytes undergo a variable number of endoreduplicative cycles consisting of S/Gap stages during terminal differentiation and cytoplasmic fragmentation (Erusalimsky and Martin 1993; Zhang et al. 1996). Endoreduplication for megakaryocytes may be important for production of large amounts of protein in platelets. Control over megakaryocytopoiesis may involve cyclin D3, as antisense oligonucleotides against cyclin D3 also suppressed endoreduplication and subsequent megakaryocyte production (Wang et al. 1995). Differentiation of megakaryocytes is also associated with sustained levels of cyclins A and E (Garcia and Cales 1996). In addition, the endoreduplicative process is associated with a reduction in cyclin B levels, but not of Cdc2 protein that results in a loss of active Cdc2 kinase complex (Wang et al. 1995; Zhang et al. 1996). Antisense targeting of cyclin B does not affect megakaryocytopoiesis (Wang et al. 1995). The lack of Cdc2 kinase activity may also be related to the downregulation of Cdc25C protein in differentiated megakaroycytes (Garcia and Cales 1996). These results suggest that megakaryocytes accomplish endoreduplication and differentiation by inhibiting the normal function of Cdc2 kinase at G2 to prevent mitosis and allow multiple rounds of DNA synthesis.

In conclusion, blocking of cells in G2 to allow for endoreduplication is a common feature in many cell types. Most cells probably accomplish this process by manipulating the amount and activity of Cdc2 kinase, its cyclin partners, and related control proteins. The mechanisms and purpose for this block vary from cell type to cell type, but all probably relate to improved function(s) of the cells. How and why each cell type accomplishes endoreduplication will be the source of future investigation. Without proper functioning of the G2 transition, none of us would be here to think about this process.

Acknowledgments. We thank Drs. Sabina Mai and Jennifer Brown Gladden for reading the manuscript. We also thank our librarian Donna Pacholok for her exhaustive efforts in tracking down references.

References

Abrieu A, Doree M, Picard A (1997) Mitogen-activated protein kinase activation down-regulates a mechanism that inactivates cyclin B-cdc2 kinase in G2-arrested oocytes. Mol Biol Cell 8:249–261

Agarwal ML, Agarwal A, Taylor WR, Stark GR (1995) p53 controls both the G2/M and the G1 cell cycle checkpoints and medicates reversible growth arrest in human fibroblasts. Proc Natl Acad Sci USA 92:8493–8497

Aloni Grinstein R, Schwartz D, Rotter V (1995) Accumulation of wild-type p53 protein upon gamma-irradiation induces a G2 arrest-dependent immunoglobulin kappa light chain gene expression. EMBO J 14:1392–1401

Alphey L, Jimenez J, White-Cooper H, Dawson I, Nurse P, Glover DM (1992) twine, a cdc25 homolog that functions in the male and female germline of *Drosophila*. Cell 69:977–988

Amon A, Surana U, Muroff I, Nasmyth K (1992) Regulation of p34CDC28 tyrosine phosphorylation is not required for entry into mitosis in *S. cerevisiae*. Nature 355:368–371

Aprelikova O, Xiong Y, Liu ET (1995) Both p16 and p21 families of cyclin-dependent kinase (CDK) inhibitors block the phosphorylation of cyclin-dependent kinases by the CDK-activating kinase. J Biol Chem 270:18195–18197

Araki K, Naito K, Haraguchi S, Suzuki R, Yokoyama M, Inoue M, Aizawa S, Toyoda Y, Sato E (1996) Meiotic abnormalities of c-mos knockout mouse oocytes: activation after first meiosis or entrance into third meiotic metaphase. Biol Reprod 55:1315–1324

Atherton-Fessler S, Parker LL, Geahlen RL, Piwnica-Worms H (1993) Mechanisms of p34cdc2 regulation. Mol Cell Biol 13:1675–1685

Barth H, Hoffmann I, Kinzel V (1996) Radiation with 1 Gy prevents the activation of the mitotic inducers mitosis-promoting factor (MPF) and cdc25-C in HeLa cells. Cancer Res 56:2268–2272

Bedi A, Barber JP, Bedi GC, el Deiry WS, Sidransky D, Vala MS, Akhtar AJ, Hilton J, Jones RJ (1995) BCR-ABL-mediated inhibition of apoptosis with delay of G2/M transition after DNA damage: a mechanism of resistance to multiple anticancer agents. Blood 86:1148–1158

Brill S, Holst P, Sigal S, Zvibel I, Fiorino A, Ochs A, Somasundaran U, Reid LM (1993) Hepatic progenitor populations in embryonic, neonatal, and adult liver. Proc Soc Exp Biol Med 204:261–269

Brizuela L, Draetta G, Beach D (1989) Activation of human CDC2 protein as a histone H1 kinase is associated with complex formation with the p62 subunit. Proc Natl Acad Sci USA 86:4362–4366

Brodsky VY, Sarkisov DS, Arefyeva AM, Panova NW, Gvasava IG (1994) Polyploidy in cardiac myocytes of normal and hypertrophic human hearts; range ofvalues. Virchows Arch 424: 429–435

Brodsky WY, Uryvaeva IV (1977) Cell polyploidy: its relation to tissue growth and function. Int Rev Cytol 50:275–332

Brodsky WY, Tsirekidze NN, Arefyeva AM (1985) Mitotic-cyclic and cycle-independent growth of cardiomyocytes. J Mol Cell Cardiol 17:445–455

Brown AJ, Jones T, Shuttleworth J (1994) Expression and activity of p40MO15, the catalytic subunit of cdk-activating kinase, during *Xenopus* oogenesis and embryogenesis. Mol Biol Cell 5:921–932

Brugarolas J, Chandrasekaran C, Gordon JI, Beach D, Jacks T, Hannon GJ (1995) Radiation-induced cell cycle arrest compromised by p21 deficiency. Nature 377:552–557

Carriere R (1967) Polyploid cell reproduction in normal adult rat liver. Exp Cell Res 46:533–540

Chen CT, Pan BT (1994) Oncogenic ras stimulates a 96-kDa histone H2b kinase activity in activated *Xenopus* egg extracts. Correlation with the suppression of p34cdc2 kinase. J Biol Chem 269:28034–28043

Chen Y, Knudsen ES, Wang JYJ (1996) Cells arrested in G1 by the v-Abl tyrosine kinase do not express cyclin A despite the hyperphosphorylation of RB. J Biol Chem 271:19637–19640

Choi TS, Fukasawa K, Zhou RP, Tessarollo L, Borror K, Resau J, Vande Woude GF (1996a) The Mos/mitogen-activated protein kinase (MAPK) pathway regulates the size and degradation of the first polar body in maturing mouse oocytes. Proc Natl Acad Sci USA 93:7032–7035

Choi T, Rulong S, Resau J, Fukasawa K, Matten W, Kuriyama R, Mansour S, Ahn N, Vande Woude GF (1996b) Mos/mitogen-activated protein kinase can induce early meiotic phenotypes in the absence of maturation-promoting factor: a novel system for analyzing spindle formation during meiosis I. Proc Natl Acad Sci USA 93:4730–4735

Clarke PR, Hoffmann I, Draetta G, Karsenti E (1993) Dephosphorylation of cdc25-C by a type-2A protein phosphatase: specific regulation during the cell cycle in *Xenopus* egg extracts. Mol Biol Cell 4:397–411

Coleman TR, Tang Z, Dunphy WG (1993) Negative regulation of the wee 1 protein kinase by direct action of the nim1/cdr1 mitotic inducer. Cell 72:919–929

Colledge WH, Carlton MB, Udy GB, Evans MJ (1994) Disruption of c-mos causes parthenogenetic development of unfertilized mouse eggs. Nature 370:65–68

Correa-Bordes J, Nurse P (1995) p25Rum1 orders S phase and mitosis by acting as an inhibitor of the p34cdc2 mitotic kinase. Cell 83:1001–1009

Courtot C, Fankhauser C, Simanis V, Lehner CF (1992) The *Drosophila* cdc25 homolog twine is required for meiosis. Development 116:405–416

Cross SM, Sanchez CA, Morgan CA, Schimke MK, Ramel S, Idzerda RL, Raskind WH, Reid BJ (1995) A p53-dependent mouse spindle checkpoint. Science 267:1353–1356

Daar I, Paules RS, Vande Woude GF (1991) A characterization of cytostatic factor activity from *Xenopus* eggs and c-mos-transformed cells. J Cell Biol 114:329–335

Dasso M, Newport JW (1990) Completion of DNA replication is monitored by a feedback system that controls the initiation of mitosis in vitro: studies in *Xenopus*. Cell 61:811–823

De Bondt HL, Rosenblatt J, Jancarik J, Jones HD, Morgan DO, Kim SH (1993) Crystal structure of cyclin-dependent kinase 2. Nature 363:595–602

Deng C, Zhang P, Harper JW, Elledge SJ, Leder P (1995) Mice lacking p21 CIP1/WAF1 undergo normal development, but are defective in G1 chechpoint control. Cell 82:675–684

Desai D, Wessling HC, Fisher RP, Morgan DO (1995) Effects of phosphorylation by CAK on cyclin binding by CDC2 and CDK2. Mol Cell Biol 15:345–350

Deshaies R (1995) The self-destructive personality of a cell cycle in transition. Curr Opin Cell Biol 7:781–789

Devault A, Martinez AM, Fesquet D, Labbe JC, Morin N, Tassan JP, Nigg EA, Cavadore JC, Doree M (1995) MAT1 (ménage à trois) a new RING finger protein subunit stabilizing cyclin H-cdk7 complexes in starfish and *Xenopus* CAK. EMBO J 14:5027–5036

Donovan JD, Toyn JH, Johnson AL, Johnston LH (1994) P40SDB25, a putative CDK inhibitor, has a role in the M/G1 transition in *Saccharomyces cerevisiae*. Genes Dev 8:1640–1653

Draetta G (1993) cdc2 activation: the interplay of cyclin binding and Thr161 phosphorylation. Trends Cell Biol 3:287–289

Draetta G, Brizuela L, Potashkin J, Beach D (1987) Identification of p34 and p13, human homologs of the cell cycle regulators of fission yeast encoded by cdc2+ and suc1+. Cell 50:319–325

Draetta GF (1997) Cell cycle: Will the real Cdk-activating kinase please stand up. Curr Biol 7:R50–R52

Ducommun B, Brambilla P, Felix MA, Franza BR Jr, Karsenti E, Draetta G (1991) cdc2 phosphorylation is required for its interaction with cyclin. EMBO J 10:3311–3319

Dunphy WG, Kumagai A (1991) The cdc25 protein contains an intrinsic phosphatase activity. Cell 67:189–196

Dunphy WG, Brizuela L, Beach D, Newport J (1988) The *Xenopus* cdc2 protein is a component of MPF, a cytoplasmic regulator of mitosis. Cell 54:423–431

Eberhart CG, Wasserman SA (1995) The pelota locus encodes a protein required for meiotic cell division: an analysis of G2/M arrest in *Drosophila* spermatogenesis. Development 121:3477–3486

Edgar BA, Datar SA (1996) Zygotic degradation of two maternal Cdc25 mRNAs terminates *Drosophila's* early cell cycle program. Genes Dev 10:1966–1977

Edgar BA, O'Farrell PH (1989) Genetic control of cell division patterns in the *Drosophila* embryo. Cell 57:177–187

Edgar BA, O'Farrell PH (1990) The three postblastoderm cell cycles of *Drosophila* embryogenesis are regulated in G2 by string. Cell 62:469–480

Edgar BA, Lehman DA, O'Farrell PH (1994) Transcriptional regulation of string (cdc25): a link betwen developmental programming and the cell cycle. Development 120:3131–3143

Elledge SJ (1996) Cell cycle checkpoints: preventing an identity crisis. Science 274:1664–1672

Enoch T, Nurse P (1990) Mutation of fission yeast cell cycle control genes abolishes dependence of mitosis on DNA replication. Cell 60:665–673

Enoch T, Carr AM, Nurse P (1992) Fission yeast genes involved in coupling mitosis to completion of DNA replication. Genes Dev 6:2035–2046

Erusalimsky JD, Martin JF (1993) The regulation of megakaryocyte polyploidization and its implications for coronary artery occlusion. Eur J Clin Invest 23:1–9

Evans T, Rosenthal ET, Youngblom J, Distel D, Hunt T (1983) Cyclin: a protein specified by maternal mRNA in sea urchin eggs that is destroyed at each cleavage division. Cell 33:389–396

Ferrell JE Jr, Wu M, Gerhart JC, Martin GS (1991) Cell cycle tyrosine phosphorylation of p34cdc2 and a microtubule-associated protein kinase homolog in *Xenopus* oocytes and eggs. Mol Cell Biol 11:1965–1971

Fesquet D, Labbe JC, Derancourt J, Capony JP, Galas S, Girard F, Lorca T, Shuttleworth J, Doree M, Cavadore JC (1993) The MO15 gene encodes the catalytic subunit of a protein kinase that activates cdc2 and other cyclin-dependent kinases (CDKs) through phosphorylation of Thr161 and its homologues. EMBO J 12:3111–3121

Fisher RP, Morgan DO (1994) A novel cyclin associates with MO15/CDK7 to form the CDK-activating kinase. Cell 78:713–724

Fogarty P, Campbell SD, Abu-Shumays R, Phalle BS, Yu KR, Uy GL, Goldberg ML, Sullivan W (1997) The *Drosophila* grapes gene is related to checkpoint gene chkl/rad27 and is required for later syncytial division fidelity. Curr Biol 7:418–426

Forsburg SL, Nurse P (1991) Cell cycle regulation in the yeasts *Saccharomyces cerevisiae* and *Schizosaccharomyces pombe*. Annu Rev Cell Biol 7:227–256

Freeman RS, Pickham KM, Kanki JP, Lee BA, Pena SV, Donoghue JD (1989) *Xenopus* homolog of the mos protooncogene transforms mammalian fibroblasts and induces maturation of *Xenopus* oocytes. Proc Natl Acad Sci USA 86:5805–5809

Fukasawa K, Vande Woude GF (1995) Mos overexpression in Swiss 3T3 cells induces meiotic-like alterations of the mitotic spindle. Proc Natl Acad Sci USA 92:3430–3434

Fukasawa K, Vande Woude GF (1997) Synergy between the Mos/mitogen-activated protein kinase pathway and loss of p53 function in transformation and chromosome instability. Mol Cell Biol 17:506–518

Fukasawa K, Choi T, Kuriyama R, Rulong S, Vande Woude GF (1996) Abnormal centrosome amplification in the absence of p53. Science 271:1744–1747

Furuno N, Nishizawa M, Okazaki K, Tanaka H, Iwashita J, Nakajo N, Ogawa Y, Sagata N (1994) Suppression of DNA replication via Mos function during meiotic divisions in *Xenopus* oocytes. EMBO J 13:2399–2410

Furuno N, Ogawa Y, Iwashita J, Nakajo N, Sagata N (1997) Meiotic cell cycle in *Xenopus* oocytes is independent of cdk2 kinase. EMBO J 16:3860–3865

Gabrielli BG, Roy LM, Maller JL (1993) Requirement for Cdk2 in cytostatic factor-mediated metaphase II arrest. Science 259:1766–1769

Gabrielli BG, De Souza CPC, Tonks ID, Clark JM, Hayward NK, Ellem KAO (1996) Cytoplasmic accumulation of cdc25B phosphatase in mitosis triggers centrosomal microtubule nucleation in HeLa cells. J Cell Sci 109:1081–1093

Galaktionov K, Beach D (1991) Specific activation of cdc25 tyrosine phosphatases by B-type cyclins; evidence for multiple roles of mitotic cyclins. Cell 67:1181–1194

Garcia P, Cales C (1996) Endoreplication in megakaryoblastic cell lines is accompanied by sustained expression of G1/S cyclins and downregulation of cdc25C. Oncogene 13:695–703

Gautier J, Norbury C, Lohka M, Nurse P, Maller J (1988) Purified maturation-promoting factor contains the product of a *Xenopus* homolog of the fission yeast cell cycle control gene cdc2+. Cell 54:433–439

Gebauer F, Richter JD (1997) Synthesis and function of Mos: the control switch of vertebrate oocyte meiosis. Bioessays 19:23–28

Gerhart J, Wu M, Kirschner M (1984) Cell cycle dynamics of an M-phase-specific cytoplasmic factor in *Xenopus laevis* oocytes and eggs. J Cell Biol 98:1247–1255

Gotoh Y, Moriyama K, Matsuda S, Okumura E, Kishimoto T, Kawasaki H, Suzuki K, Yahara I, Sakai H, Nishida E (1991) *Xenopus* M phase MAP kinase: isolation of its cDNA and activation by MPF. EMBO J 10:2661–2668

Gotoh Y, Masuyama N, Dell K, Shirakabe K, Nishida E (1995) Initiation of *Xenopus* oocyte maturation by activation of the mitogen-activated protein kinase cascade. J Biol Chem 270:25898–25904

Gould KL, Nurse P (1989) Tyrosine phosphorylation of the fission yeast cdc2+ protein kinase regulates entry into mitosis. Nature 342:39–45

Gould KL, Moreno S, Owen DJ, Sazer S, Nurse P (1991) Phosphorylation at Thr167 is required for *Schizosaccharomyces pombe* p34cdc2 function. EMBO J 10:3297–3309

Grana X, Reddy EP (1995) Cell cycle control in mammalian cells: role of cyclins, cyclin-dependent kinases (CDKs), growth suppressor genes and cyclin-dependent kinase inhibitors (CKIs). Oncogene 11:211–219

Graves JD, Campbell JS, Krebs EG (1995) Protein serine threonine kinases of the MAPK cascade. Ann NY Acad Sci 766:320–343

Gu Y, Rosenblatt J, Morgan DO (1992) Cell cycle regulation of CDK2 activity by phosphorylation of Thr160 and Tyr15. EMBO J 11:3995–4005

Guadagno TM, Newport JW (1996) Cdk2 kinase is required for entry into mitosis as a positive regulator of Cdc2-cyclin B kinase activity. Cell 84:73–82

Harper JW, Elledge SJ (1996) Cdk inhibitors in development and cancer. Curr Opin Genet Dev 6:56–64

Harper JW, Adami GR, Wei N, Keyomarsi K, Elledge SJ (1993) The p21 Cdk-interacting protein Cip1 is a potent inhibitor of G1 cyclin-dependent kinases. Cell 75:805–816

Harper JW, Elledge SJ, Keyomarsi K, Dynlacht B, Tsai LH, Zhang P, Dobrowolski S, Bai C, Connell Crowley L, Swindell E et al. (1995) Inhibition of cyclin-dependent kinases by p21. Mol Biol Cell 6:387–400

Hartley RS, Rempel RE, Maller JL (1996) In vivo regulation of the early embryonic cell cycle in Xenopus. Dev Biol 173:408–419

Hartwell LH, Weinert TA (1989) Checkpoints: controls that ensure the order of cell cycle events. Science 246:629–634

Harvey M, Sands AT, Weiss RS, Hegi ME, Wiseman RW, Pantazis P, Giovanella BC, Tainsky MA, Bradley A, Donehower LA (1993) In vitro growth characteristics of embryo fibroblasts isolated from p53-deficient mice. Oncogene 8:2457–2467

Hashimoto N, Watanabe N, Furuta Y, Tamemoto H, Sagata N, Yokoyama M, Okazaki K, Nagayoshi M, Takeda N, Ikawa Y et al. (1994) Parthenogenetic activation of oocytes in c-mos-deficient mice. Nature 370:68–71

Hayles J, Fisher D, Woollard A, Nurse P (1994) Temporal order of S phase and mitosis in fission yeast is determined by the state of the p34cdc2-mitotic B cyclin complex. Cell 78:813–822

Heald R, McLoughlin M, McKeon F (1993) Human wee1 maintains mitotic timing by protecting the nucleus from cytoplasmically activated Cdc2 kinase. Cell 74:463–474

Herzinger T, Funk JO, Hillmer K, Eick D, Wolf DA, Kind P (1995) Ultraviolet B irradiation-induced G2 cell cycle arrest in human keratinocytes by inhibitory phosphorylation of the cdc2 cell cycle kinase. Oncogene 11:2151–2156

Hicks GG, Egan SE, Greenberg AH, Mowat M (1991) Mutant p53 tumor suppressor alleles release ras-induced cell cycle growth arrest. Mol Cell Biol 11:1344–1352

Hirakawa T, Ruley HE (1988) Rescue of cells from ras oncogene-induced growth arrest by a second, complementing, oncogene. Proc Natl Acad Sci USA 85:1519–1523

Hoffman LH, Wooding FB (1993) Giant and binucleate trophoblast cells of mammals. J Exp Zool 266:559–577

Hoffmann I, Clarke PR, Marcote MJ, Karsenti E, Draetta G (1993) Phosphorylation and activation of human cdc25-C by cdc2-cyclin B and its involvement in the self-amplification of MPF at mitosis. EMBO J 12:53–63

Hoffmann I, Draetta G, Karsenti E (1994) Activation of the phosphatase activity of human cdc25A by a cdk2-cyclin E dependent phosphorylation at the G1/S transition. EMBO J 13:4302–4310
Holloway SL, Glotzer M, King RW, Murray AW (1993) Anaphase is initiated by proteolysis rather than by the inactivation of maturation-promoting factor. Cell 73:1393–1402
Honigberg SM, McCarroll RM, Esposito RE (1993) Regulatory mechanisms in meiosis. Curr Opin Cell Biol 5:219–225
Izumi T, Maller JL (1993) Elimination of cdc2 phosphorylation sites in the cdc25 phosphatase blocks initiation of M-phase. Mol Biol Cell 4:1337–1350
Izumi T, Maller JL (1995) Phosphorylation and activation of the *Xenopus* Cdc25 phosphatase in the absence of Cdc2 and Cdk2 kinase activity. Mol Biol Cell 6:215–226
Izumi T, Walker DH, Maller JL (1992) Periodic changes in phosphorylation of the *Xenopus* cdc25 phosphatase regulate its activity. Mol Biol Cell 3:927–939
James SW, Mirabito PM, Scacheri PC, Morris NR (1995) The *Aspergillus nidulans* bimE (blocked-in-mitosis) gene encodes multiple cell cycle functions involved in mitotic checkpoint control and mitosis. J Cell Sci 108:3485–3499
Jeffrey PD, Russo AA, Polyak K, Gibbs E, Hurwitz J, Massague J, Pavletich NP (1995) Mechanism of CDK activation revealed by the structure of a cyclinA-CDK2 complex. Nature 376:313–320
Jin P, Gu Y, Morgan DO (1996) Role of inhibitory CDC2 Phosphorylation in radiation-induced G2 arrest in human cells. J Cell Biol 134:963–970
Jinno S, Suto K, Nagata A, Igarashi M, Kanaoka Y, Nojima H, Okayama H (1994) Cdc25A is a novel phosphatase functioning early in the cell cycle. EMBO J 13:1549–1556
Kakizuka A, Sebastian B, Borgmeyer U, Hermans-Borgmeyer I, Bolado J, Hunter T, Hoekstra MF, Evans RM (1992) A mouse cdc25 homolog is differentially and developmentally expressed. Genes Dev 6:578–590
Kanki JP, Donoghue DJ (1991) Progression from meiosis I to meiosis II in *Xenopus* oocytes requires de novo translation of the mosxe protooncogene. Proc Natl Acad Sci USA 88:5794–5798
Kao GD, McKenna WG, Maity A, Blank K, Muschel RJ (1997) Cyclin B1 availability is a rate-limiting component of the radiation-induced G2 delay in HeLa cells. Cancer Res 57:753–758
Kato JY, Matsuoka M, Polyak K, Massague J, Sherr CJ (1994) Cyclic AMP-induced G1 phase arrest mediated by an inhibitor (p27Kip1) of cyclin-dependent kinase 4 activation. Cell 79:487–496
Kharbanda S, Saleem A, Datta R, Yuan ZM, Weichselbaum R, Kufe D (1994a) Ionizing radiation induces rapid tyrosine phosphorylation of p34cdc2. Cancer Res 54:1412–1414
Kharbanda S, Yuan ZM, Rubin E, Weichselbaum R, Kufe D (1994b) Activation of Src-like p56/p53lyn tyrosine kinase by ionizing radiation. J Biol Chem 269:20739–20743
King RC, Riley SF, Cassidy JD, White PE, Paik YK (1981) Giant polytene chromosomes from the ovaries of a *Drosophila* mutant. Science 212:441–443
Kinoshita N, Yamano H, Niwa H, Yoshida T, Yanagida M (1993) Negative regulation of mitosis by the fission yeast protein phosphatase ppa2. Genes Dev 7:1059–1071
Knoblich JA, Sauer K, Jones L, Richardson H, Saint R, Lehner CF (1994) Cyclin E controls S phase progression and its down-regulation during *Drosophila* embryogenesis is required for the arrest of cell proliferation. Cell 77:107–120
Kosako H, Gotoh Y, Nishida E (1994) Requirement for the MAP kinase kinase/MAP kinase cascade in *Xenopus* oocyte maturation. EMBO J 13:2131–2138
Krek W, Nigg EA (1991a) Differential phosphorylation of vertebrate p34cdc2 kinase at the G1/S and G2/M transitions of the cell cycle: identification of major phosphorylation sites. EMBO J 10:305–316
Krek W, Nigg EA (1991b) Mutations of p34cdc2 phosphorylation sites induce premature mitotic events in HeLa cells: evidence for a double block to p34cdc2 kinase activation in vertebrates. EMBO J 10:3331–3341
Kubiak JZ, Weber M, de Pennart H, Winston NJ, Maro B (1993) The metaphase II arrest in mouse oocytes is controlled through microtubule-dependent destruction of cyclin B in the presence of CSF. EMBO J 12:3773–3778

Kudryavtsev BN, Kudryavtseva MV, Sakuta GA, Stein GI (1993) Human hepatocyte polyploidization kinetics in the course of the life cycle. Virchows Arch B Cell Pathol Incl Mol Pathol 64:387–393

Kuerbitz SJ, Plunkett BS, Walsh WV, Kastan MB (1992) Wild-type p53 is a cell-cycle checkpoint determinant following irradiation. Proc Natl Acad Sci USA 89:7491–7495

Kumagai A, Dunphy WG (1991) The cdc25 protein controls tyrosine dephosphorylation of the cdc2 protein in a cell-free system. Cell 64:903–914

Kumagai A, Dunphy WG (1992) Regulation of the cdc25 protein during the cell cycle in *Xenopus* extracts. Cell 70:139–151

Kumagai A, Dunphy WG (1995) Control of the Cdc2/cyclin B complex in *Xenopus* egg extracts arrested at a G2/M checkpoint with DNA synthesis inhibitors. Mol Biol Cell 6:199-213

Kumagai A, Dunphy WG (1996) Purification and molecular cloning of Plx1, a Cdc25-regulatory kinase from *Xenopus* egg extracts. Science 273:1377–1380

Labib K, Moreno S (1996) rum1: A CDK inhibitor regulating G1 progression in fission yeast. Trends Cell Biol 6:62–66

Lee MG, Nurse P (1987) Complementation used to clone a human homologue of the fission yeast cell cycle control gene cdc2. Nature 327:31–35

Lee MS, Ogg S, Xu M, Parker LL, Donoghue DJ, Maller JL, Piwnica-Worms H (1992) cdc25+ encodes a protein phosphatase that dephosphorylates p34cdc2. Mol Biol Cell 3:73–84

Lee TH, Kirschner MW (1996) An inhibitor of p34cdc2 cyclin B that regulates the G2/M transition in *Xenopus* extracts. Proc Natl Acad Sci USA 93:352–356

Lew DJ, Reed SI (1995) A cell cycle checkpoint monitors cell morphogenesis in budding yeast. J Cell Biol 129:739–749

Liu F, Stanton JJ, Wu ZQ, Piwnica-Worms H (1997) The human Myt1 kinase preferentially phosphorylates Cdc2 on threonine 14 and localizes to the endoplasmic reticulum and Golgi complex. Mol Cell Biol 17:571–583

Llamazares S, Moreira A, Tavares A, Girdham C, Spruce BA, Gonzalez C, Karess RE, Glover DM, Sunkel CE (1991) polo encodes a protein kinase homolog required for mitosis in *Drosophila.* Genes Dev 5:2153–2165

Lundgren K, Walworth N, Booher R, Dembski M, Kirschner M, Beach D (1991) mik1 and wee1 cooperate in the inhibitory tyrosine phosphorylation of cdc2. Cell 64:1111–1122

Maity A, McKenna WG, Muschel RJ (1995) Evidence for post-transcriptional regulation of cyclin B1 mRNA in the cell cycle and following irradiation in HeLa cells. EMBO J 14:603–609

Makela TP, Tassan JP, Nigg EA, Frutiger S, Hughes GJ, Weinberg RA (1994) A cyclin associated with the CDK-activating kinase MO15. Nature 371:254–257

Matsuoka M, Kato JY, Fisher RP, Morgan DO, Sherr CJ (1994) Activation of cyclin-dependent kinase 4 (cdk4) by mouse MO15-associated kinase. Mol Cell Biol 14:7265–7275

Matten W, Daar I, Vande Woude GF (1994) Protein kinase A acts at multiple points to inhibit *Xenopus* oocyte maturation. Mol Cell Biol 14:4419–4426

Matten WT, Copeland TD, Ahn NG, Vande Woude GF (1996) Positive feedback between MAP kinase and Mos during *Xenopus* oocyte maturation. Dev Biol 179:485–492

McGowan CH, Russell P (1993) Human Wee1 kinase inhibits cell division by phosphorylating p34cdc2 exclusively on Tyr15. EMBO J 12:75–85

Mendenhall MD (1993) An inhibitor of p34CDC28 protein kinase activity from *Saccharomyces cerevisiae.* Science 259:216–219

Milne DM, Campbell DG, Caudwell FB, Meek DW (1994) Phosphorylation of the tumor suppressor protein p53 by mitogen-activated protein kinases. J Biol Chem 269:9253–9260

Minshull J, Sun H, Tonks NK, Murray AW (1994) A MAP kinase-dependent spindle assembly checkpoint in *Xenopus* egg extracts. Cell 79:475–486

Moreno S, Nurse P (1994) Regulation of progression through the G1 phase of the cell cycle by the rum1 + gene. Nature 367:236–242

Mueller PR, Coleman TR, Kumagai A, Dunphy WG (1995) Myt1: A membrane-associated inhibitory kinase that phosphorylates Cdc2 on both threonine-14 and tyrosine-15. Science 270:86–90

Murray A (1995) Cyclin ubiquitination: the destructive end of mitosis. Cell 81:149–152
Murray AW, Solomon MJ, Kirschner MW (1989) The role of cyclin synthesis and degradation in the control of maturation promoting factor activity. Nature 339:280–286
Murray AW (1992) Creative blocks: cell-cycle checkpoints and feedback controls. Nature 359:599–604
Nagata A, Igarashi M, Jinno S, Suto K, Okayama H (1991) An additional homolog of the fission yeast cdc25+ gene occurs in humans and is highly expressed in some cancer cells. New Biol 3:959–968
Nebreda AR, Hunt T (1993) The c-mos proto-oncogene protein kinase turns on and maintains the activity of MAP kinase, but not MPF, in cell-free extracts of *Xenopus* oocytes and eggs. EMBO J 12:1979–1986
Nebreda AR, Gannon JV, Hunt T (1995) Newly synthesized protein(s) must associate with p34cdc2 to activate MAP kinase and MPF during progesterone-induced maturation of *Xenopus* oocytes. EMBO J 14:5597–5607
Nigg EA (1993) Targets of cyclin-dependent protein kinases. Curr Opin Cell Biol 5:187–193
Nigg EA (1996) Cyclin-dependent kinase 7: At the cross-roads of transcription, DNA repair and cell cycle control? Curr Opin Cell Biol 8:312–317
Nishizawa M, Furuno N, Okazaki K, Tanaka H, Ogawa Y, Sagata N (1993) Degradation of Mos by the N-terminal proline (Pro2)-dependent ubiquitin pathway on fertilization of *Xenopus* eggs: possible significance of natural selection for Pro2 in Mos. EMBO J 12:4021–4027
Norbury C, Nurse P (1992) Animal cell cycles and their control. Annu Rev Biochem 61:441–470
Norbury C, Blow J, Nurse P (1991) Regulatory phosphorylation of the p34cdc2 protein kinase in vertebrates. EMBO J 10:3321–3329
Nugroho TT, Mendenhall MD (1994) An inhibitor of yeast cyclin-dependent protein kinase plays an important role in ensuring the genomic integrity of daughter cells. Mol Cell Biol 14:3320–3328
Oberpriller JO, Ferrans VJ, Carroll RJ (1983) Changes in DNA content, number of nuclei and cellular dimensions of young rat atrial myocytes in response to left coronary artery ligation. J Mol Cell Cardiol 15:31–42
O'Connell MJ, Raleigh JM, Verkade HM, Nurse P (1997) Chk1 is a wee1 kinase in the G2 DNA damage checkpoint inhibiting cdc2 by Y15 phosphorylation. EMBO J 16:545–554
O'Connor PM, Ferris DK, Hoffmann I, Jackman J, Draetta G, Kohn KW (1994) Role of the cdc25C phosphatase in G2 arrest induced by nitrogen mustard. Proc Natl Acad Sci USA 91:9480–9484
Osmani SA, Ye XS (1996) Cell cycle regulation in *Aspergillus* by two protein kinases. Biochem J 317:633–641
Page AW, Orr Weaver TL (1997) Stopping and starting the meiotic cell cycle. Curr Opin Genet Dev 7:23–31
Pan BT, Chen CT, Lin SM (1994) Oncogenic Ras blocks cell cycle progression and inhibits p34cdc2 kinase in activated *Xenopus* egg extracts. J Biol Chem 269:5968–5975
Pan BT, Zhang Y, Brott B, Chen DH (1997) The 96 kDa protein kinase activated by oncogenic Ras in *Xenopus* egg extracts is also activated by constitutively active Mek: activation requires serine/threonine phosphorylation. Oncogene 14:1653–1660
Papa S, Capitani S, Matteucci A, Vitale M, Santi P, Martelli AM, Maraldi NM, Manzoli FA (1987) Flow cytometric analysis of isolated rat liver nuclei during growth. Cytometry 8:595–601
Parker LL, Piwnica-Worms H (1992) Inactivation of the p34cdc2-cyclin B complex by the human WEE 1 tyrosine kinase. Science 257:1955–1957
Parker LL, Walter SA, Young PG, Piwnica-Worms H (1993) Phosphorylation and inactivation of the mitotic inhibitor Wee 1 by the nim1/cdr1 kinase. Nature 363:736–738
Patra D, Dunphy WG (1996) Xe-p9, a *Xenopus* Suc1/Cks homolog, has multiple essential roles in cell-cycle control. Genes Dev 10:1503–1515
Peters JM, King RW, Hoog C, Kirschner MW (1996) Identification of BIME as a subunit of the anaphase-promoting complex. Science 274: 1199–1201
Pines J (1993) Cyclins and cyclin-dependent kinases: take your partners. Trends Biochem Sci 18:195–197

Poon RY, Yamashita K, Adamczewski JP, Hunt T, Shuttleworth J (1993) The cdc2-related protein p40MO15 is the catalytic subunit of a protein kinase that can activate p33cdk2 and p34cdc2. EMBO J 12:3123–3132

Poon RYC, Hunter T (1995) Dephosphorylation of Cdk2 Thr160 by the cyclin-dependent kinase-interacting phosphatase KAP in the absence of cyclin. Science 270:90–93

Poon RYC, Jiang W, Toyoshima H, Hunter T (1996) Cyclin-dependent kinases are inactivated by a combination of p21 and Thr-14/Tyr-15 phosphorylation after UV-induced DNA damage. J Biol Chem 271:13283–13291

Posada J, Yew N, Ahn NG, Vande Woude GF, Cooper JA (1993) Mos stimulates MAP kinase in *Xenopus* oocytes and activates a MAP kinase in vitro. Mol Cell Biol 13:2546–2553

Reed BH (1995) Drosophila development pulls the strings of the cell cycle. Bioessays 17:553–556

Rempel RE, Sleight SB, Maller JL (1995) Maternal Xenopus Cdk2-cyclin E complexes function during meiotic and early embryonic cell cycles that lack a G1 phase. J Biol Chem 270:6843–6855

Ridley AJ, Paterson HF, Noble M, Land H (1988) Ras-mediated cell cycle arrest is altered by nuclear oncogenes to induce Schwann cell transformation. EMBO J 7:1635–1645

Rosenberg B, Pfitzer P (1983) Ploidy in the hearts of elderly patients. Virchows Arch B Cell Pathol Incl Mol Pathol 42:19–24

Roszell JA, Fredi JL, Irving CC (1978) The development of polyploidy in two classes of rat liver nuclei. Biochim Biophys Acta 519:306–316

Roy LM, Haccard O, Izumi T, Lattes BG, Lewellyn AL, Maller JL (1996) Mos proto-oncogene function during oocyte maturation in *Xenopus*. Oncogene 12:2203–2211

Roy R, Adamczewski JP, Seroz T, Vermeulen W, Tassan JP, Schaeffer L, Nigg EA, Hoeijmakers JH, Egly JM (1994) The MO15 cell-cycle kinase is associated with the TFIIH transcription-DNA repair factor. Cell 79:1093–1101

Russo AA, Jeffrey PD, Pavletich NP (1996) Structural basis of cyclin-dependent kinase activation by phosphorylation. Nat Struct Biol 3:696–700

Sagata N (1996) Meiotic metaphase arrest in animal oocytes: its mechanisms and biological significance. Trends Cell Biol 6:22–28

Sagata N (1997) What does Mos do in oocytes and somatic cells? Bioessays 19:13–21

Sagata N, Oskarsson M, Copeland T, Brumbaugh J, Vande Woude GF (1988) Function of c-mos proto-oncogene product in meiotic maturation in Xenopus oocytes. Nature 335:519–525

Sagata N, Daar I, Oskarsson M, Showalter SD, Vande Woude GF (1989a) The product of the mos proto-oncogene as a candidate "initiator" for oocyte maturation. Science 245:643–646

Sagata N, Watanabe N, Vande Woude GF, Ikawa Y (1989b) The c-mos proto-oncogene product is a cytostatic factor responsible for meiotic arrest in vertebrate eggs. Nature 342:512–518

Sauer K, Knoblich JA, Richardson H, Lehner CF (1995) Distinct modes of cyclin E/cdc2c kinase regulation and S-phase control in mitotic and endoreduplication cycles of *Drosophila* embryogenesis. Genes Dev 9:1327–1339

Schwob E, Bohm T, Mendenhall MD, Nasmyth K (1994) The B-type cyclin kinase inhibitor p40SIC1 controls the G1 to S transition in *S. cerevisiae*. Cell 79:233–244

Seger R, Krebs EG (1995) The MAPK signaling cascade. FASEB J 9:726–735

Serizawa H, Makela TP, Conaway JW, Conaway RC, Weinberg RA, Young RA (1995) Association of Cdk-activating kinase subunits with transcription factor TFIIH. Nature 374:280–282

Serrano M, Lin AW, McCurrach ME, Beach D, Lowe SW (1997) Oncogenic ras provokes premature cell senescence associated with accumulation of p53 and p16INK4a. Cell 88:593–602

Sherr CJ (1993) Mammalian G1 cyclins. Cell 73:1059–1065

Shiekhattar R, Mermelstein F, Fisher RP, Drapkin R, Dynlacht B, Wessling HC, Morgan DO, Reinberg D (1995) Cdk-activating kinase complex in a component of human transcription factor TFIIH. Nature 374:283–287

Sia RA, Herald HA, Lew DJ (1996) Cdc28 tyrosine phosphorylation and the morphogenesis checkpoint in budding yeast. Mol Biol Cell 7:1657–1666

Sigal SH, Gupta S, Gebhard CF Jr, Holst P, Neufeld D, Reid LM (1995) Evidence for a terminal differentiation process in the rat liver. Differentiation 59:35–42

Smythe C, Newport JW (1992) Coupling of mitosis to the completion of S phase in Xenopus occurs via modulation of the tyrosine kinase that phosphorylates p34cdc2. Cell 68:787–797

Solomon MJ, Glotzer M, Lee TH, Philippe M, Kirschner NW (1990) Cyclin activation of p34cdc2. Cell 63:1013–1024

Solomon MJ, Lee T, Kirschner MW (1992) Role of phosphorylation in p34cdc2 activation: identification of an activating kinase. Mol Biol Cell 3:13–27

Solomon MJ, Harper JW, Shuttleworth J (1993) CAK, the p34cdc2 activating kinase, contains a protein identical or closely related to p40MO15. EMBO J 12:3133–3142

Sorger PK, Murray AW (1992) S-phase feedback control in budding yeast independent of tyrosine phosphorylation of p34cdc28. Nature 355:365–368

Stern B, Nurse P (1996) A quantitative model for the cdc2 control of S phase and mitosis in fission yeast. Trends Genet 12:345–350

Stewart E, Enoch T (1996) S-phase and DNA-damage checkpoints: a tale of two yeasts. Curr Opin Cell Biol 8:781–787

Stewart N, Hicks GG, Paraskevas F, Mowat M (1995) Evidence for a second cell cycle block at G2/M by p53. Oncogene 10:109–115

Strausfeld U, Fernandez A, Capony JP, Girard F, Lautredou N, Derancourt J, Labbe JC, Lamb NJ (1994) Activation of p34cdc2 protein kinase by microinjection of human cdc25C into mammalian cells. Requirement for prior phosphorylation of cdc25C by p34cdc2 on sites phosphorylated at mitosis. J Biol Chem 269:5989–6000

Stukenberg PT, Lustig KD, McGarry TJ, King RW, Kuang J, Kirschner NW (1997) Systematic identification of mitotic phosphoproteins. Curr Biol 7:338–348

Tsukada T, Tomooka Y, Takai S, Ueda Y, Nishikawa S, Yagi T, Tokunaga T, Takeda N, Suda Y, Abe S (1993) Enhanced proliferative potential in culture of cells from p53-deficient mice. Oncogene 8:3313–3322

Tuel-Ahlgren L, Jun X, Waddick KG, Jin J, Bolen J, Uckun FM (1996) Role of tyrosine phosphorylation in radiation-induced cell cycle-arrest of leukemic B-cell precursors at the G2-M transition checkpoint. Leuk Lymphoma 20:417–426

Uckun FM, Tuel-Ahlgren L, Waddick KG, Jun X, Jin JH, Myers DE, Rowley RB, Burkhardt AL, Bolen JB (1996) Physical and functional interactions between Lyn and p34cdc2 kinases in irradiated human B-cell precursors. J Biol Chem 271:6389–6397

Varmuza S, Prideaux V, Kothary R, Rossant J (1988) Polytene chromosomes is mouse trophoblast giant cells. Development 102:127–134

Verlhac MH, Kubiak JZ, Weber M, Geraud G, Colledge WH, Evans MJ, Maro B (1996) Mos is required for MAP kinase activation and is involved in microtubule organization during meiotic maturation in the mouse. Development 122:815–822

Vikhanskaya F, Erba E, D'Incalci M, Broggini M (1994) Introduction of wild-type p53 in a human ovarian cancer cell line not expressing endogenous p53. Nucleic Acids Res 22:1012–1017

Wang Z, Zhang Y, Kamen D, Lees E, Ravid K (1995) Cyclin D3 is essential for megakaryocytopoiesis. Blood 86:3783–3788

Watanabe N, Vande Woude GF, Ikawa Y, Sagata N (1989) Specific proteolysis of the c-mos proto-oncogene product by calpain on fertilization of *Xenopus* eggs. Nature 342:505–511

Watanabe N, Broome M, Hunter T (1995) Regulation of the human WEE1Hu CDK tyrosine 15-kinase during the cell cycle. EMBO J 14:1878–1891

Weber M, Kubiak JZ, Arlinghaus RB, Pines J, Maro B (1991) c-mos proto-oncogene product is partly degraded after release from meiotic arrest and persists during interphase in mouse zygotes. Dev Biol 148:393–397

Wu L, Russell P (1993) Nim1 kinase promotes mitosis by inactivating Wee 1 tyrosine kinase. Nature 363:738–741

Wu L, Russell P (1997) Nif1, a novel mitotic inhibitor in *Schizosaccharomyces pombe*. EMBO J 16:1342–1350

Xiong Y, Hannon GJ, Zhang H, Casso D, Kobayashi R, Beach D (1993) p21 is a universal inhibitor of cyclin kinases. Nature 366:701–704

Ye XS, Fincher RR, Tang A, O'Donnell K, Osmani SA (1996) Two S-phase checkpoint systems, one involving the function of both BIME and Tyr15 phosphorylation of p34cdc2, inhibit NIMA and prevent premature mitosis. EMBO J 15:3599–3610

Yew N, Oskarsson M, Daar I, Blair DG, Vande Woude GF (1991) mos gene transforming efficiencies correlate with oocyte maturation and cytostatic factor activities. Mol Cell Biol 11:604–610

Yew N, Mellini ML, Vande Woude GF (1992) Meiotic initiation by the mos protein in *Xenopus*. Nature 355:649–652

Yew N, Strobel M, Vande Woude GF (1993) Mos and the cell cycle: the molecular basis of the transformed phenotype. Curr Opin Genet Dev 3:19–25

Zhang Y, Wang ZG, Ravid K (1996) The cell cycle in polyploid megakaryocytes is associated with reduced activity of cyclin B1-dependent Cdc2 kinase. J Biol Chem 271:4266–4272

Zhao X, Singh B, Arlinghaus RB (1991) Inhibition of c-mos protein kinase blocks mouse zygotes at the pronuclei stage. Oncogene 6:1423–1426

Zybina EV, Zybina TG (1996) Polytene chromosomes in mammalian cells. Int Rev Cytol 165:53–119

Mechanisms of Interferon Action

Douglas W. Leaman[1]

1 Introduction

The interferons (IFNs) are a family of cytokines that elicit pleiotropic biological effects (Pestka et al. 1987; DeMaeyer and DeMaeyer-Guignard 1988; Sen and Lengyel 1992). Although best known (and named) for their ability to inhibit viral replication in treated cells, IFNs are also capable of regulating cellular proliferation, differentiation, and immunological responses (Lengyel 1982; Pestka et al. 1987; Baron et al. 1992; Vilcek and Sen 1996). Over the past 40 years, studies of IFN expression, structure, and function have provided a number of important scientific breakthroughs. The IFNs were the first cytokines to be cloned and characterized, and the first to be used therapeutically. The study of how the transcription of the IFN genes is regulated has provided insight into the complexities of inducible gene expression in eukaryotic organisms, and the analysis of IFN-induced proteins has uncovered a number of important regulators of many physiological processes. Perhaps the most far-reaching discovery to have emerged from analysis of the IFN system has been the identification of the JAK and STAT signaling components, a finding that has ushered in a new era of cytokine and growth factor signal transduction research.

This chapter focuses on the biochemistry of IFN action, emphasizing recent advances in understanding how the IFN receptors and signal transduction components function. It also covers some of the mechanisms by which the IFNs regulate cellular proliferation and inhibit viral replication, features that have led to the utilization of IFNs for therapeutic purposes. Finally, this report ends with a description of a novel IFN subtype which has an unexpected biological function in the normal physiology of pregnancy, illustrating that we have by no means uncovered all of the mysteries surrounding these important cytokines.

[1]Gemini Technologies Inc., 11,000 Cedar Ave., Suite 140, Cleveland, OH 44106, USA

Progress in Molecular and Subcellular Biology, Vol. 20
A. Macieira-Coelho (Ed.)

2 The Interferon Proteins and Genes

2.1 Classification and Biological Properties

The IFNs were first purified on the basis of their ability to interfere with virus replication (Isaacs and Lindenmann 1957), a common characteristic that has resulted in the grouping together of both related and unrelated proteins (see below). IFNs are synthesized and secreted by cells following exposure to virus or other stimuli, and they act upon neighboring cells to inhibit the replication of many different DNA and RNA viruses (DeMaeyer and DeMaeyer-Guignard 1988). IFNs also affect the function of the immune system, and have potent effects on cell proliferation and differentiation. They increase the abundance or affect the activities of specific effector proteins, many of which are encoded by subsets of early response genes known collectively as IFN-stimulated genes (ISGs) (Sen 1991; Stark et al. 1997). The protein products of these genes then mediate the physiological responses of the IFNs (Hovanessian 1991; Williams 1995; Silverman 1996; Stark et al. 1997).

Mammalian IFNs are broadly classified into two structurally discrete categories. The type I IFNs include the ubiquitous IFNα, -β, and -ω subtypes, and the species-restricted IFNτ (see below and Sect. 7). The only known type II IFN subtype is IFNγ. Type I and II IFNs differ in their primary amino acid sequences, evolutionary relationships, receptor cross-reactivities, sites of production, and inducibility. Nonetheless, type I and II IFNs induce overlapping sets of responsive genes (Sect. 5), and activate remarkably similar signaling cascades (Sect. 4), two features that contribute to the similarities in type I and II IFN function.

The type I IFNs are part of the body's first line of defense against viruses, foreign cells (including tumor cells), microbes, and other pathogens (DeMaeyer and DeMaeyer-Guignard 1988). Although low levels of type I IFNs are expressed constitutively, they are produced by cells in substantial quantities in response to virus, bacteria, and certain cytokines or growth factors (Sen and Lengyel 1992). The exact IFN subtypes produced depends on the inducer and the cell type (see Sect. 2.3). Unlike acquired immune responses, such as the production of epitope-specific antibodies, the type I IFN response is very rapid, usually peaking within several hours of the initial infection (Baron et al. 1992). Type I IFNs can also activate components of the immune system, and can enhance T cell cytotoxicity against virus-infected or allogeneic cells, mediated, to a large extent, by upregulation of Class I major histocompatibility complex (MHC) antigens which present viral or tumor antigens (Fleischmann and Fleischmann 1992).

IFNγ, produced only by T cells and natural killer cells in response to foreign antigens or T cell mitogens (Hardy and Young 1992; Sen and Lengyel 1992), is more an immunomodulatory cytokine than an antiviral agent, since its primary physiological role appears to be to regulate macrophage function (Talmadge et al. 1986). Among the biochemical changes observed in IFNγ-activated macrophages are the induced expression of cytokines (interleukin-1 or tumor necrosis factor-α), Class I and Class II MHC, cell adhesion molecules (LFA-1), Fc receptors, and components of the complement cascade (C3 receptors) (Hardy and Young 1992). These features contribute to the antimicrobial and/or antitumoricidal actions of macrophages, and highlight the importance of IFNγ in the host defense system. More extensive summaries of the biological and biochemical functions of type I and II IFNs can be found elsewhere (Lengyel 1982; Pestka et al. 1987; Dron and Tovey 1992; Hardy and Young 1992; Stark et al. 1997).

2.2 Evolution of the IFN Genes

Many fundamental differences in the functions of IFN subsets can be traced to the evolutionary relationships among the different genes. The type I genes, for example, have all evolved from a common primordial progenitor (Wilson et al. 1983) and are arrayed on the short arm of chromosome 9 in humans (Weissman and Weber 1986; DeMaeyer and DeMaeyer-Guignard 1988). The IFNα and IFNβ subtypes evolved first, presumably as a result of the duplication of this progenitor gene about 400 million years ago, prior to the divergence of mammals (Wilson et al. 1983; Gillespie et al. 1984). Since that time, IFNβ has remained as a single-copy gene in most species, whereas the IFNα prototypic gene has subsequently amplified to high copy numbers, comprising up to 30 distinct loci in certain mammals (Weissman et al. 1982; DeMaeyer and DeMaeyer-Guignard 1988). Some of these copies represent nonfunctional pseudogenes, but at least 13 of the 18 IFNα alleles in humans are known to encode fully functional proteins (Pestka et al. 1987; DeMaeyer and DeMaeyer-Guignard 1988; Allen and Diaz 1996). The relative antiviral and antiproliferative activities of different IFNα subtypes can differ (Fish et al. 1983; Foster et al. 1996), as can the relative inducibilities of their genes (Hiscott et al. 1984), suggesting that there may be distinct physiological circumstances under which one or more of the subtypes plays a predominant role in warding off viral infection. Nonetheless, it is not currently known why multiple IFNα genes have been retained in all mammalian species throughout evolution.

IFNω evolved from IFNα ~ 100 million years ago (Henco et al. 1985), before the radiation of mammalian species and probably before the amplification of the prototypic IFNα gene (Gillespie et al. 1984). Like the IFNα genes, the IFNω genes have duplicated to high copy numbers in many species, such as ruminants (Capon et al. 1985; Leaman and Roberts 1992), but not in humans (Capon et al. 1985). A major difference between the IFNω and other type I IFNs is the length of the mature protein. IFNω proteins are 172 amino acid residues long, as compared to 166 residues for the IFNβ and most IFNα proteins (although some IFNαs are 165 residues long) (DeMaeyer and DeMaeyer-Guignard 1988). The biological significance, if any, of the additional 6 amino acids is unclear. The IFNω proteins elicit many of the same biological responses as other type I IFNs, including inhibiting viral replication and cellular proliferation (Capon et al. 1985; DeMaeyer and DeMaeyer-Guignard 1988). Whether they also have unique functions is not known.

A fourth class of type I IFN has evolved exclusively in species belonging to the *Ruminantia* suborder of mammals. This unusual IFN, now called IFNτ, has a unique pattern of expression in trophoblast cells of the developing trophectoderm (the presumptive placenta), where it is proposed to function in the establishment of pregnancy (Roberts et al. 1992). IFNτ possesses biological activities characteristic of other type I IFNs, including antiviral, antiproliferative and immunomodulatory activities (see Roberts et al. 1992), although it may have additional properties that make it well suited for its role in early pregnancy. Like the IFNω, from which they evolved, mature IFNτ proteins are 172 residues in length, and they interact with a common type I IFN receptor (Hansen et al. 1989). A more detailed description of IFNτ regulation and function is provided in Section 7 of this chapter.

The type II IFNγ gene evolved from a distinct progenitor gene that appeared more recently than the prototypic type I IFN gene (DeMaeyer and DeMaeyer-Guignard 1988). Unlike the intronless type I IFN genes, the IFNγ gene is more typical of higher eukaryotic genes, consisting of four exons separated by three introns, the first of which contains sequences involved in regulating IFNγ gene expression (Young 1996). The single IFNγ gene in humans resides on chromosome 12 (Gray and Goeddel 1982). The mature IFNγ protein is 143 amino acids in length, and shares very little sequence similarity with the type I IFNs (Pestka et al. 1987; DeMaeyer and DeMaeyer-Guignard 1988). The expression of the IFNγ gene is subject to both developmental and lineage-specific constraints not involved in the regulation of most type I IFNs (Sect. 2.3.1).

2.3 Tissue Distribution and Regulation of IFN Gene Expression

2.3.1 Type I IFN Genes

Although normally quiescent, the expression and synthesis of type I IFNs can be induced in almost all cell types (Sen and Lengyel 1992). IFNβ is the predominant subtype expressed by most somatic cells, including fibroblasts and epithelial cells (Havell et al. 1978), whereas IFNα and IFNω are the major species induced in cells of hematopoeitic lineage (DeMaeyer and DeMaeyer-Guignard 1988). The primary inducers are viruses, but other organisms, including bacteria and protozoa, can also evoke an IFN response (Lengyel and Sen 1992). The double-stranded (ds) RNA intermediates that accumulate during the replicative cycles of many viruses appear to be the major inducers of IFN gene transcription (Lai and Joklik 1973; Levy and Salazar 1992; Welsh and Sen 1997). Indeed, if dsRNA is introduced directly into cells or animals, it can effectively mimic the virus-dependent induction of some type I IFN genes. Regulation of IFN synthesis occurs predominantly at the transcriptional level, and although some of the transcription factors involved in regulating IFN gene expression are known, questions persist as to how virus and other stimuli generate the biological cues necessary to activate these transcriptional effectors. Transcription of the IFN genes is transient, rarely lasting more than a few hours (DeMaeyer and DeMaeyer-Guignard 1988), which can be attributed to the transient nature of the transcription factors required for IFN gene induction, to the upregulation of repressor molecules that suppress transcription after several hours (Taniguchi 1989; Maniatis et al. 1992), and to the presence of AU-rich destabilizing signals in the 3′-untranslated portions of IFN transcripts which determine the short half-life of the IFN mRNA (Whittemore and Maniatis 1990).

The IFNβ gene is more amenable to transcriptional analyses than the other type I IFN genes, in part because its expression can be induced in many different cell types, but also because of the relative ease in studying a single-copy gene versus the multi-copy genes typical of other type I IFNs. Human IFNβ promoter sequences from position −125 to −38 (relative to the transcription start site) are sufficient to confer full virus inducibility upon heterologous promoters (Taniguchi 1989; Maniatis et al. 1992). This region of the promoter is composed of four overlapping positive regulatory domains (PRDs I–IV), and at least one negative regulatory domain (NRD) (Goodbourn and Maniatis 1988). Unfortunately, these designations disguise the fact that several of the PRD also contribute to the downregulation of IFNβ expression following induction. The PRDI element, for example, binds members of the IFN-regulatory factor (IRF) family, including the transactivator IRF-1, and the transcriptional repressor IRF-2 (Miyamoto et al. 1988; Harada et al. 1989). PRDI also binds a poorly characterized repressor PRDI-BFI, and the myeloid cell-specific repressor ICSBP (Keller and Maniatis 1991; Nelson et al. 1993). Thus, it seems that while PRDI plays an obligate role in transcriptional induction of the IFNβ gene, its predominant role is to maintain transcriptional quiescence in the absence of inducer. An additional IRF-1 binding site is found within the PRDIII element which, because of its low affinity for IRF-1, may actually bind other IRF family members in vivo (Fujita et al. 1988). PRDII binds the nuclear factor κB (NFκB) transcription factor, and PRDIV binds an ATF-2-containing factor (Lenardo et al. 1989; Visvanathan and Goodbourne 1989; Thanos and Maniatis 1992; Du et al. 1993). Both factors are activated by virus or dsRNA and play important roles in IFNβ gene induction. The binding and activities of all factors that interact with the IFNβ promoter are influenced by the high mobility group protein, HMGI(Y), which induces specific bends in the promoter DNA, augmenting proper protein-protein interactions (Thanos and Maniatis 1992; Falvo et al. 1995). The functional promoter assembly, complete with associated transcription factors, is known collectively as an enhanceosome, and optimal activation of IFNβ expression requires the presence of all PRD sites in their proper spatial orientation (Falvo et al. 1995; Thanos and Maniatis 1995). Indeed, individual PRD sites are poorly responsive to virus when placed in front of a heterologous promoter, either alone or in multiple copies (Thanos and Maniatis 1992), highlighting the importance of interactions between the different PRD-specific transcription factors.

Despite these advances in our understanding of IFNβ gene regulation, many questions persist. As mentioned earlier, it is unclear how the transcription factors that regulate IFNβ expression are modified in response to virus infection or by stimulation with dsRNA. The dsRNA-dependent serine/threonine kinase PKR appears to play a major role in activating some of the factors (Kumar et al. 1994; Maran et al. 1994), but is not sufficient to explain all inducing effects, since IFN genes can be activated by virus in PKR-null cells (Yang et al. 1995). The identification of all factors that regulate IFNβ promoter activity, such as the NRD-binding factors, remains incomplete. In addition, the results of transcriptional studies carried out with the human promoter do not always apply to promoters from other species. The mouse IFNβ promoter, for example, shares roughly 70% nucleotide sequence identity with the human promoter, yet differs markedly in its inducibility following exposure of cells to virus or dsRNA (Dirks et al. 1989). The reasons for this difference are unknown. Finally, although both IRF-1 and IRF-2 have been implicated in IFNβ gene regulation, mice carrying deletions of the IRF-1 or IRF-2 genes showed no defects in either basal or virus-induced IFNβ gene expression (Matsuyama et al. 1993), suggesting that currently unknown factors contribute to IFNβ regulation, either normally or as surrogate regulators in the absence of the IRFs.

Much less is known about the transcriptional regulation of the IFNα or IFNω genes. Unlike the highly conserved IFNβ gene promoter, IFNα genes share only moderate promoter sequence identity, even within a given species (Henco et al. 1985; Houle and Santoro 1996), and the individual genes often differ significantly in their responsiveness to different virus or other stimuli (Hiscott et al. 1984; Dent et al. 1996). A molecular explanation for these differences is still largely unknown. Only about 100 base pairs upstream of the site of transcriptional initiation is usually required for optimal IFNα gene inducibility by virus (Weidle and Weissmann 1983). Although some of the same promoter elements involved in IFNβ gene regulation are present within IFNα promoters, including PRDI-like elements that bind IRF-1 and IRF-2 (Miyamoto et al. 1988), others, such as NFκB elements, are generally missing, and IFNα promoters tend to lack the overall organization of regulatory elements that is characteristic of the IFNβ promoter. At least one IFNα promoter-specific sequence motif (GGAAATG) has been identified through functional studies of the IFNα2 promoter (MacDonald et al. 1990). To date, however, the so-called TG factor that presumably binds this site has not been cloned, and the roles of other putative promoter elements remain poorly understood. IFNω gene promoters possess considerable sequence identity with the IFNα gene promoters, and IFNs ω, α and β are induced by many of the same stimuli (reviewed in Roberts et al. 1992). Few studies have addressed the transcriptional control of IFNω genes, and more work in this area is clearly warranted.

2.3.2
The IFNγ Gene

The major inducers of IFNγ synthesis in T cells are factors involved in T cell stimulation, such as infectious agents that induce monocytes to produce interleukin (IL)-12, antigens to which the body is presensitized, polypeptide growth factors, IL-1, IL-2, and the newly identified IFNγ-inducing factor (Hardy and Young 1992; Okamura et al. 1995; Young 1996). The T cell-restricted nature of IFNγ expression is regulated, in part, by methylation of IFNγ promoter sequences. In $CD4^+$ memory T cells expressing IFNγ, the promoter is not methylated, whereas in nonexpressing cells, a CpG motif located near position −54 relative to the transcription start site is constitutively methylated, resulting in the repression of promoter activity. Treatment of nonexpressing TH2 cells with 5-azacytidine, an agent that blocks DNA methylation, overcomes the repression and permits IFNγ induction in those cells (reviewed in Young 1996).

IFNγ induction requires the activation of factors that bind to regulatory elements located both in the 5′-flanking region and within the first intron (Hardy et al. 1985, 1987; Penix et al. 1996; Young 1996). Recent studies have demonstrated that regulation of the IFNγ gene by cytokines, including IL-2 and IL-12, may be mediated by sequence elements in the first intron of the gene that bind STATs (signal transducers and activators of transcription, Sect. 4.1.2). Multiple STAT recognition sites, tandemly arranged within this intron, have specificities for different STAT family members. The types and numbers of STAT complexes that are bound to these elements may dictate how strongly IFNγ is induced by a variety of cytokines (Xu et al. 1996). Importantly, the same studies also provided the first clear example of cooperative interaction between STATs bound to tandem DNA recognition sequences (Xu et al. 1996), a feature of STAT transcriptional regulation that most likely extends to other genes as well. In addition to STAT-binding sites, a variety of other transcription factor recognition sequences have been identified, including those that bind NFκB, nuclear factor-activated T cell (NFAT), and AP-1 sites (Hardy and Young 1992; Young 1996). Negative regulatory regions of the promoter have also been localized (Young 1996), some of which bind the YY-1 repressor that may act to inhibit binding of positive regulatory factors (such as AP1) to adjacent elements (Ye et al. 1996). As with the type I IFN genes, much remains to be learned about IFNγ gene transcriptional control.

3 The Interferon Receptors

3.1 The Cytokine Receptor Superfamily

As prototypical cytokines, IFNs elicit their biological effects through interactions with high affinity cell surface receptors (Rubinstein and Orchansky 1986; Langer and Pestka 1988; Uzé et al. 1995). The presence of such receptors was proposed over 30 years ago (Friedman 1967), but it was not until 1980 that ^{125}I-labeled mouse IFNα was used to demonstrate specific binding to IFN-sensitive, but not IFN-resistant mouse cells (Auget 1980). This binding displayed all of the hallmarks of standard ligand/receptor interactions, and subsequent studies of receptor specificity demonstrated that all type I IFNs competed for the same cell surface-binding sites (Branca and Baglioni 1981). IFNγ, on the other hand, could not compete for the same sites, but instead bound to a distinct receptor that did not bind type I IFNs (Celada et al. 1984). IFN receptors are now classified into two distinct receptor types: type I IFN receptors bind all type I IFNs and the type II receptor binds IFNγ (Farrar and Schreiber 1993; Uzé et al. 1995). Both IFN receptor types are ubiquitously expressed, usually at about 1000 per cell (Rubinstein and Orchansky 1986), and both are multisubunit complexes, as described below.

The individual receptor subunits comprising the type I and II IFN receptors have recently been identified through molecular cloning studies. Components of both IFN receptor types belong to the larger family of cytokine receptors, typified by the presence of ligand-binding domains that resemble immunoglobin constant regions (known as D200 domains) (Uzé et al. 1995). Cytokine receptors are further subclassified into either class I or class II (Bazan 1990; Cosman et al. 1990; Kishimoto et al. 1994: Uzé et al. 1995). Most class I receptors possess a characteristic WSXWS motif located at variant positions within the receptor chains, and all have four highly conserved cysteine residues which form disulfide bonds within the D200 domain (Uzé et al. 1995).

Class II cytokine receptors are distinguished from class I receptors by the absence of the WSXWS motif and variant positioning of the four conserved cysteines within the D200 domain (Uzé et al. 1995). All IFN receptor components belong to the class II family (Uzé et al. 1995) and thus are distinct from the interleukin receptors which, with the exception of the IL-10 receptor, are classified as class I receptors (Cosman et al. 1990; Stahl and Yancopolous 1993; Kishimoto et al. 1994). As with other cytokine receptors, the individual IFN receptor subunits are devoid of intrinsic kinase catalytic function, requiring receptor-associated cytoplasmic kinases to activate downstream signaling pathways, as described in Section 4.1.1.

3.2
The Type I IFN Receptor Complex

The type I IFN receptor (IFNAR) is capable of binding all known type I IFN subtypes (Uzé et al. 1995). The cloning of cDNAs encoding different IFNAR subunits revealed that each of these exists in multiple splice variant forms, only some of which are functional (Uzé et al. 1990; Novick et al. 1994; Domansky et al. 1995; Lutfalla et al. 1995). For this reason, the type I receptor has been more complex to study than the type II IFN receptor, and the precise makeup of the prototypical type I IFN receptor remains somewhat controversial. What follows is a brief description of the current understanding of the type I receptor complex. For consistency, the nomenclature used will adhere to the suggested terminology put forth by the International Society for IFN and Cytokine Society nomenclature committee, although it should be noted that alternative designations can be found in the literature.

The IFNAR-1 subunit was isolated by using a functional screen to identify human cDNAs capable of conferring responsiveness to human IFNα8 upon mouse cells (Uzé et al. 1990). Interestingly, subsequent analyses demonstrated that ectopic IFNAR-1 expression in mouse cells affects their responsiveness to the human IFNα8 subspecies, but not other human IFNαs, due to low affinity cross-reactivity of most human IFNα subtypes with the endogenous mouse receptors (Uzé et al. 1990). Indeed, the binding of these other human IFNα subtypes to murine cells is not augmented by expression of human IFNAR-1, resulting in early speculation that IFNAR-1 is not required for IFN binding. Since that time, however, IFNAR-1 knockout experiments in mice have demonstrated conclusively that IFNAR-1 is necessary for optimal binding of all type I IFNs, and for transduction of IFN-dependent signals (Muller et al. 1994; Hwang et al. 1995). IFNAR-1 is a 110-kDa transmembrane protein which contains tandem D200 domains within its extracellular domain (Uzé et al. 1995). The intracellular domain contains a number of conserved tyrosines that are phosphorylated in response to IFN binding and serve to couple the receptor to signaling molecules (Uzé et al. 1995). The best characterized of these is tyrosine 466 which, once phosphorylated, appears to be the primary site of STAT interaction (Yan et al. 1996; see Sect. 4.3). Tyrosines 527 and 538 have been implicated in binding other molecules (Pfeffer et al. 1997), as described in Section 4.4.

A soluble form of the other type I receptor component, IFNAR-2, was purified initially from human urine (Novick et al. 1994). A membrane-bound form of the same protein was cloned by reverse genetics using the amino-acid information derived from the soluble protein to screen a cDNA library (Novick et al. 1994). Although this cDNA encoded a nonfunctional protein, an alternatively spliced variant of the same gene has since been identified (Domansky et al. 1995; Lutfalla et al. 1995). The longer IFNAR-2 protein, designated IFNAR-2c, complements the defect in a mutant human cell line (U5A) that is deficient in type I IFN responsiveness (Lutfalla et al. 1995), thereby demonstrating its functional role in IFN signaling. The other IFNAR2 forms, on the other hand, cannot complement U5A cells (Lutfalla et al. 1995). IFNAR-2c is a 489-amino acid, 100-kDa transmembrane protein. Its cytoplasmic domain is 251 residues in length, and contains subdomains that participate in binding JAKs (Janus kinases). Interestingly, phosphorylation of IFNAR-2c can be observed following IFNβ, but not IFNα binding (Abramovich et al. 1994), suggesting that these two IFN subtypes may interact with the IFNAR complex somewhat differently. The functional significance of this observation is currently unknown. The roles, if any, of the shorter splice variant forms of the IFNAR-2 gene, including the soluble receptor (IFNAR-2a) and the shorter membrane-bound protein (IFNAR-2b), also remain unclear. It is worth noting that additional splice variants of IFNAR-1 have also been identified (Cook et al. 1996), raising the possibility that the assembly of these different IFNAR forms into distinct receptor complexes may contribute to the pleiotropic effects associated with type I IFN. For now, the simplest view of type I IFN signaling involves the heterodimerization of the full-length IFNAR1 and IFNAR-2c, leading to activation of receptor-associated JAKs and initiation of IFN-dependent signaling events (Fig. 1).

3.3 The Type II IFN Receptor Complex

Early studies that made use of murine/human somatic cell hybrids suggested that at least two different human gene products, encoded on chromosomes 6 and 21, were required to confer responsiveness to human IFNγ upon hamster cells (Jung et al. 1987). The ligand-binding chain of the receptor, IFNGR-1, was cloned by screening cDNA expression libraries with an antiserum directed against affinity-purified human IFNγ receptor protein (Auget et al. 1988). IFNGR-1 is a 475-amino acid glycoprotein with a molecular mass of 90kDa. The first 231 residues make up the extracellular domain, which includes

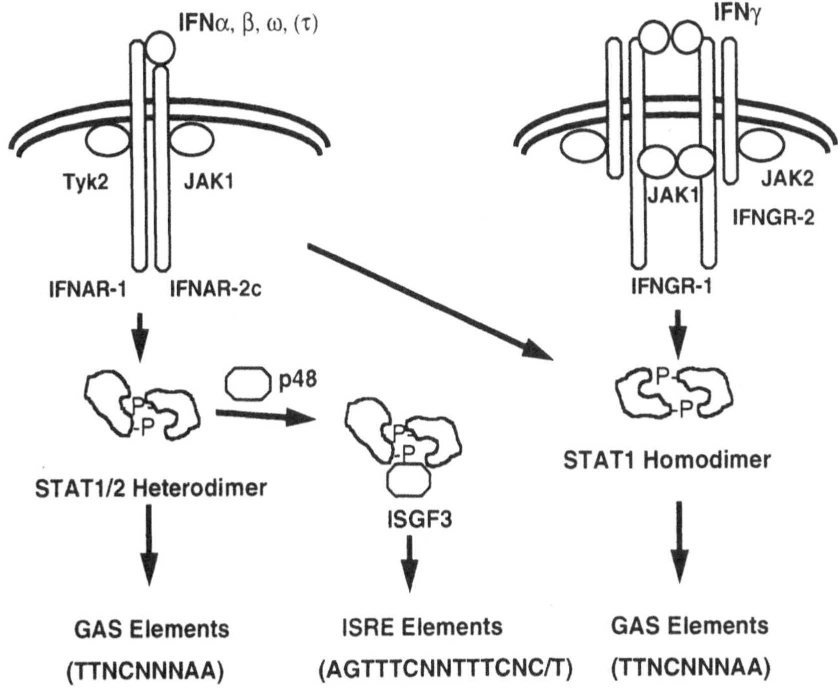

Fig. 1. Components of the type I and II IFN signaling pathways. The type I IFNs bind the type I receptor, comprised of at least two subunits, *IFNAR-1* and *IFNAR-2c*. Ligand-induced dimerization of receptor components leads to activation of *JAK1* and *Tyk2*, phosphorylation of receptor tyrosines and recruitment of *STAT1* and *STAT2*. Once phosphorylated, the STATs form STAT1/2 heterodimers or STAT1/1 homodimers that migrate to the nucleus to bind *GAS* elements and promote gene transcription. *STAT1/2* heterodimers can also bind to a third component, p48, to form ISGF3, which recognizes *ISRE* elements found in the regulatory regions of certain ISGs. *IFNγ* binds as a dimer, inducing dimerization of IFNGR-1 subunits, followed by recruitment of IFNGR-2 subunits to the supercomplex. The receptor-associated *JAK1* and *JAK2* catalyze the phosphorylation of tyrosine 440 of *IFNGR-1*, and tyrosine 701 of the recruited STAT1. Activated *STAT1* forms homodimers and binds to GAS elements in the nucleus

characterisitic immunoglobulin-like structures and several putative N-linked glycosylation sites (Auget et al. 1988; Uzé et al. 1995). The intracellular domain of IFNGR-1 is 221 residues in length and contains sequences required for coupling the receptor to signaling, components (Farrar et al. 1991, 1992; Greenlund et al. 1994; Kaplan et al. 1996). Stoichiometric analyses have revealed that IFNγ binds as a dimer that induces homodimerization of individual IFNGR-1 chains (Farrar and Schreiber 1993). Expression of human IFNGR-1, which is encoded on chromosome 6, in murine cells is sufficient to confer high affinity binding to Human IFNγ, but not to confer full biological responsiveness (Fischer et al. 1990). Conversely, expression of murine IFNGR-1 in human cells is not sufficient to confer full responsiveness to murine IFNγ, further demonstrating the need for a second receptor component to transduce IFNγ-specific signals (Hemmi et al. 1994).

The cloning of the second human IFNγ receptor subunit took advantage of the earlier somatic cell genetic studies. Ectopic expression of the IFNGR-1 chain in murine cells harboring human chromosome 21 reconstituted full responsiveness to human IFNγ in these cells (Fischer et al. 1990), permitting isolation of the accessory chain of the IFNγ receptor, IFNGR-2, by positional cloning (Soh et al. 1994). Murine IFNGR-2 was isolated independently by expression cloning in COS cells (Hemmi et al. 1994). Human IFNGR-2 is a 310-residue transmembrane protein with only 63 cytoplasmic residues (Soh et al. 1994). This short cytoplasmic tail contains a sequence motif resembling the box2 region of certain cytokine receptors, required for the association of JAK family tyrosine Kinases (Sect. 4.1.1). IFNγ binding promotes IFNGR-1 dimerization, and each IFNGR-1 subunit associates with individual IFNGR-2 subunits, the sole function of which appears to be the recruitment of JAK2 (Bach et al. 1996). The assembled supercomplex promotes activation of the associated JAKs and initiates signaling (Sect. 4.2). Phosphorylation of a single tyrosine residue, at position 440 of IFNGR-1, is required to recruit STAT1 to the activated receptor (Greenlund et al. 1994). Other sequence motifs within the cytoplasmic portions of the receptor chains are required for dimerization and internalization (Farrar et al. 1992; Kaplan et al. 1996). The assembled type II receptor complex is depicted in Fig. 1.

4 IFN-Dependent Signaling Pathways

4.1 Identification of IFN-Dependent Signaling Factors

The initial cloning of the type I and II IFN receptor subunits revealed little information about the signaling mechanism. Although numerous studies had implicated both tyrosine and serine/threonine phosphorylation in the early stages of IFN responses (Levy and Darnell 1990; Pfeffer and Tan 1990), the cytoplasmic tails of the receptors contained few recognizable sequence motifs, such as tyrosine kinase domains characteristic of growth factor receptors or serine/threonine kinase domains typical of the TGFβ/activin family of receptors (Ullrich and Schelessinger 1990; Helden 1995). By the early 1990s, experiments exploring the transcriptional regulation of ISGs had provided a general model of type I IFN signaling: IFN binding to IFNAR activated several latent cytoplasmic transcription factors that migrated to the nucleus and bound to specific promoter elements to induce ISG expression (Levy and Darnell 1990; Stark and Kerr 1992). Ultimately, the identities of the major signaling factors were uncovered by the convergence of two complementary approaches. A biochemical approach was used to purify the transcription factors, followed by cloning of the corresponding genes through reverse genetics (Schindler et al. 1992a,b; Veals et al. 1992), and a somatic cell genetic approach was used to isolate mutant human cell lines deficient in signaling, which were then used to define the defective components functionally through complementation (reviewed in Darnell et al. 1994). The combined results of these two approaches defined a new mechanism of signaling, collectively called JAK/STAT pathways, which is now known to transmit signals initiated by a wide variety of growth factors and cytokines. An exhaustive review of JAK/STAT signaling is beyond the scope of this chapter and is available elsewhere (Darnell et al. 1994; Ihle and Kerr 1995; Schindler and Darnell 1995; Taniguchi 1995; Ihle 1996; Leaman et al. 1996a; Darnell 1997). However, a brief overview is provided, since it was the study of IFN signaling that revealed for the first time the presence of this important signaling mechanism.

4.1.1 The JAKs

The JAKs are tyrosine kinases that reside in the cytoplasm and associate with the intracellular domains of cytokine and growth factor receptors, where they couple ligand binding to the phosphorylation of effector molecules (Ihle and Kerr 1995; Schindler and Darnell 1995; Leaman et al. 1996a). The four known JAK family members in mammals are JAK1, JAK2, JAK3, and Tyk2 (see Schindler and Darnell 1995), and a related protein from *Drosophila* has also been identified (Hou and Perrimon 1997). The mammalian JAKs contain a number of shared structural domains, including two tandem tyrosine kinase-like domains, of which only the second (C-proximal) is catalytically active (Ihle et al. 1994). Detailed descriptions of the ligand-dependent activation and functions of specific JAKs can be found in a number of recent reviews (Ihle et al. 1994; Schindler and Darnell 1995; Leaman et al. 1996a). In the context of IFN signaling, mutant cell lines deficient in individual JAKs have been used to demonstrate that JAKs are functionally required for all known IFN-dependent effects, including the phosphorylation of receptor chains and the activation of STATs (Velazquez et al. 1992; Müller et al. 1993a; Watling et al. 1993). Interestingly, some JAKs have also been implicated in maintaining the structural integrity of the IFN receptors themselves (Velazquez et al. 1995; Briscoe et al. 1996). The identities of the individual JAKs involved in IFN signaling are described in Sections 4.2 and 4.3, and are shown in Fig. 1.

4.1.2 The STATs

The STATs comprise an interesting family of signaling molecules that mediate responses to numerous extracellular signaling proteins (Darnell et al. 1994; Schindler and Darnell 1995; Ihle 1996; Leaman et al. 1996a; Darnell 1997).

STATs have several conserved structural motifs, including a DNA-binding domain and a C-terminal transactivation domain, but their *src* homology region-2 (SH2) domains contribute most significantly to the unique mechanism of STAT action (see Schindler and Darnell 1995; Ihle 1996; Darnell 1997). Latent STATs are recruited to liganded receptor complexes, usually by docking to specific phosphotyrosine-containing recognition sequences, a process mediated by their SH2 domains (Shuai et al. 1994; Improta et al. 1994). Once associated with receptors, the STATs are phosphorylated on a tyrosine residue located near position 700 (Shuai et al. 1993). Depending on the receptor system this phosphorylation can be catalyzed by intrinsic receptor tyrosine kinases, as in responses to some growth factors (Leaman et al. 1996b; Vignais et al. 1996), or by receptor-associated JAKs, as in responses to most cytokines, including the IFNs (Darnell et al. 1994; Ihle et al. 1994; Schindler and Darnell 1995). Phosphorylated STATs dimerize via reciprocal SH2-phosphotyrosine interactions to form homo- or heterodimers, and then migrate to the nucleus, where they bind to specific DNA response elements to activate the transcription of target genes (Darnell et al. 1994; Schindler and Darnell 1995; Ihle and Kerr 1995; Decker et al. 1997). There are currently seven known STAT family members in mammals, STATs1–4, 5a and 5b (encoded by different genes) and 6 (Ihle 1996; Darnell 1997). Most STATs also exist in shorter splice-variant forms that have distinct, and in some cases antagonistic, functions when compared to the full-length proteins (Schaefer et al. 1995; see also Darnell 1997). Furthermore, STAT dimers can interact with other proteins to create novel transcription factors with unique DNA sequence specificities (see Sect. 4.3). Overall, specificity in STAT signaling can be viewed as being regulated at two levels: (1) by the types of STATs activated in response to a given ligand, and (2) by the types of STAT dimers or multimers that are formed as a consequence of this activation. The IFNs provide clear examples of both levels of regulation, as described in the following sections.

4.2 Type II IFN Signaling

Although the details of type I IFN signaling were uncovered first, the IFNγ pathway is less complex, and thus provides a good introduction to a prototypical JAK/STAT signaling pathway. As described in Section 3.2, IFNγ binding promotes dimerization of the IFNGR-1 chains and recruitment of IFNGR-2 to the supercomplex, resulting in activation of JAK1 and JAK2, which associate constitutively with the IFNGR-1 and IFNGR-2 chains, respectively (Kaplan et al. 1996; Muller et al. 1994; Watling et al. 1993; Bach et al. 1996). Although the precise mechanism of JAK activation is not known, one possibility is that receptor oligomerization brings the JAKs into juxtaposition, allowing cross-phosphorylation. Once activated, one or both JAKs phosphorylate tyrosine 440

(Y^{440}) of the IFNGR-1 chain(s), providing a recruitment sites for latent STAT1 molecules (Greenlund et al. 1994). The STAT1 SH2 domain recognizes Y^{440} and its flanking sequences, and is the major determinant of STAT1 specificity for the IFNγ receptor (Heim et al. 1995). The recruited STAT1 monomers are phosphorylated on tyrosine by the JAKs, and form STAT1 homodimers through reciprocal SH2 domain/phosphotyrosine interactions (Shuai et al. 1993). Activated STAT1 homodimer, known also as gamma-activated factor (GAF), migrates to the nucleus and binds to gamma activated sequence (GAS, consensus TTNCNNNAA) elements, located in the 5′-regulatory regions of IFNγ-responsive genes such as IRF-1, GBP, and FcγRI (see Darnell et al. 1994; Ihle and Kerr 1995; Schindler and Darnell 1995; Leaman et al. 1996a; Decker et al. 1997). Although GAS elements are necessary for gene induction by IFNγ, they respond poorly to IFNγ when placed ahead of minimal promoters, implicating the importance of STAT interactions with other transcription factors, only a few examples of which are currently available (Perez et al. 1994; see also Leaman et al. 1996a). Other putative second messengers of the IFNγ pathway are described in Section 4.4.

4.3 Type I IFN Signaling

Binding of IFNs to the type I receptor leads to dimerization of the receptor components and activation of tyk2 and JAK1, associated with IFNAR-1 and IFNAR-2c, respectively (Colamonici et al. 1994; Novick et al. 1994). As in the case of IFNγ signaling, each JAK is probably activated by the reciprocal kinase, and together they phosphorylate the receptor components, the STATs and possibly other downstream signaling molecules. Both STAT1 and 2 are required for transmission of type I IFN signals (Schindler et al. 1992a,b; Müller et al. 1993b; Leung et al. 1995). Phosphorylated STAT1 and 2 heterodimerize and migrate to the nucleus where they can bind to a subset of GAS elements to activate transcription (Li et al. 1996) or associate with a third component, p48, to form the heterotrimeric transcription factor IFN-stimulated gene factor 3 (ISGF3) (Fu et al. 1992; Schindler et al. 1992a,b; Darnell et al. 1994), which binds to DNA sequences known as IFN-stimulated response elements (ISREs, consensus AGTTTCNNTTTCNC/T) that are found in the 5′-regulatory regions of most type I IFN-reponsive genes (Darnell et al. 1994; Schindler and Darnell 1995; Stark et al. 1997). Type I IFNs also induce formation of STAT1 homodimer, leading to the transcriptional induction of some of the same genes induced by IFNγ (Decker et al. 1997). Interestingly, IFNγ can also induce the expression of at least one ISRE-containing gene, 9–27, through STAT1 homodimer association with p48 (Bluyssen et al. 1995). These latter two features provide a molecular explanation for some of the similarities in type I and II IFN function that were recognized many years ago.

Studies conducted in the mutant human cell lines U3A and U6A, lacking STAT1 and STAT2, respectively, have suggested that the activation of STATs by type I IFNs is an ordered process (Improta et al. 1994; Leung et al. 1995). STAT2 tyrosine phosphorylation in IFNα-treated U3A cells is essentially normal (Improta et al. 1994), whereas STAT1 phosphorylation in U6A cells is weak or absent (Leung et al. 1995), indicating that tyrosine-phosphorylated STAT2 provides an essential docking site for STAT1. This mechanism of STAT activation appears to be unique to the type I IFN system, since there are currently no other examples of legends that require the ordered activation of STATs, even when multiple STATs are activated (Guschin et al. 1995; Stahl et al. 1995; Han et al. 1996; Leaman et al. 1996b; Vignais et al. 1996). At least one study has suggested that STAT1 and 2 preassociate with the IFNAR-2c receptor chain prior to IFN binding (Li et al. 1997). This interaction is probably phosphotyrosine-independent, although the precise receptor sequence requirements have not been defined. The significance of this preassociation is unknown, but it may provide the specificity needed for STAT2 activation, since the STAT2 SH2 domains do not appear to confer specificity for type I receptor phosphotyrosines (Li et al. 1997).

While it is clear that STATs must be phosphorylated on tyrosine to be activated, several studies have suggested that they must also undergo serine phosphorylation to attain maximal transcriptional activity (Wen et al. 1995; Zhang et al. 1995). A single serine residue, at position 727 of the mature STAT1 protein, is phosphorylated both constitutively and in response to certain ligands, such as IFNγ and platelet-derived growth factor (Wen et al. 1995). Conversion of this serine to alanine decreases the ability of STAT1 to activate transcription of GAS-containing genes (Wen et al. 1995), but has less effect on the induction of ISGF3-responsive genes, presumably because STAT2 is the major transcriptional effector within this trimeric complex (Müller et al. 1993b). The identities of the kinases required for STAT serine phosphorylation are unknown. The mitogen-activated protein kinases (MAPKs) have been implicated in STAT serine phosphorylation (David et al. 1995b; Wen et al. 1995), but their involvement in this process remains controversial.

An important aspect of IFN-dependent signaling that has received little attention is the downregulation of the pathways that follows the initial induction. IFN responses are transient, even in cells continuously exposed to IFNs (Friedman et al. 1984; Larner et al. 1984). Exposure of cells to tyrosine phosphatase inhibitors can prolong an IFN response (Igarashi et al. 1993; David et al. 1993; Haque et al. 1995), suggesting that dephosphorylation of the receptor components, the JAKs, and/or the STATs is required to downregulate the pathway. STAT1, for example, appears to be dephosphorylated in the nucleus and recycled back to the cytoplasm (Haspel et al. 1996), although the nuclear phosphatase involved in this process has not been identified. Furthermore, in murine cells lacking the myeloid cell-specific tyrosine phosphatase HCP (known also as SHPTP1), the IFN-dependent phosphorylation of JAK1 is augmented (David et al. 1995a), implicating HCP in JAK1-deactivation in these cells. Unfortunately, the identities of phosphatases acting in nonlymphoid cell types are unknown. It should be noted that dephosphorylation may not be the only mechanism of STAT downregulation. Inhibitors of proteosome activity also sustain STAT1-dependent responses (Kim and Maniatis 1996), suggesting that degradation of phosphorylated STAT1 via ubiquitin-dependent pathways contributes to the inactivation of STAT signals. Clearly, more work in this area is needed.

4.4 Other IFN-Induced Signaling Components

Although STATs are required for many of the early events associated with ISG induction, additional signaling mechanisms are needed to mediate the full range of IFN-dependent transcription responses. STAT-containing complexes disappear before ISG transcription ceases (Friedman et al. 1984; Levy et al. 1989), suggesting that secondary mechanisms of transcriptional regulation must come into play. Indeed, numerous studies have implicated a variety of second messenger systems in potentiating of IFN responses (Pfeffer and Tan 1991; Pfeffer and Strulovici 1992), although the functions of many of these proteins remains unclear.

Early reports demonstrated an IFN-dependent increase in cellular diacylglycerol (Pfeffer and Tan 1991), an activator of serine/threonine protein kinase-C (PKC) isoforms which have been implicated in transducing signals initiated by hormones that utilize G-protein-linked receptors, or by growth factors that utilize tyrosine kinase-containing receptors (Housey et al. 1988; Ullrich and Schlessinger 1990). IFNα induces the rapid activation of PKC isoforms β and ε (Pfeffer et al. 1991; Pfeffer and Strulovici 1992), and inhibitors of PKC, such as H-7 or staurosporin, can inhibit IFN actions in fibroblasts (see Pfeffer and Strulovici 1992). H-7 can inhibit the DNA-binding activity of certain STAT complexes (Zhang et al. 1995), leading to speculation that PKC may play a direct or indirect role in regulating STAT serine phosphorylation. However, other data conflict with this assessment (Wen and Darnell 1997), and a clear role for PKC in IFN signaling has yet to emerge.

A second serine/threonine kinase, PKR, has also been implicated in type I and type II IFN signaling (Williams 1995). Studies carried out with fibroblasts devoid of PKR (PKR −/−) demonstrated that induction of the GBP gene was deficient in response to either type I or II IFNs (Yang et al. 1995; Kumar et al. 1997). Furthermore, IFNγ-dependent antiviral responses were reduced or deficient in the PKR −/− cells as compared to PKR +/+ cells (Yang et al. 1995). PKR is thought to regulate the activity of IFN-dependent transcription factors such as IRF-1 and NFκB, presumably through serine or threonine modifications (Watanabe et al. 1991; Kumar et al. 1994, 1997; Kirchoff et al. 1995). Although not yet proven, these factors (IRF-1 in particular) are thought to cooperate with STATs in the induction, and perhaps in the maintenance, of ISG expression (Harada et al. 1989; Imam et al. 1990; Reis et al. 1992). Whether PKR has other functions, such as modifying STATs, remains to be determined.

Arachidonic acid is released from IFN-treated cells, implicating phospholipase A2 (PLA2) in some aspect of IFN signaling (Hannigan and Williams 1991; Flati et al. 1996). PLA2 activation by IFNα is rapid, and specific PLA2 inhibitors can prevent either the formation or the release of ISGF3 from the receptor complex complex (Flati et al. 1996). JAK1 is required in some aspect of PLA2 activation, most likely in the recruitment of PLA2 to the receptor (Flati et al. 1996). Unfortunately, the relative importance and overall function of PLA2 in IFN signaling remains uncertain.

Phosphatidylinositol 3-kinase (PI3K) has recently been implicated in type I IFN responses (Pfeffer et al. 1997). PI3K is best known for its role in phosphorylating lipids in the phosphatidylinositol (4,5)-bisphosphate pathway to generate different signals in response to specific ligands (Zvelebil et al. 1996). Recent data have suggested that PI3K can also act as a protein kinase, although few substrates are known. Interestingly, STAT3 has been implicated in the IFN-dependent activation of PI3K, acting as an adapter to couple PI3K to the IFNAR-1 chain of the activated type I IFN receptor (Pfeffer et al. 1997). In the model proposed, STAT3 binds to the phosphorylated tyrosines at positions 527 and 538 of the IFNAR-1 chain, is itself phosphorylated on tyrosine, and serves as a docking site for the p85 regulatory subunit of PI3K (Pfeffer et al. 1997). Activated PI3K can then presumably initiate additional signaling cascades, either through its lipid or protein kinase activities. Identification of specific substrates for PI3K action in the IFN pathway will certainly increase our understanding of ISG regulation.

In summary, it is clear that JAK/STAT signaling is necessary to initiate IFN responses, but other pathways are likely to contribute to the full range of IFN-dependent effects. The identities of these supporting pathways are beginning to emerge, but functional analyses are still in their early stages.

5 IFN-Stimulated Genes

5.1 Kinetics of ISG Induction

An important advancement in the study of IFN action resulted from the identification of genes that were transcriptionally regulated by treatment of cells with IFN (reviewed in Darnell et al. 1994; Sen and Ransohoff 1997; Stark et al.

1997). As mentioned, these early studies had indicated that the expression of a large number of normally quiescent genes is upregulated by treatment of cells with IFN. The induction of ISG expression is largely protein synthesis-independent, and relies, at least initially, on the activation of STAT-containing transcription factors. ISG expression is transient, declining to basal levels within 48 h of stimulation even in the continued presence of IFN, during which time cells are unresponsive to newly added IFN for many hours after the initial induction (De Maeyer and De Maeyer-Guignard 1988). Although receptor internalization may contribute to this refractory state (Pfeffer and Donner 1990), cells treated with protein synthesis inhibitors such as cycloheximide showed extended ISG responses (Friedman et al. 1984), suggesting that IFN-induced transcriptional repressors also contribute to the suppression of ISG expression (see Sect. 5.2.4).

There is little doubt that the protein products of the ISGs mediate IFN-dependent effects, including antiviral and antiproliferative responses. What is not always clear is precisely which gene products are involved in the different IFN activities. Many ISGs were identified initially as differentially regulated mRNA species, rather than on the basis of their functions (see Sen 1991; Stark and Kerr 1992). Others were identified as enzymes that showed increased activity in virus-treated or IFN-stimulated cells, and were only later shown to be regulated at the transcriptional level as well (Staeheli 1990; Sen 1991). What follows is a brief description of some of the better-characterized IFN-stimulated proteins, and of their respective roles in mediating the antiviral and antiproliferative effects of IFNs. It should be noted that over 40 ISGs have been identified to date, and more complete lists of their identities and known functions can be found elsewhere (De Maeyer and De Maeyer-Guignard 1988; Staeheli 1990; Sen and Ransohoff 1997).

5.2 Mediators of Antiviral Responses

5.2.1 The 2′,5′-Oligoadenylate Synthetases

Among the earliest identified IFN-induced enzymes were the 2′,5′-oligoadenylate (2–5 A) synthetases (Revel and Chebath 1986; De Maeyer and De Maeyer-Guignard 1988; Hovanessian 1991). The 2–5 A synthetases are a family of enzymes, encoded by different genes, that exist in small, medium, and large forms, and reside in different subcellular compartments (Sen and Ransohoff 1997). Expression of 2–5 A synthetases is induced by IFN and the latent proteins are activated by dsRNA to polymerize ATP molecules into a series of 2′,5′-linked oligoadenylates. The newly synthesized 2–5 A oligomers activate a latent cellular ribonuclease, RNase L, which targets mRNAs and rRNAs for degradation (Silverman 1996). Although RNase L is present in unstimulated cells, its expression is induced by IFNs. Thus, the IFN-dependent increase in 2–5 A synthetase and RNase L protein levels primes cells to respond quickly to a subsequent virus infection, whereupon the induced enzymes are activated by the viral dsRNA (De Maeyer and De Maeyer-Guignard 1988; Sen and Ransohoff 1997). Although RNase L will target mRNA indiscriminantly, 2–5 A oligomers are relatively unstable, and so their synthesis in the proximity of viral dsRNA structures targets RNase L activity preferentially toward the viral RNA, rather than cellular mRNA (Staeheli 1990). Overall, this system is involved both in inhibiting viral replication and, to a lesser extent, in suppressing cellular proliferation until the viral challenge is eliminated. Among the types of viruses whose replication is inhibited by the 2–5 A synthetase/RNase L system are the picornaviruses, mengo virus, and encephalomyocarditis virus (Staeheli 1990; Silverman 1996). Whether the different forms of the 2–5 A synthetase have different biological specificities remains to be determined. Interestingly, studies in which the RNase L gene has been knocked out in mice have demonstrated that RNase L is absolutely required for antiviral responses, and that it also regulates apoptosis in certain cell types (Zhou et al. 1997).

Recently, by chemically synthesizing 2–5 A-linked antisense oligonuleotides containing sequences that are complementary to specific cellular messages, Silverman and collegues have demonstrated that they can target specific mRNA species for degradation, presumably by RNase L (see Maran et al. 1994). This approach provides an alternative to the traditional antisense strategies in which antisense DNA/mRNA hybrids are degraded primarily by RNase H (Walder and Walder 1988).

5.2.2 The Mx Proteins

In 1963, Lindenmann et al. reported the identification of a genetic locus in the mouse, Mx, that was required for resistance to influenza virus (Lindenmann et al. 1963). The protein product of this locus, Mx1, a 75-kDa nuclear protein, is induced in cells treated with type I IFNs or infected with virus. Mx1 has been implicated in the protection of cells against influenza virus, particularly strains A and B (Revel and Chebath 1986; Staeheli 1990). In cells overexpressing Mx1, the adsorption, penetration, and transport of influenza virus nucleocapsid into the nucleus is unaffected, but viral transcription is inhibited (Staeheli et al. 1986). A human homolog, MxA, is capable of inhibiting both influenza and vesicular stomatitus virus (VSV) replication (Staeheli 1990). Unlike Mx1, MxA is localized to the cytoplasm, and it appears to affect differentially the replication of the two virus types (Staeheli and Pavlovic 1991). The inhibition of VSV by MxA is at the level of viral transcription, whereas the inhibition of influenza virus, while not yet fully defined, does not appear to be related to transcriptional inhibition. The overall mechanisms of action of the human or murine Mx proteins are unknown, but may require their GTP binding or GTPase activities, since MxA proteins in which the GTP/GDP-binding elements are destroyed are ineffective in inhibiting virus replication (Pavlovic and Staeheli 1991). Mx may complex with certain viral proteins, perhaps RNA polymerases, to inhibit their functions (Staeheli 1990; Stark et al. 1997) Alternatively, since the Mx proteins resemble a yeast protein, Vps1, that is involved in vacuolar sorting (Rothman et al. 1990), they may inhibit the replication of some viruses by redirecting the intracellular trafficking of viral proteins. Whether Mx has roles other than in the antiviral response is unknown.

5.2.3 The dsRNA-Induced Kinase PKR

The dsRNA-dependent serine/threonine kinase PKR, another IFN-induced enzyme involved in protecting cells against virus (Samuel 1993; Williams 1995), resides in the cytoplasm as a dimer of identical 68-kDa subunits (Patel et al. 1995). Although it is normally present at low constitutive levels, expression of PKR is upregulated by either IFN, virus, or dsRNA (Samuel 1993). Activation of PKR's catalytic function requires the binding of a cofactor, such as dsRNA, heparin, or other polyanionic molecule (Samuel 1993; Williams 1995). Once activated, PKR phosphoryates cellular substrates, including the α subunit of the eukaryotic translation initiation factor eIF-2, leading to its inactivation and a global inhibition of protein synthesis (see Samuel 1993). PKR activation can also inhibit cell growth, and ectopic overexpression of PKR in mammalian or yeast cells is generally toxic (see Williams 1995).

PKR functions in several cellular signaling cascades. It is absolutely required for the dsRNA-dependent activation of the transcription factor NFκB, which activates transcription of the IFNβ gene and some ISGs (Maran et al. 1994; Kumar et al. 1997). Although recombinant PKR can catalyze the in vitro phosphorylation of IκB, the inhibitor of NFκB (Kumar et al. 1994), the in vivo relationship between these two proteins has not been established. A mutant cell line deficient in the dsRNA-dependent activation of NFκB has recently been isolated (Leaman et al. 1997). Interestingly, these P2.1 cells have wild-type PKR that is activated by dsRNA in vivo, implying that the PKR-dependent activation of NFκB is not direct. Although the identity of the defective component in P2.1 cells is unknown, it is likely to be a kinase or other protein involved in transferring PKR-dependent signals to effector proteins such as NFκB, perhaps through the potentiation of a phosphorylation cascade.

As mentioned previously, PKR has also been implicated in mediating responses to IFN (Williams 1995; Kumar et al. 1997) and other ligands, including platelet-derived growth factor (Mundschau and Faller 1995). How PKR activation is controlled by these extracellular signaling ligands is unclear. Clearly, PKR must have multiple mechanisms of activation, and many different functions in the normal or ligand-dependent regulation of cell growth, only some of which are currently understood. The availability of cells from PKR knockout mice should now provide a means of addressing its functional relevance in a variety of physiological processes (Yang et al. 1995).

5.2.4 IFN-Regulatory Factors (IRFs)

The IRFs are a family of transcription factors that mediate multiple aspects of IFN responses, some of which have been described above. Included within this rather diverse family are IRF-1, -2, -3, and -4, p48, ICSBP, and ICSAT/Pip (Harada et al. 1989; Veals et al. 1992; Au et al. 1995; Nelson et al. 1993; Eisenbeis et al. 1995; Yamagata et al. 1996; Mittruecker et al. 1997). Some of these family members are transcriptional activators that act either alone (IRF-1; Harada et al. 1989; Nelson et al. 1993) or as components of higher-order complexes (p48; Veals et al. 1992). Others act as transcriptional repressors (IRF-2, ICSBP), and at least one family member, ICSAT/Pip, can act as either a transactivator or repressor, depending on the target gene (Brass et al. 1996; Yamagata et al. 1996). The subcellular distributions of the IRFs also vary, with p48 found both in the nucleus and in the cytoplasm, and the rest predominantly in the nucleus. Some family mambers display a cell-type-restricted pattern of expression in cells of myeloid origin (e.g., ICSBP, ICSAT/Pip, and IRF-4) while others are widely expressed in adult tissues (Harada et al. 1989; Veals et al. 1992; Au et al. 1995; Nelson et al. 1993; Eisenbeis et al. 1995; Yamagata et al. 1996).

IRF-1 is the most extensively studied member of the family. Identified originally as a mediator of IFNβ gene expression (Miyamoto et al. 1988), its most critical function may be in T cell maturation since specific subsets of T cells are reduced or missing in IRF-1-deficient mice (Matsuyama et al. 1993). IRF-1 has also been implicated as a tumor suppressor since the genetic locus that includes the gene is frequently deleted in certain human leukemias (Willman et al. 1993). The IFNγ-dependent induction of several ISGs, including GBP and inducible nitric oxide synthetase, are defective in IRF-1-minus cells (Briken et al. 1995) and, although IRF-1 has been implicated in sustaining the IFN-induced expression of certain ISGs, the kinetics of ISG induction in IRF-1 knockout mice are essentially identical to those observed in wild-type control animals (Matsuyama et al. 1993). IFNβ induction by virus was also normal in the IRF-1 knockout animals, although the dsRNA-dependent induction of IFN genes was defective (Matsuyama et al. 1993). These data suggest that redundant mechanisms of transcriptional regulation come into play when IRF-1 is missing.

A similar uncertainty surrounds the role of IRF-2 in IFN and ISG regulation. Although IRF-2 has been implicated in the silencing of certain ISGs and IFN genes in wild-type cells, these genes displayed normal transcriptional responses to virus and IFN in IRF-2 knockout mice (Matsuyama et al. 1993). Although secondary repressors may function in the absence of IRF-2, the identities and normal roles of such factors are currently unknown. IRF-2-null mice had pronounced defects in hematopoiesis and B lymphopoiesis (Matsuyama et al. 1993), as did IRF-4 knockout mice (Mittruecker et al. 1997). ICSBP-deficient mice developed a chronic myelogenous leukemia-like disease, and were severely immunodeficient (Holtschke et al. 1996). Thus, it seems that many IRF family members have major roles in hematopoietic cell growth and differentiation.

5.3 IFN Regulation of Cell-Cycle Components

Some of the same proteins that contribute to the IFN-dependent inhibtion of viral replication also mediate the antiproliferative effects of IFN. PKR, for example, is transcriptionally induced by IFNs and, once activated, can lead to a global down-regulation of protein synthesis, as can the 2–5 A synthetase/RNase L system (see above). In both cases, the end result is general inhibition of cell growth. However, both systems also require an activator (usually dsRNA) that is often missing in uninfected cells, suggesting that other IFN-induced proteins mediate antiproliferative effects when virally derived products are absent. The IFN-dependent regulation of several proto-oncogenes has been linked to the antiproliferative effects of IFNs. Treatment of human Daudi Burkitt lymphoma cells with IFNα causes a decrease in *c-myc* mRNA levels (Einat and Kimchi 1985), and ectopic overexpression of *c-myc* prior to treatment with IFNα partially overcomes the IFN-induced growth arrest (Einat and Kimchi 1988). Although these data suggest that reduced *c-myc* expression contributes to the antiproliferative effects of IFNα, the *c-myc*-responsive genes that are downregulated as a result of IFN treatment are not known. Indeed, at least one *c-myc*-regulated gene, p48 (Weihua et al. 1997), is upregulated by IFN, suggesting that the effect of *c-myc* downregulation must affect only a subset of responsive genes. Regardless, the incomplete restoration of cell proliferation in cells with deregulated *c-myc* suggests that other pathways must also contribute to the growth-inhibitory function of IFNα (Einat and Kimchi 1988).

As antiproliferative agents, IFNs must either force cells to exit from the cell cycle, or they must prolong or arrest one or more phases of the cell cycle. Early work suggested that the latter was the case, and the IFNs prolonged all phases of the cell cycle, particularly G1 and G2 (see Rubinstein and Orchansky 1986). Although much has been learned since that time about the identities of cell-cycle regulators, including the numerous cyclins, cyclin-dependent kinases (cdk), and phosphatases, scant information regarding the specific control of these factors by IFNs has been forthcoming. What follows is a brief description of the few papers that have addressed the issue of IFN-dependent cell-cycle control. Although still rather rudimentary, these results provide an interesting glimpse into the potential complexities of IFN regulation of cell growth.

A focal point for the inhibition of cell-cycle progression by IFNs appears to be the retinoblastoma protein, pRB (Burke et al. 1992), which acts as an inhibitor of the cell cycle by binding to various transcription factors, notably E2F, and inhibiting their transcriptional activities (Sherr 1994). pRB is normally phosphorylated in late G1 by cyclin-cdk complexes, and the hyperphosphorylated form releases E2F, which, in turn, activates genes that are required for DNA replication (Heichman and Roberts 1994). Treatment of cells with IFNα has recently been shown to inhibit the cell-cycle kinases that regulate pRb phosphorylation, thereby suppressing pRB phosphorylation and slowing progression into S phase (Burke et al. 1992; Tiefenbrun et al. 1996). Several putative targets for IFN actions have been identified. Expression of cyclin D3, for example, is strongly downregulated by IFNα, resulting in a decrease in cyclin D-associated cdk4 and cdk6 kinase activities (Tiefenbrun et al. 1996). In addition, the expression and activities of both cyclin E and cyclin A are reduced in IFN-treated cells (Zhang and Kumar 1994; Tiefenbrun et al. 1996). These cyclins have been implicated in the activation of cdk2 near the start of S phase, and both are essential for the initiation of DNA replication (Heichman and Roberts 1994). The IFN-dependent decrease in cyclin-cdk activity has also been attributed to a corresponding decrease in cdc25A gene expression (Tiefenbrun et al. 1996). cdc25A is a phosphatase that contributes to cdk2 activation by removing a phosphate from tyrosine 15 (King et al. 1994). In the absence of cdc25A, cdk2 remains inactive, thereby preventing activation of cdk2-cyclin A and cdk2-cyclin E complexes (Doree and Galas 1994). The overall reduction in expression of both cyclins, coupled with the decrease in cdk2 activation, results in decreased pRB phsophorylation, thereby preventing cellular progression into S phase.

IFNγ has been proposed to upregulate the expression of $p21^{WAF1/CIP1/CAP1}$ (Chin et al. 1996). p21 functions by binding to cyclin-cdk2 complexes, thereby inhibiting their kinase activites and preventing G1 to S phase transition (Sherr 1994). This inhibition is thought to be directly related to the relative levels of p21 and cyclin-cdk complexes, e.g., higher p21 levels increase the likelihood of binding to cyclin-cdk2 complexes to inhibit their function. The IFNγ-dependent upregulation of p21 is at the transcriptional level, and is mediated by STAT1 homodimer (GAF) (Chin et al. 1996). IFNα has also been shown to induce p21 expression in Daudi cells, resulting in an observable increase in p21 association with cdk2 that correlates with an arrest of these cells in G1

(Subramanian and Johnson 1997). Thus, IFN-dependent regulation of cdk inhibitors may represent an additional level of cell-cycle control by the IFNs, although confirmation of this possibility is needed.

The expression of the gene 200 cluster of ISGs on mouse chromosome 1 has been implicated in regulating cellular proliferation (Choubey et al. 1996; Min et al. 1996). One gene product in particular, p202, has been shown to modulate growth through direct interactions with a variety of transcription factors, including E2F, *fos*, *jun*, and the p50 and p65 subunits of NFκB, in many cases inhibiting their transcriptional activites toward target genes (Choubey et al. 1996; Min et al. 1996). p202 can also bind to the hyperphosphorylated form of pRB, inhibiting its role in G1 to S progression (Choubey and Lengyel 1995). Whether IFNs also directly regulate cyclin-cdk complexes in mouse cells, or whether proteins analogous to the mouse p200 proteins function in human cells is not known.

Little else is known about the IFN-dependent regulation of cell-cycle components. For example, the role, if any, of p53 in these cell-cycle regulatory events is unclear. Although p21 induction by IFNγ appears to be independent of p53 upregulation (since p53 protein levels are not affected by IFN), a supporting role for p53 in this process has not been ruled out. p53 is an important regulator of p21 transcription (el-Deiry et al. 1993), and we have found that IFN cannot upregulate p21 in p53-null cell, although the growth-inhibtory effects of IFN are intact in these same cells (D.W.L., G. Stark, unpubl. observation). Interestingly, the murine p202 protein described above can influence p53 function by binding to the p53-binding protein 53BP1, resulting in decreased p53-dependent transcriptional regulation (Datta et al. 1996). How, or if, this contributes to growth arrest is unclear.

6 Therapeutic Potentials of the IFNs

IFNs exhibit potent antiviral and antitumor activities, two properties that make them very attractive candidates for therapeutic agents. Early attempts to test the efficacy of purified IFNs in the treament of a variety of viral illnesses failed to demonstrate a significant reduction in the term or severity of the viral infections (Dianzani 1992a). Treatment of human cancers with IFNs were somewhat more promising, but suffered from many of the same drawbacks as the antiviral studies. In all cases, the studies were hampered by the lack of substantial quantities of highly purified IFN. In addition, IFN treatment was accompanied by undesirable side effects, primarily fever, chills, fatigue, and in some cases anemia and/or thrombocytopenia (Dianzani 1992b), and it was believed that impurities in the IFN preparations were to blame for many of these toxic effects (Dianzani 1992b). With the advent of large-scale production of biologically active recombinant IFNs in the 1980s came the hope that higher doses and pure preparations would improve the therapeutic efficacy of IFNs. Unfortunately, patients continued to exhibit adverse therapeutic reactions, which intensified with higher doses (Dianzani 1992a; Quesada 1992; Gutterman 1994). In addition, a substantial proportion of patients developed anti-IFN antibodies, particularly those patients undergoing chronic therapy with high IFN doses. Although these drawbacks are substantial, IFNs have, nonetheless, become an important tool for combating a variety of human diseases. What follows is a brief listing of diseases for which IFNs have been approved as therapeutic agents in the US. It should be noted that the accepted use of IFNs in other countries is more extensive (Tyring 1992).

6.1 Cancers

It is perhaps surprising that most of the successful therapeutic applications of IFNs have been against cancers, and not viral infections (Gutterman 1994). Hairy cell leukemia (HCL) is one such example in which IFNs have shown promise. HCL is a rare B-cell neoplasm characterized by the presence of "hairy cells" in the peripheral blood and in bone marrow. As the disease progresses, patients often develop enlarged spleens (splenomegaly) and are susceptible to recurrent oportunistic infections (Platanias and Golomb 1992). Prior to the introduction of IFNα therapy, splenectomy was the most common treatment. HCL pateints undergoing IFNα therapy show a high (~90%) response rate, of which 5–10% go into complete remission (Gutterman 1994). Although the mechanism of IFNα action against HCL is incompletely understood, it most likely relates to IFNα-induced B-cell differentiation, the inhibition of hairy cell responsiveness to B cell growth factors, or the activation of antineoplastic immune cell function (Platanias and Golomb 1992).

Chronic myelogenous leukemia (CML) is a well-characterized human cancer that frequently involves a chromosomal translocation between chromosomes 9 and 22 [t(9:22)], leading to the generation of the Philadelphia Chromosome (Leibowitz and Young 1989). This translocation results in the transfer of the c-abl protooncogene from chromosome 9 to the breakpoint cluster region (bcr) on chromosome 22, forming the hybrid bcr/abl gene encoding an abnormal, deregulated tyrosine kinase (Gale and Butturini 1993). The disease is characterized by both a chronic phase, in which patients exhibit thrombocytosis, leukocytosis, and splenomegaly, followed by an accelerated phase during which patients quickly enter into a fatal myeloid or lymphoid blast crisis (Platanias and Golomb 1992). IFNα has been used effectively to treat CML patients, and 80–90% respond positively to the therapy, with roughly 20% going into remission (Platanias and Golomb 1992). As in the case of HCL, the mechanism of IFNα action against CML is not clear, and may relate to the antiproliferative, antiviral (viruses have been implicated in the animal CML model) or immunomodulatory functions of IFNα (Platanias and Golomb 1992).

IFNs have shown promise in the treatment of a variety of solid tumors. Among the tumors that have been targeted are papilloma virus-associated tumors, malignant melanomas, bladder carcinomas, ovarian carcinomas, renal cell carcinomas and a variety of others (Strander and Oberg 1992). Since many tumors have been targeted, the responses on a whole can be considered quite variable. IFNs (primarily IFNα) are often used in combination with other antitumor treatments, such as chemotherapy, surgery, or radiation therapy. Indeed, IFN monotherapy is not likely to provide many additional breakthroughs, but the use of IFNs as adjuvants may find greater acceptance as new combinatorial therapies are developed.

Karposi's sarcoma is a malignant cutaneous neoplasm that is a frequent component of the acquired immunodeficiency syndrome (AIDS). Although the AIDS-associated Karposi's sarcoma is seldom life-threatening, the disfigurement and discomfort associated with later stages of the disease justify the need for effective treatments. Systemic IFNα therapy can lead to tumor regression in 40–50% of patients, with 20–30% showing a complete response (de Wit 1992). The mechanism of action is not completely clear, but may relate to increased tumor antigen presentation, or decreased cellular proliferation in IFN-treated individuals (de Wit 1992).

6.2 Viral Diseases

In the US, the only viral disease for which IFNα is an approved treatment is hepatitis B virus (HBV), although IFNα treatment of hepatitis C and Herpes virus is also approved in many other countries (Tyring 1992). In patients with chronic HBV infections, treatment with IFNα for 6 months can induce a sustained remission in 40–50% of patients (Hoofnagle 1992). In responding patients, liver function is improved, and in a low percentage of these individuals (~15%), HBV antigens are eliminated, and the patients are no longer considered as carriers (Hoofnagel 1992; Gutterman 1994). The poor responsiveness to IFNα therapy in the majority of patients is thought to be due to viral mutations and virus-inhibited cellular responsiveness to IFN, probably mediated by the HBV terminal protein which blocks IFN signaling in infected cells (Gutterman 1994). Combinatorial therapies that address these problems are needed to enhance the responses observed with IFN monotherapy.

6.3 Other Diseases

Multiple sclerosis (MS) is the most recent addition to the list of diseases for which IFNs have been approved for clinical use in the US. MS is a neurological disorder that strikes young adults and is characterized by discrete lesions in the brain and spinal cord, leading to progressive neurological impairment (Panitch and Bever 1992). Patients with relapsing-remitting MS exhibit periods of attacks or exacerbations that affect daily function (Knobler et al. 1993). Although the cause of MS is unknown, it is believed to occur in genetically predisposed individuals, and its effects may be associated with occasional viral infections (Sibley et al. 1984). IFNα is largely ineffective in combating MS symptoms, but subcutaneous IFNβ treatment three times a week has been found to lessen the severity and/or frequency of MS exacerbations (Knobler et al. 1993). IFNγ, on the other hand, worsens these symptoms (Panitch and Bever 1992). Indeed, it is believed that IFNβ acts in treated patients to inhibit the actions of IFNγ, either by reducing the IFN-induced expression of MHC class II in microglial cells of the CNS, or by directly inhibiting IFNγ expression (Panitch and Bever 1992). The disparity between IFNα and IFNβ effectiveness in controlling MS exacerbations is not fully understood. Although these two IFN subtypes share a common receptor, they also have subtle differences in their effects on cells, including differences in the ligand-induced phosphorylation of IFNAR subunits described earlier (Abramovich et al. 1994). Identification of the specific mode of IFNβ action may permit the generation of more effective therapies.

Chronic granulomatous disease (CGD) is an inherited immune deficiency disease (Malech 1992). Patients with CGD display defects in the ability of their neutrophiles, monocytes, and eosinophiles to produce superoxide, hydrogen peroxide, and other antimicrobial oxidants, resulting in an increased susceptability to bacterial and fungal infections. CGD patients must take daily prophylactic antibiotics, and IFNγ has been approved for prophylactic use to further reduce the risk of infection. IFNγ is administered by subcutaneous injection on alternative days, and the mode of action is believed to be the upregulation of host defense mechanisms (Malech 1992).

In addition to the diseases listed here, IFNs have shown efficacy in treating other clinical conditions, such as childhood angiomatous diseases, leprosy, leishmaniasis, lupus erythematosus, basal cell carcinomas, and others.

Reviews that address these additional applications are available elsewhere (Tyring 1992; Gutterman 1994).

7 Interferons as Hormones of Pregnancy

Although best known for regulating immune function or mediating cellular responses to virus infection, several surprising biological roles for IFNs have been revealed over the past 10 years. The most dramatic example has been the discovery that, in some species, IFNs play an obligatory role in the establishment of pregnancy. In 1982 it was demonstrated that explanted ovine conceptuses (the embryo proper plus associated fetal membranes) secreted large amounts of a low molecular weight protein into the culture medium. This protein was subsequently purified to homogeneity, and named ovine trophoblast protein-1 (oTP-1) to reflect its production by trophoblast cells of the developing trophectoderm (Godkin et al. 1982). Expression of oTP-1 in vivo coincided with a critical period in embryonic development known as the time of maternal recognition of pregnancy (Short 1969). During this period in early pregnancy, the conceptus must signal its presence to the mother in order to maintiain a functional corpus luteum and prevent the initiation of a subsequent estrous cycle (see Roberts et al. 1992). Purified oTP-1 was shown, in sheep, to extend estrous cycle length when infused into the uteri of nonpregnant animals (Godkin et al. 1984), whereas purified conceptus secretory proteins from which oTP-1 had been removed by immunoadsorption were ineffective in this regard (see Roberts et al. 1992). These and other studies led to the designation of oTP-1 as the major antiluteolysin of sheep.

The identification of oTP-1, and the bovine homolog bTP-1, as type I IFNs came about through protein-sequencing studies and the molecular cloning of their corresponding cDNAs (Imakawa et al. 1987; Stewart et al. 1987; Roberts et al. 1992). These results were quite unexpected. IFN production by conceptuses and placental tissues of other species had been described previously (see Roberts et al. 1992), but the magnitude of oTP-1 and bTP-1 production by preimplantation ovine and bovine conceptuses (about 100 μg of protein per conceptus in a 24-h culture period), coupled with their proposed role in maternal recognition of pregnancy, revealed a novel physiological role for IFNs in ruminant species. The trophoblast IFNs have since been recognized as a distinct subset of type I IFN, designated IFNτ, in view of their unique structure, function, and pattern of expression in early pregnancy.

The IFNτ proteins share around 70% amino acid sequence similarity with the IFNω, from which they are presumed to have evolved (see Sect. 2.2). As with other type I IFNs, IFNτ can protect many cell types from lysis by a range of viruses at a potency roughly equivalent to that exhibited by IFNα or IFNω, and IFNτ also has antiproliferative and immunomodulatory effects that are similar to other type I IFNs (Roberts et al. 1992; Alexenko et al. 1997). What sets IFNτ apart is its ability to extend estrous cycle length in domestic ruminant species, and its unusual pattern of expression in early embryonic development (Roberts et al. 1992), implicating a unique mode of gene regulation. Whereas IFNα, -β and -ω genes are strongly upregulated by viruses or dsRNA, the IFNτ genes are only slightly induced under the same conditions (Cross and Roberts 1991). Instead, the IFNτ genes appear to be developmentally regulated in a cell-specific manner, although the exact transcriptional regulators have yet to be identified (Cross and Roberts 1991; Leaman et al. 1994).

IFNτ can compete with other type I IFNs for the same cell surface binding sites, and can also activate the same repertoires of JAKs and STATs (Hansen et al. 1989; Alexenko et al. 1997). In many instances, the same ISGs are induced, both in vivo and in vitro (D.W.L. and R.M. Roberts, unpubl.). However, it is clear that IFNτ must also have specific effects on cells of the uterine endometrium since it is roughly tenfold more potent than other type I IFNs in extending estrous cycle length when infused into the uteri of nonpregnant animals (see Roberts et al. 1992). It seems likely, therefore, that IFNτ activates additional signaling pathways, either exclusively or preferentially, that are involved in regulating the physiological processes that govern luteal lifespan. The identification of these pathways should provide critical insight into the mechanism of IFNτ action.

Although IFNτ appears to be restricted to species within the *Artiodactyla* suborder of mammals, recent work has demonstrated that other IFNs may function to regulate reproductive parameters in other species. An atypical porcine IFNω subspecies that is distinct from the IFNω and IFNτ subtypes identified in other mammals is produced by trophoblast cells of periimplantation pig conceptuses (Lefevre and Boulay 1993). Interestingly, pig conceptuses also secrete IFNγ (Roberts et al. 1992; Lefevre and Boulay 1993), providing an unusual circumstance in which both type I and II IFNs are expressed concomitantly. Although the biological function of this porcine IFNω is not yet known, it may represent the first example of a new type I IFN subtype that, like IFNτ, does not act primarily as an antiviral agent. Whether other species also express this or related IFNs during pregnancy remains to be determined.

8 Conclusion

Although it has been 40 years since IFNs were first described, the level of scientific and clinical interest in these multifunctional cytokines has not diminished. Rapid progress has been made recently in the identification of the biochemical pathways that mediate IFN actions through the cloning of IFN receptor and signaling components. So what does the future hold in store? Although the possibilities are limitless, the areas with the greatest potential for progress pertain to the identification of factors that regulate IFN gene expression following exposure of cells to virus or other stimuli, and the further characterization of factors that mediate IFN-dependent regulation of cell growth and differentiation. Advances in these areas would be of interest not only to the basic researcher, but also to those working to improve the therapeutic effectiveness of IFNs. Scientific progress in related fields, such as the study of cytokine and growth factor signaling, cell cycle regulation, and control of apoptosis (to name a few) is advancing rapidly, and it is likely that work in these areas will impinge on future IFN studies, much as the identification of IFN signaling and antiviral components has advanced our understanding of how other extracellular signaling proteins mediate cellular responses. Finally, as described in the previous section, the demonstration that IFNs play an important role in early embryonic development in some species suggests that IFNs may function in other areas of mammalian physiology, including those that are not directly related to the immune system.

Acknowledgments. I would like to thank Drs. George Stark, Robert Silverman and Clayton Huntley for critical reading of this manuscript. I apologize for not having referenced many important publications; in the space available, it is impossible to cite all of the significant contributions published in this field.

References

Abramovich C, Shulman LM, Ratovitski E, Harroch S, Tovey M, Eid P, Revel M (1994) Differential tyrosine phosphorylation of the IFNAR chain of the type I interferon receptor and of an associated surface protein in response to IFN-alpha and IFN-beta. EMBO J 13:5871–5877

Alexenko AP, Leaman D, Li J, Roberts RM (1997) The antiproliferative and antiviral activities of interferon-tau variants on human cells. J Interferon Cytokine Res (in press)

Allen G, Diaz MO (1996) Nomenclature of the human interferon proteins. J Interferon Cytokine Res 16:181–184

Au WC, Moore PA, Lowther W, Juang YT, Pitha PM (1995) Identification of a member of the interferon regulatory factor family that binds to the interferon-stimulated response element and activates expression of interferon-induced genes. Proc Natl Acad Sci USA 92:11657–11661

Auget M (1980) High-affinity binding of ^{125}I-labelled mouse interferon to a specific cell-surface receptor. Nature 284:459–461

Auget M, Dembic Z, Merlin G (1988) Molecular cloning and expression of the human interferon-γ receptor. Cell 55:273–280

Bach EA, Tanner JW, Marsters S, Ashkenazi A, Aguet M, Shaw AS, Schreiber RD (1996) Ligand-induced assembly and activation of the gamma interferon receptor in intact cells. Mol Cell Biol 16:3214–3221

Baron S, Coppenhaver DH, Dianzani F, Fleischmann WR Jr, Hughes TK Jr, Klimpel GR, Niesel DW, Stanton GJ, Tyring SK (1992) Introduction to the interferon system. In: Baron S (ed) Interferon: principles and medical applications. University of Texas at Galveston Medical Branch, Galveston, pp 1–15

Bazan JF (1990) Structural design and molecular evolution of a cytokine receptor superfamily. Proc Natl Acad Sci USA 87:6934–6938

Bluyssen HAR, Muzaffar R, Vlieststra RJ, van der Made ACJ, Leung S, Stark GR, Kerr IM, Trapman J, Levy DE (1995) Combinatorial association and abundance of components of interferon-stimulated gene factor 3 dictate the selectivity of interferon responses. Proc Natl Acad Sci USA 92:5645–5649

Branca AA, Baglioni C (1981) Evidence that types I and II interferons have different receptors. Nature 294:768–771

Brass AL, Kehrli E, Eisenbeis CF, Storb U, Singh H (1996) Pip, a lymphoid-restricted IRF, contains a regulatory domain that is important for autoinhibition and ternary complex formation with the Ets factor PU.1. Genes Dev 10:2335–2347

Briken V, Ruffner H, Schultz U, Schwarz A, Reis LFL, Strehlow I, Decker T, Staeheli P (1995) Interferon regulatory factor 1 is required for mouse Gbp gene activation by gamma interferon. Mol Cell Biol 15:975–982

Briscoe J, Rogers NC, Witthuhn BA, Watling D, Harpur AG, Wilks AF, Stark GR, Ihle JN, Kerr IM (1996) Kinase-negative mutants of JAK1 can sustain interferon-gamma-inducible gene expression but not an antiviral state. EMBO J 15:799–809

Burke LC, Bybee A, Thomas NS (1992) The retinoblastoma protein is partially phosphorylated during early G1 in cycling cells but not in G1 cells arrested with alpha-interferon. Oncogene 7:783–788

Capon DJ, Shepard HM, Goeddel DV (1985) Two distinct families of human and bovine interferon-α genes are coordinately expressed and encode functional polypeptides. Mol Cell Biol 5:768–779

Celada A, Gray PW, Rinderknecht E, Screiber RD (1984) Evidence for a gamma-interferon receptor that regulates macrophage tumoricidal activity. J Exp Med 160:55–62

Chin YE, Kitagawa M, Su WC, You ZH, Iwamoto Y, Fu XY (1996) Cell growth arrest and induction of cyclin-dependent kinase inhibitor p21 WAF1/CIP1 mediated by STAT1. Science 272:719–722

Choubey D, Lengyel P (1995) Binding of an interferon-inducible protein (p202) to the retinoblastoma protein. J Biol Chem 270:6134–6140

Choubey D, Li S-J, Datta B, Gutterman JU, Lengyel P (1996) Inhibition of E2F-mediated transcription by p202. EMBO J 15:5668–5678

Colamonici O, Yan H, Domanski P, Handa R, Smalley D, Mullersman J, Witte M, Krishnan K, Krolewski J (1994) Direct binding to and tyrosine phosphorylation of the alpha subunit of the type I interferon receptor by p135tyk2 tyrosine kinase. Mol Cell Biol 14:8133–8142

Cook JR, Cleary CM, Mariano TM, Izotova L, Pestka S (1996) Differential responsiveness of a splice variant of the human type I interferon receptor to interferons. J Biol Chem 271:13448–13453

Cosman D, Lyman SD, Idzerda RL, Beckmann MP, Park LS, Goodwin RG, March CJ (1990) A new cytokine receptor superfamily. Trends Biochem Sci 15:265–270

Cross JC, Roberts RM (1991) Constitutive and trophoblast-specific expression of a class of bovine interferon genes. Proc Natl Acad Sci USA 88:3817–3821

Darnell JE Jr (1997) STAT molecules and gene regulation. Science 277:1630–1635

Darnell JE Jr, Kerr IM, Stark GR (1994) Jak-STAT pathways and transcriptional activation in response to IFNs and other extracellular signaling proteins. Science 264:1415–1421

Datta B, Li B, Choubey D, Nallur G, Lengyel P (1996) p202, an interferon-inducible modulator of transcription, inhibits transcriptional activation by the p53 tumor suppressor protein, and a segment from the p53-binding protein 1 that binds to p202 overcomes this inhibition. J Biol Chem 271:27544–27555

David M, Grimley PM, Finbloom DS, Larner AC (1993) A nuclear tyrosine phosphatase downregulates interferon-induced gene expression. Mol Cell Biol 13:7515–7521

David M, Chen HE, Goelz S, Larner AC, Neel BG (1995a) Differential regulation of the alpha/beta interferon-stimulated JAK/STAT pathway by the SH2 domain-containing tyrosine phosphatase SHPTP1. Mol Cell Biol 15:7050–7058

David M, Petricoin E III, Benjamin C, Pine R, Weber MJ, Larner AC (1995b) Requirement for MAP kinase (ERK2) activity in interferon α- and interferon β-stimulated gene expression through STAT proteins. Science 269:1721–1723

Decker T, Kovarik P, Meinke A (1997) GAS elements: a few nucleotides with a major impact on cytokine-induced gene expression. J Interferon Cytokine Res 17:121–134

De Maeyer E, De Maeyer-Guignard J (1988) Interferons and other regulatory cytokines. John Wiley, New York

Dent CL, Macbride SJ, Sharp NA, Gewert DR (1996) Relative transcriptional inducibility of the human interferon-alpha subtypes conferred by the virus-responsive enhancer sequence. J Interferon Cytokine Res 16:99–107 de Wit R (1992) Karposi's sarcoma and AIDS. In: Baron S (ed) Interferon: principles and medical applications. University of Texas at Galveston Medical Branch, Galveston, pp 475–485

Dianzani F (1992a) Biological basis for therapy and for side effects. In: Baron S (ed) Interferon: principles and medical applications. University of Texas at Galveston Medical Branch, Galveston, pp 409–416

Dianzani F (1992b) How to use an endogenous system as a therapeutic agent. J Interferon Res Special Issue:100–118

Dirks W, Mittnacht S, Rentrop M, Hayser H (1989) Isolation and functional characterization of the murine interferon-β1 promoter. J Interferon Res 9:125–133

Domanski P, Witte M, Kellum M, Rubinstein M, Hackett R, Pitha P, Colamonici OR (1995) Cloning and expression of a long form of the beta subunit of the interferon alpha beta receptor that is required for signaling. J Biol Chem 270:21606–21611

Doree M, Galas S (1994) The cyclin-dependent protein kinases and the control of cell division. FASEB J 8:1114–1121

Dron M, Tovey M (1992) Interferon α/β, gene structure and regulation. In: Baron S (ed) Interferon: principles and medical applications. University of Texas at Galveston Medical Branch, Galveston, pp 33–43

Du W, Thanos D, Maniatis T (1993) Mechanisms of transcriptional synergism between distinct virus- inducible enhancer elements. Cell 74:887–898

Einat M, Kimchi A (1985) Close link between reduction of *c-myc* expression by interferon and G1/G0 arrest. Nature 313:597–600

Einat M, Kimchi A (1988) Transfection of fibroblasts with activated *c-myc* confers resistance to antigrowth effects of interferon. Oncogene 2:485–491

Eisenbeis CF, Singh H, Storb U (1995) Pip, a novel IRF family member, is a lymphoid-specific, PU.1-dependent transcriptional activator. Genes Dev 9:1377–1387

el-Deiry WS, Tokino T, Velculescu VE, Levy DB, Parsons R, Trent JM, Lin D, Mercer WE, Kinzler KW, Vogelstein B (1993) WAF1, a potential mediator of p53 tumor suppression. Cell 75:817–825

Falvo JV, Thanos D, Maniatis T (1995) Reversal of intrinsic DNA bends in the IFNβ gene enhancer by transcription factors and the architectural protein HMG I(Y). Cell 83:1101–1111

Farrar MA, Schreiber RD (1993) The molecular cell biology of interferon-gamma and its receptor. Annu Rev Immunol 11:571–611

Farrar MA, Campbell JD, Schreiber RD (1992) Identification of a functionally important sequence in the C terminus of the interferon-gamma receptor. Proc Natl Acad Sci USA 89:11706–11710

Farrar MA, Fernandez-Luna J, Schreiber RD (1991) Identification of two regions within the cytoplasmic domain of the human interferon-gamma receptor required for function. J Biol Chem 266:19626–19635

Fischer T, Rehm A, Aguet M, Pfizenmaier K (1990) Human chromosome 21 is necessary and sufficient to confer human IFNγ receptor gene. Cytokine 2:157–161

Fish EN, Banerjee K, Stebbing N (1983) Human leukocyte interferon subtypes have different antiproliferative and antiviral activities on human cells. Biochem Biophys Res Commun 112:537–546

Flati V, Haque SJ, Williams BRG (1996) Interferon-α-induced phosphorylation and activation of cytosolic phospholipase A_2 is required for the formation of interferon-stimulated gene factor three. EMBO J 15:1566–1571

Fleischmann WR Jr, Fleischmann CM (1992) Mechanisms of interferons' antitumor actions. In: Baron S (ed) Interferon: principles and medical applications. University of Texas at Galveston Medical Branch, Galveston, pp 299–309

Foster GR, Rodriques O, Ghouze F, Schulte-Frohlinde E, Testa D, Liao MJ, Stark GR, Leadbeater L, Thomas HC (1996) Different relative activities of human cell-derived interferon-α subtype: IFN-α8 has very high antiviral potency. J Interferon Cytokine Res 16:1027–1033

Friedman RL, Manly SP, McMahon M, Kerr IM, Stark GR (1984) Transcriptional posttranscriptional regulation of interferon-induced gene expression in human cells. Cell 38:745–755

Friedman RM (1967) Interferon binding, the first step in establishment of antiviral activity. Science 156:1760–1762

Fu X-Y, Schindler C, Improta T, Aebersold R, Darnell JE Jr (1992) The proteins of ISGF-3, the interferon α-induced transcriptional activator, define a gene family involved in signal transduction. Proc Natl Acad Sci USA 89:7840–7843

Fujita T, Sakakibara J, Sudo Y, Miyamoto M, Kimura Y, Taniguchi T (1988) Evidence for a nuclear factor(s), IRF-1, mediating induction and silencing properties to human IFN-beta gene regulatory elements. EMBO J 7:3397–3405

Gale RP, Butturini A (1993) Recent progress in understanding chronic myelogenous leukemia. Leuk Lymphoma 11 Suppl 1:3–5

Gillespie D, Pequignot E, Carter WE (1984) Evolution of interferon genes. In: Carne PE, Carter WA (eds) Handbook of experimental pharmacology, vol 71. Springer, Berlin Heidelberg New York, pp 45–70

Godkin JD, Bazer FW, Moffat J, Sessions F, Roberts RM (1982) Purification and properties of a major, low molecular weight protein released by the trophoblast of sheep blastocysts at day 13–21. J Reprod Fertil 65:141–149

Godkin JD, Bazer FW, Thatcher WW, Roberts RM (1984) Proteins released by cultured day 15–16 conceptuses prolong luteal maintenance when introduced into the uterin lumen of cyclic ewes. J Reprod Fertil 71:57–65

Goodbourn S, Maniatis T (1988) Overlapping positive and negative regulatory domains of the human β-interferon gene. Proc Natl Acad Sci USA 85:1447–1451

Gray PW, Goeddel DV (1982) Structure of the human immune interferon gene. Nature 298:859–863

Gray PW, Leong S, Fennie EH, Farrar MA, Pingel JT, Fernandez-Luna J, Schreiber RD (1989) Cloning and expression of the cDNA for the murine interferon γ receptor. Proc Natl Acad Sci USA 86:8479–8501

Greenlund AC, Farrar MA, Viviano BL, Schreiber RD (1994) Ligand-induced IFNγ receptor tyrosine phosphorylaton couples the receptor to its signal transduction system (p91). EMBO J 13:1591–1600

Guschin D, Rogers N, Briscoe J, Witthuhn B, Watling D, Horn F, Pellegrini S, Yasukawa K, Heinrich P, Stark GR, Ihle JN, Kerr IM (1995) A major role for the protein tyrosine kinase JAK1 in the JAK/STAT signal transduction pathway in response to interleukin-6. EMBO J 14:1421–1429

Gutterman JU (1994) Cytokine therapeutics: lessons from interferon α. Proc Natl Acad Sci USA 91:1198–1209

Han Y, Leaman DW, Watling D, Rogers N, Groner B, Kerr IM, Wood WI, Stark GR (1996) Participation of JAK and STAT proteins in growth hormone-induced signaling. J Biol Chem 271:5947–5952

Hannigan GE, Williams BRG (1991) Signal transduction by interferon-α through arachidonic acid metabolism. Science 251:204–207

Hansen TR, Kazemi M, Keisler DH, Malathy PV, Imakawa K, Roberts RM (1989) Complex binding of the embryonic interferon, ovine trophoblast protein-1, to endometrial receptors. J Interferon Res 9:215–225

Haque SJ, Flati V, Deb A, Williams BRG (1995) Roles of protein-tyrosine phosphatases in Statlalpha-mediated cell signaling. J Biol Chem 270:25709–25714

Harada H, Fujita T, Miyamoto M, Kimurak Y, Murayama M, Furia A, Miyata T, Taniguchi T (1989) Structurally similar but functionally distinct factors, IRF-1 and IRF-2, bind to the same regulatory elements of IFN and IFN-inducible genes. Cell 58:729–739

Hardy KJ, Young HA (1992) IFNγ, gene structure and regulation. In: Baron S (ed) Interferon: principles and medical applications. University of Texas at Galveston Medical Branch, Galveston, pp 47–64

Hardy KJ, Peterlin BM, Atchison RE, Stobo JD (1985) Regulation of expression of the human interferon gamma gene. Proc Natl Acad Sci USA 82:8173–8177

Hardy KJ, Manger B, Newton M, Stobo JD (1987) Molecular events involved in regulating human interferon-gamma gene expression during T cell activation. J Immunol 138:2353–2358

Haspel RL, Salditt-Georgieff M, Darnell JE Jr (1996) The rapid inactivation of nuclear tyrosine phosphorylated Statl depends upon a protein tyrosine phosphatase. EMBO J 15: 6262–6270

Heichman KA, Roberts JM (1994) Rules to replicate by. Cell 79:557–562

Heim MH, Kerr IM, Stark GR, Darnell JE Jr (1995) Contribution of STAT SH2 groups to specific interferon signaling by the Jak-STAT pathway. Science 267:1347–1349

Heldin C-H (1995) Dimerization of cell surface receptors in signal transduction. Cell 80:213–223

Hemmi S, Bohni R, Stark G, Di Marco F, Aguet M (1994) A novel member of the interferon receptor family complements functionality of the murine interferon gamma receptor in human cells. Cell 76:803–810

Henco K, Brosium J, Fujisawa A, Fujisawa J-I, Haynes JR, Hochstadt J, Kovacic T, Pasek M, Schambock A, Schmid J, Todokoro K, Walchli M, Nagata S, Weissman CJ (1985) Structural relationship of human interferon alpha genes and pseudogenes. Mol Biol 185:227–260

Hiscott J, Cantell K, Weissman C (1984) Differential expression of human interferon genes. Nucleic Acids Res 12:3727–3749

Holtschke T, Lohler J, Kanno Y, Fehr T, Giese N, Rosenbauer F, Lou J, Knobeloch KP, Gabriele L, Waring JF, Bachmann MF, Zinkernagel RM, Morse HC 3rd, Ozato K, Horak I (1996) Immunodeficiency and chronic myelogenous leukemia-like syndrome in mice with a targeted mutation of the ICSBP gene. Cell 87:307–317

Hoofnagle JH (1992) Interferon therapy of viral hepatitis. In: Baron S (ed) Interferon: principles and medical applications. University of Texas at Galveston Medical Branch, Galveston, pp 433–462

Hou XS, Perrimon N (1997) The JAK-STAT pathway in *Drosophila*. Trends Genet 13:105–110

Houle JL, Santoro N (1996) Analysis of human interferon-alpha gene promoters by multiple sequence alignment. J Interferon Cytokine Res 16:93–98

Housey GM, Johnson MD, Hsiao WL, O'Brian CA, Weinstein IB (1988) Structural and functional studies of protein kinase C. Adv Exp Med Biol 234:127–140

Hovanessian AG (1991) Interferon-induced and double-stranded RNA-activated enzymes: a specific protein kinase and 2′-5′-oligoadenylate synthetases. J Interferon Res 11:199–205

Hwang SY, Hertzog PJ, Holland KA, Sumarsono SH, Tymms MJ, Hamilton JA, Whitty G, Bertoncello I, Kola I (1995) A null mutation in the gene encoding a type I interferon receptor component eliminates antiproliferative and antiviral responses to interferons α and β and alters macrophage responses. Proc Natl Acad Sci USA 92:11284–11288

Igarashi K-I, David M, Larner AC, Finbloom DS (1993) In vitro activation of a transcription factor by gamma interferon requires a membrane-associated tyrosine kinase and is mimicked by vanadate. Mol Cell Biol 13:3984–3989

Ihle JN (1996) STATs: signal transducers and activators of transcription. Cell 84:331–334

Ihle JN, Kerr IM (1995) Jaks and Stats in signaling by the cytokine receptor superfamily. Trends Genet 11:69–74

Ihle JN, Witthuhn BA, Quelle FW, Yamamoto K, Thierfelder WE, Kreider B, Silvennoinen O (1994) Signaling by the cytokine receptor superfamily: JAKs and STATs. TIBS 19:222–227

Imakawa K, Anthony RV, Kazemi M, Marotti KR, Polites HG, Roberts RM (1987) Interferon-like sequence of ovine trophoblast protein secreted by embryonic trophectoderm. Nature 330:377–379

Imam AMA, Ackrill AM, Dale TC, Kerr IM, Stark GR (1990) Transcription factors induced by interferons alpha and gamma. Nucleic Acids Res 18:6573–6580

Improta T, Schindler C, Horvath CM, Kerr IM, Stark GR, Darnell JE Jr (1994) Transcription factor ISGF-3 formation requires phosphorylated Stat91 protein, but Stat113 protein is phosphorylated independently of Stat91 protein. Proc Natl Acad Sci USA 91:4776–4780

Isaacs A, Lindenmann J (1957) Virus interference. 1. The interferon. Proc R Soc Lond Ser B Biol Sci 147:258–267

Jung V, Rashidbaigi A, Jones C, Tischfield JA, Shows TB, Pestka S (1987) Human chromosomes 6 and 21 are required for sensitivity to human IFNγ. Proc Natl Acad Sci USA 84:4151–4155

Kaplan DH, Greenlund AC, Tanner JW, Shaw AS, Schreiber RD (1996) Identification of an interferon-gamma receptor alpha chain sequence required for JAK-1 binding. J Biol Chem 271:9–12

Keller AD, Manaitis T (1991) Identification and characterization of a novel repressor of b-interferon gene expression. Genes Dev 5:868–879

Kim TK, Maniatis T (1996) Regulation of interferon-gamma-activated STAT1 by the ubiquitin-proteasome pathway. Science 273:1717–1719

King RW, Jackson PK, Kirschner MW (1994) Mitosis in transition. Cell 79:563–571

Kirchhoff S, Koromilas AE, Schaper F, Grashoff M, Sonenberg N, Hauser H (1995) IRF-1 induced cell growth inhibition and interferon induction requires the activity of the protein kinase PKR. Oncogene 11:439–445

Kishimoto T, Taga T, Akira S (1994) Cytokine signal transduction. Cell 76:253–262

Knobler RL, Geeenstein JI, Johnson KP, Lublin FD, Panitch HS, Conway K, Grant-Gorsen SV, Muldoon J, Marcus SG, Wallenberg JC, Williams GJ, Yoshizawa CN (1993) Systemic recombinant human interferon-β treatment of relapsing-remitting multiple sclerosis: pilot study analysis and 6-year follow-up. J Interferon Res 13:333–340

Kumar A, Haque J, Lacoste J, Hiscott J, Williams BRG (1994) Double-stranded RNA-dependent protein kinase activates transcription factor NF-kappaB by phosphorylating IkB. Proc Natl Acad Sci USA 91:6288–6292

Kumar A, Yang Y-L, Flati V, Der S, Kadereit S, Deb A, Haque J, Reis L, Weissmann C, Williams BRG (1997) Deficient cytokine signaling in mouse embryo fibroblasts with a targeted deletion in the PKR gene: role of IRF-1 and NF-kappaB. EMBO J 16:406–416

Lai MHT, Joklik K (1973) The induction of interferon by temperature-sensitive mutants of reovirus, UV-irradiated reovirus and subviral particles. Virology 51:191–204

Langer JS, Pestka S (1988) Interferon receptors. Immunol Today 9:393–400

Larner AC, Jonak G, Cheng Y-SE, Korant B, Knight E, Darnell JE Jr (1984) Transcriptional induction of two genes in human cells by beta interferon. Proc Natl Acad Sci USA 81:6733–6737

Leaman DW, Roberts RM (1992) Genes for the trophoblast interferons in sheep, goat and musk ox, and distribution of related genes among mammals. J Interferon Res 12:1–11

Leaman DW, Cross JC, Roberts RM (1994) Multiple regulatory elements are required to direct trophoblast interferon gene expression in choriocarcinoma cells and trophectoderm. Mol Endocrinol 8:456–468

Leaman DW, Leung S, Li X, Stark GR (1996a) Regulation of STAT-dependent pathways by growth factors and cytokines. FASEB J 10:1578–1588

Leaman DW, Pisharody S, Flickinger TW, Commane MA, Schlessinger J, Kerr IM, Levy DE, Stark GR (1996b) Roles of JAKs in activation of STATs and stimulation of *c-fos* gene expression by epidermal growth factor. Mol Cell Biol 16:369–375

Leaman DW, Salvekar A, Patel R, Sen G, Stark GR (1997) Isolation of a mutant human cell line defective in interferon-stimulated gene silencing and in responsiveness to double-stranded RNA. EMBO J (submitted)

Lefevre F, Boulay V (1993) A novel and atypical type one interferon gene expressed by trophoblast during early pregnancy. J Biol Chem 268:19760–19768

Leibowitz D, Young KS (1989) The molecular biology of CML: a review. Cancer Invest 7:195–203

Lenardo MJ, Fan CM, Maniatis T, Baltimore D (1989) The involvement of NFκB in β-interferon gene regulation reveals its role as a widely inducible mediator of signal transduction. Cell 57:287–294

Lengyel P (1982) Biochemistry of interferons and their actions. Annu Rev Biochem 51:251–282

Leung S, Qureshi SA, Kerr IM, Darnell JE Jr, Stark GR (2995) Role of STAT2 in the alpha interferon signaling pathway. Mol Cell Biol 15:1312–1317

Levy D, Darnell JE Jr (1990) Interferon-dependent transcriptional activation: signal transduction without second messenger involvement? New Biol 2:923–928

Levy DE, Kessler DS, Pine R, Darnell JE Jr (1989) Cytoplasmic activation of ISGF3, the positive regulator of interferon-α-stimulated transcription, reconstituted in vitro. Genes Dev 3:1362–1371

Levy HB, Salazar AM (1992) Interferon inducers. In: Baron S (ed) Interferon: principles and medical applications. University of Texas at Galveston Medical Branch, Galveston, pp 65–76

Li X, Leung S, Qureshi S, Darnell JE Jr, Stark GR (1996) Formation of STAT1-STAT2 heterodimers and their role in the activation of IRF-1 gene transcription by interferon-α. J Biol Chem 271:5790–5794

Li X, Leung S, Kerr IM, Stark GR (1997) Functional subdomains of STAT2 required for preassociation with the alpha interferon receptor for signaling. Mol Cell Biol 17:2048–2056

Lindenmann J, Lane CA, Hobson D (1963) The resistance of A2G mice to myxoviruses. J Immunol 90:942–950

Lutfalla G, Holland SJ, Cinato E, Monneron D, Reboul J, Rogers NC, Smith JM, Stark GR, Gardiner K, Mogensen KE, Kerr IM, Uzé G (1995) Mutant U5A cells are complemented by an interferon-αβ receptor subunit generated by alternative processing of a new member of a cytokine receptor gene cluster. EMBO J 14:5100–5108

MacDonald NJ, Kuhl D, Maguire D, Naf D, Gallant P, Goswamy A, Hug H, Bueler H, Chaturvedi M, de la Fuente J, Ruffner H, Meyer F, Weissman C (1990) Different pathways mediate virus inducibility of the human IFN-α1 and IFN-β genes. Cell 60:767–779

Malech HL (1992) Interferon gamma as infection prophylaxis in chronic granulomatous disease. In: Baron S (ed) Interferon: principles and medical applications. University of Texas at Galveston Medical Branch, Galveston, pp 563–573

Maniatis T, Whittemore L-A, Du W, Fan C-M, Keller AD, Palombella VJ, Thanos D (1992) Positive and negative control of human interferon-β gene expression. In: McKnight SL, Yamamoto KR (eds) Transcriptional regulation, part 2, Cold Spring Harbor Laboratory Press, Cold Spring Harbor, New York, pp 1193–1220

Maran A, Maitra RK, Kumar A, Dong B, Xiao W, Li G, Williams BR, Torrence PF, Silverman RH (1994) Blockage of NFκB signaling by selective ablation of an mRNA target by 2-5 (A) antisense chimeras. Science 265:789–792

Matsuyama T, Kimura T, Kitagawa M, Pfeffer K, Kawakami T, Watanabe N, Kundig TM, Amakawa R, Kishihara K, Wakeham A, Potter J, Furlonger CL, Narendran A, Suzuki H, Ohashi PS, Paige CJ, Taniguchi T, Mak TW (1993) Targeted disruption of IRF-1 or IRF-2 results in abnormal type I IFN gene induction and aberrant lymphocyte development. Cell 75:83–97

Min W, Ghosh S, Lengyel P (1996) The interferon-inducible p202 protein as a modulator of transcription: Inhibition of NFκB, *c-Fos*, and *c-Jun* activities. Mol Cell Biol 16:359–368

Mittruecker H-W, Matsuyama T, Grossman A, Kundig TM, Potter J, Shahinian A, Wakeham A, Patterson B, Ohashi PS, Mak TW (1997) Requirement for the transcription factor LSIRF/IRF4 for mature B and T lymphocyte function. Science 275:540–543

Miyamoto M, Fujita T, Kimura Y, Maruyama M, Harada H, Sudo Y, Miyata T, Taniguchi T (1988) Regulated expression of a gene encoding a nuclear factor, IRF-1, that specifically binds to IFN-β gene regulatory elements. Cell 54:903–913

Müller M, Briscoe J, Laxton C, Guschin D, Ziemiecki D, Silvennoinen O, Harpur AG, Barbieri G, Witthuhn BA, Schindler C, Pelligrini S, Wilks AF, Ihle JN, Stark GR, Kerr IM (1993a) The protein tyrosine kinase JAK1 complements defects in interferon-alpha/beta and -gamma signal transduction. Nature 366:129–135

Müller M, Laxton C, Briscoe J, Schindler C, Improta T, Darnell JE Jr, Stark GR, Kerr IM (1993b) Complementation of a mutant cell line: central role of the 91 kDa polypeptide of ISGF3 in the interferon-alpha and -gamma signal transduction pathways. EMBO J 12:4221–4228

Muller U, Steinhoff U, Reis LF, Hemmi S, Pavlovic J, Zinkernagel RM, Aguet M (1994) Functional role of type I and type III interferons in antiviral defense. Science 264:1918–1912

Mundschau LJ, Faller DV (1995) Platelet-derived growth factor signal transduction through the interferon-inducible kinase PKR. Immediate early gene induction. J Biol Chem 270:3100–3106

Nelson N, Marks MS, Driggers PH, Ozato K (1993) Interferon consensus sequence-binding protein, a member of the interferon regulatory factor family, suppresses interferon-induced gene transcription. Mol Cell Biol 13:588–599

Novick D, Cohen B, Rubinstein M (1994) The human interferon α/β receptor: characterization and molecular cloning. Cell 77:391–400

Okamura H, Tsutsui H, Komatsu T, Yutsudo M, Hakura A, Tanimoto T, Torigoe K, Okura T, Nukada Y, Hattori K, Akita K, Nambe M, Tanabe F, Konishi K, Fuguta S, Kurimoto M (1995) Cloning of a new cytokine that induces IFN-γ production by T cells. Nature 378:88–91

Panitch HS, Bever C Jr (1992) Clinical use of interferons in multiple sclerosis. In: Baron S (ed) Interferon: principles and medical applications. University of Texas at Galveston Medical Branch, Galveston, pp 581–587

Patel RC, Stanton P, McMillan NM, Williams BRG, Sen GC (1995) The interferon-inducible double-stranded RNA-activated protein kinase self-associates in vitro and in vivo. Proc Natl Acad Sci USA 92:8283–8287

Pavlovic J, Staeheli P (1991) The antiviral potentials of Mx proteins. J Interferon Res 11:215–219

Penix LA, Sweetser MT, Weaver WM, Hoeffler JP, Kerppola TK, Wilson CB (1996) The proximal regulatory element of the interferon-γ promoter mediates selective expression in T cells. J Biol Chem 271:31964–31972

Perez C, Coeffier E, Moreau-Gachelin F, Wietzerbin J, Benech PD (1994) Involvement of the transcription factor PU1/Spi-1 in myeloid cell-restricted expression of an interferon-inducible gene encoding the human high-affinity Fcγ receptor. Mol Cell Biol 14:5023–5031

Pestka S, Langer JA, Zoon KC, Samuel CE (1987) Interferons and their actions. Annu Rev Biochem 56:727–777

Pfeffer LM, Donner DB (1990) The down-regulation of alpha-interferon receptors in human lymphoblastoid cells: relation to cellular responsiveness to the antiproliferative action of alpha-interferon. Cancer Res 50:2654–2657

Pfeffer LM, Strulovici B (1992) Transmembrane second messengers for IFN α/β. In: Baron S (ed) Interferon: principles and medical applications. University of Texas at Galveston Medical Branch, Galveston, pp 150–160

Pfeffer LM, Tan YH (1991) Do second messengers play a role in interferon signal transduction? Trends Biochem Sci 16:321–323

Pfeffer LM, Eisenkraft BL, Reich NC, Improta T, Baxter G, Daniel-Issakani S, Strulovici B (1991) Transmembrane signaling by interferon alpha involves diacylglycerol production and activation of the epsilon isoform of protein kinase C in Daudi cells. Proc Natl Acad Sci USA 88:7988–7992

Pfeffer LM, Mullersman JE, Pfeffer SR, Murti A, Shi W, Yang CH (1997) STAT3 as an adapter to couple phosphatidylinositol 3-kinase to the IFNAR1 chain of the type I interferon receptor. Science 276:1418–1420

Platanias LC, Golomb HM (1992) Clinical use of interferon: Hairy cell, chronic myelogenous and other leukemias. In: Baron S (ed) Interferon: principles and medical applications. University of Texas at Galveston Medical Branch, Galveston, pp 487–499

Quesada JR (1992) Toxicity and side effects of interferons. In: Baron S (ed) Interferon: principles and medical applications. University of Texas at Galveston Medical Branch, Galveston, pp 427–432

Reis LFL, Harada H, Wolchok JD, Taniguchi T, Vilcek J (1992) Critical role of a common transcription factor, IRF-1, in the regulation of IFN-β and IFN-inducible genes. EMBO J 11: 185–193

Revel M, Chebath J (1986) Interferon activated genes. Trends Biochem Sci 11:166–170

Roberts RM, Cross JC, Leaman DW (1992) Interferons as hormones of pregnancy. Endocrin Rev 13:432–452

Rothman JH, Raymond CK, Gilbert T, O'Hara PJ, Stevens TH (1990) A putative GTP binding protein homolous to interferon-inducible Mx proteins performs an essential function in yeast protein sorting. Cell 61:1063–1074

Rubinstein M, Orchansky P (1986) The interferon receptors. CRC Crit Rev Biochem 21:249–277

Samuel CE (1993) The eIF-2α protein kinases, regulators of translation in eukaryotes from yeasts to humans. J Biol Chem 268:7603–7606

Schaefer TS, Sanders, LK, Nathans D (1995) Cooperative transcriptional activity of Jun and Stat3β, a short form of Stat3. Proc Natl Acad Sci USA 92:9097–9101

Schindler C, Darnell JE Jr (1995) Transcriptional responses to polypeptide ligands: the JAK-STAT pathway. Annu Rev Biochem 64:621–651

Schindler C, Fu X-Y, Improta T, Aebersold R, Darnell JE Jr (1992a) Proteins of transcription factor ISGF-3: one gene encodes the 91- and 84-kDa ISGF-3 proteins that are activated by interferon α. Proc Natl Acad Sci USA 89:7836–7839

Schindler C, Shuai K, Prezioso VR, Darnell JE Jr (1992b) Interferon-dependent tyrosine phosphorylation of a latent cytoplasmic transcription factor. Science 257:809–813

Sen GC (1991) Transcriptional regulation of interferon-inducible genes. In: Cohen P, Foulkes JC (eds) Hormonal regulation of transcription. Elsevier, Amsterdam, pp 349–374

Sen GC, Lengyel P (1992) The interferon system. A bird's eye view of its biochemistry. J Biol Chem 8:5017–5020

Sen GC, Ransohoff RM (1997) Transcriptional regulation in the interferon system. Landes Bioscience, Georgetown, Texas

Sherr CJ (1994) G1 phase progression: cycling on cue. Cell 79:551–555

Short RV (1969) Implantationand the maternal recognition of pregnancy. In: Wolstenholme GEW, O'Connor M (eds) Foetel autonomy: Ciba Foundation Symposium J&A Churchill, London, pp 2–26

Shuai K, Stark GR, Kerr IM, Darnell JE Jr (1993) A single phosphotyrosine residue of Stat91 required for gene activation by interferon-γ. Science 261:1744–1746

Shuai K, Horvath CM, Huang LHT, Qureshi SA, Cowburn D, Darnell JE Jr (1994) Interferon activation of the transcription factor Stat91 involves dimerization through SH2-phosphotyrosyl peptide interactions. Cell 76:821–828

Sibley WA, Bamford CR, Clark K (1984) Clinical viral infections and multiple sclerosis. Lancet 1:1313–1315

Silverman RH (1996) 2-5 A dependent RNase L: a regulated endoribonuclease in the interferon system. In: D'Alessio G, Riordan JF (eds) Ribonucleases: structure and function. Academic Press, New York, pp 515–551

Soh J, Donnelly RJ, Kotenko S, Mariano TM, Cook JR, Wang N, Emanuel S, Schwartz B, Miki T, Pestka S (1994) Identification and sequence of an accessory factor required for activation of the human interferon gamma receptor. Cell 76:793–802

Staeheli P (1990) Interferon-induced proteins and the antiviral state. Adv Virus Res 38:147–200

Staeheli P, Pavlovic J (1991) Inhibition of vesicular stomatitis virus mRNA synthesis by human MxA protein. J Virol 65:4498–4501

Staeheli P, Haller O, Boll W, Lindenmann J, Weissmann C (1986) Mx protein: constitutive expression in NIH 3T3 cells transformed with Mx cDNA confers selective resistance to influenza virus. Cell 44:147–158

Stahl N, Yancopoulos GD (1993) The alphas, betas, and kinases of cytokine receptor complexes. Cell 74:587–590

Stahl N, Farruggella TJ, Boulton TG, Zhong Z, Darnell JE Jr, Yancopoulos GD (1995) Choice of STATs and other substrates specified by modular tyrosine-based motifs in cytokine receptors. Science 267:1349–1353

Stark GR, Kerr IM (1992) Interferon-dependent signaling pathways: DNA elements, transcription factors, mutations, and effects of viral proteins. J Interferon Res 12:147–151

Stark GR, Kerr IM, Williams BRG, Silverman RH, Schreiber RD (1997) How cells respond to interferons. Annu Rev Biochem (in press)

Stewart HJ, McCann SHE, Barker PJ, Lee KE, Lamming GE, Flint APF (1987) Interferon sequence homology and receptor binding activity of ovine trophoblast antileuteolytic protein. J Endocrinol 115:R13–15

Strander H, Oberg K (1992) Clinical uses of interferons: solid tumors. In: Baron S (ed) Interferon: principles and medical applications. University of Texas at Galveston Medical Branch, Galveston, pp 533–561

Subramaniam PS, Johnson HM (1997) A role for the cyclin-dependent kinase inhibitor p21 in the G1 cell cycle arrest mediated by the type I interferons. J Interferon Cytokine Res 17:11–15

Talmadge KW, Gallati H, Sinigaglia F, Walz A, Garotta G (1986) Identity between human interferon-gamma and "macrophage-activating factor" produced by human T lymphocytes. Eur J Immunol 16:1471–1477

Taniguchi T (1989) Regulation of interferon-β gene: structure and function of cis-elements and trans-factors. J Interferon Res 9:633–640

Taniguchi T (1995) Cytokine signaling through nonreceptor protein tyrosine kinases. Science 268:251–255

Thanos D, Maniatis T (1992) The high mobility group protein HMG I(Y) is required for NFκB-dependent virus induction of the human IFN-beta gene. Cell 71:777–789

Thanos D, Maniatis T (1995) Virus induction of human IFNβ gene expression requires the assembly of an enhanceosome. Cell 83:1091–1100

Tiefenbrun N, Melamed D, Levy N, Resnitzky D, Hoffman I, Reed SI, Kimchi A (1996) Alpha interferon suppresses the cyclin D3 and cdc25A genes, leading to a reversible G0-like arrest. Mol Cell Biol 16:3934–3944

Tyring SK (1992) Introduction to clinical uses of interferons. In: Baron S (ed) Interferon: principles and medical applications. University of Texas at Galveston Medical Branch, Galveston, pp 399–408

Ullrich A, Schlessinger J (1990) Signal transduction by receptors with tyrosine kinase activity. Cell 61:203–212

Uzé G, Lutfalla G, Gresser J (1990) Genetic transfer of a functional human interferon α receptor into mouse cells: cloning and expression of its cDNA. Cell 60:225–232

Uzé G, Lutfalla G, Mogensen KE (1995) α and β interferons and their receptor and their friends and relations. J Interferon Cytokine Res 15:3–26

Veals SA, Schindler C, Leonard D, Fu X-Y, Aebersold R, Darnell JE Jr, Levy DE (1992) Subunit of an alpha-inteferon-responsive transcription factor is related to interferon regulatory factor and Myb families of DNA-binding proteins. Mol Cell Biol 12:3315–3324

Velazquez L, Fellous M, Stark GR, Pellegrini S (1992) A protein tyrosine kinase in the interferon α/β signaling pathway. Cell 70:313–322

Velazquez L, Mogensen KE, Barbieri G, Fellous M, Uzé G, Pellegrini S (1995) Distinct domains of the protein tyrosine kinase tyk2 required for binding of interferon-α/β and for signal transduction. J Biol Chem 270:3327–3334

Vignais M-L, Sadowski HB, Watling D, Rogers NC, Gilman M (1996) Platelet-derived growth factor induces phosphorylation of multiple JAK family kinases and STAT protein. Mol Cell Biol 16:1759–1769

Vilcek J, Sen GC (1996) Interferons and other cytokines. In: Field BN, Knipe DM, Howley PM, et al. (eds) Fields virology, 3rd edn. Lippincott-Raven, Philadelphia, pp 375–399

Visvanathan KV, Goodbourne S (1989) Double-stranded RNA activates binding of NF-kappaB to an inducible element in the human β-interferon promoter. EMBO J 8:1129–1138

Walder RY, Walder JA (1988) Role of RNase H in hybrid-arrested translation by antisense oligonucleotides. Proc Natl Acad Sci USA 85:5011–5015

Watanabe N, Sakakibara J, Hovanessian AG, Taniguchi T, Fujita T (1991) Activation of IFN-β element by IRF-1 requires a post-translational event in addition to IRF-1 synthesis. Nucleic Acids Res 19:4421–4428

Watling D, Guschin D, Muller M, Silvennoinen O, Witthuhn BA, Quelle FW, Rogers NC, Schindler C, Stark GR, Ihle JN, Kerr IM (1993) Complementation by the protein tyrosine kinase JAK2 of a mutant cell line defective in the interferon-gamma signal transduction pathway. Nature 366:166–170

Weidel U, Weissman C (1983) The 5′-flanking region of a human IFN-α gene mediates viral induction of transcription. Nature 303:442–445

Weihua X, Linder DJ, Kalvakolanu DV (1997) The interferon-inducible murine p48 (ISGF3γ) gene is regulated by protooncogene *c-myc*. Proc Natl Acad Sci USA 94:7227–7232

Weissmann C, Weber H (1986) The interferon genes. Prog Nucleic Acid Res Mol Biol 33:251–300

Weissmann C, Nagata S, Boll W, Fountoulakis M, Fujisawa A, Fujisawa JI, Haynes J, Henco K, Mantei N, Ragg H, Schein C, Schmid J, Shaw G, Streuli M, Taira H, Todokoro K, Weidle U (1982) Structure and expression of human IFN-alpha genes. Philos Trans R Soc Lond Biol 299:7–28

Welsh RM, Sen GC (1997) Nonspecific host responses to viral infection. In: Nathanson N (ed) Viral pathogenesis. Lippincott-Raven, Philadelphia, pp 109–141

Wen Z, Darnell JE Jr (1997) Mapping of STAT3 serine phosphorylation to a single residue (727) and evidence that serine phosphorylationhas no influence on DNA binding of STAT1 and STAT3. Nucleic Acids Res 25:2062–2067

Wen Z, Zhong Z, Darnell JE Jr (1995) Maximal activation of transcription by Stat1 and Stat3 requires both tyrosine and serine phosphorylation. Cell 82:241–250

Whittemore LA, Maniatis T (1990) Postinduction turnoff of beta-interferon gene expression. Mol Cell Biol 10:1329–1337

Williams BRG (1995) The role of the dsRNA-activated kinase, PKR, in signal transduction. Semin Virol 6:191–202

Willman CL, Sever CE, Pallavicini MG, Harada H, Tanaka N, Slovak ML, Yamamoto H, Harada K, Meeker TC, List AF et al. (1993) Deletion of IRF-1, mapping to chromosome 5q31.1, in human leukemia and preleukemic myelodysplasia. Science 259:968–971

Wilson V, Jeffreys AJ, Barrie PA, Boseley PG, Slocombe PM, Easton A, Burke DC (1983) A comparison of vertebrate interferon gene families detected by hybridization with human interferon DNA. J Mol Biol 166:457–475

Xu X, Sun YL, Hoey T (1996) Cooperative DNA binding and sequence-selective recognition conferred by the STAT amino-terminal domain. Science 273:794–797

Yamagata T, Nishida J, Tanaka T, Sakai R, Mitani K, Yoshida M, Taniguchi Yazaki Y, Hirai H (1996) A novel interferon regulatory factor family transcription factor, ICSAT/Pip/LSIRF, that negatively regulates the activity of interferon-regulated genes. Mol Cell Biol 16:1283–1294

Yan H, Krishnan K, Greenlund AC, Gupta S, Lim J TE, Schreiber RD, Schindler CW, Krolewski JJ (1996) Phosphorylated interferon-α receptor 1 subunit (IFNaR1) acts as a docking site for the latent form of the 113 kDa STAT2 protein. EMBO J 15:1064–1074

Yang CH, Shi W, Basu L, Murti A, Constantinescu SN, Blatt L, Croze E, Mullersman JE, Pfeffer LM (1996) Direct association of STAT3 with the IFNAR-1 chain of the human type I interferon receptor. J Biol Chem 271:8057–8061

Yang Y-L, Reis LF, Pavlovic A, Aguzzi A, Schafer R, Kumar A, Williams BRG, Aguet M, Weissmann C (1995) Deficient signaling in mice devoid of double-stranded RNA-dependent protein kinase. EMBO J 14:6095–6106

Ye J, Cippitelli M, Dorman L, Ortaldo JR, Young HA (1996) The nuclear factor YY1 suppresses the human gamma interferon promoter through two mechanisms: inhibition of AP1 binding and activation of a silencer element. Mol Cell Biol 16:4744–4753

Young HA (1996) Regulation of interferon-gamma gene expression. J Interferon Cytokine Res 16:563–568

Zhang X, Blenis J, Li H-C, Schindler C, Chem-Kiang S (1995) Requirement of serine phosphorylation for formation of STAT-promoter complexes Science 267:1990–1994

Zhang K, Kumar R (1994) Interferon-alpha inhibits cyclin E- and cyclin D1-dependent CDK-2 kinase activity associated with RB protein and E2F in Daudi cells. Biochem Biophys Res Commun 200:522–528

Zhou A, Paranjape J, Brown TL, Nie H, Naik S, Dong B, Chang A, Trapp B, Fairchild R, Colmenares C, Silverman RH (1997) Interferon action and apoptosis are defective in mice devoid of 2′,5′-oligoadenylate dependent RNase L. EMBO J 16:6355–6363

Zvelebil MJ, MacDougall L, Leevers S, Volinia S, Vanhaesebroeck B, Gout I (1996) Structural and functional diversity of phosphoinositide 3-kinases. Philos Trans R Soc Lond B Biol Sci 351:217–223

Growth-Inhibiting N-Substituted Endogenous Peptides

Kjell Elgjo[1], Karl Ludwig Reichelt[2], and David S. Gembitsky[1]

1 Introduction

Considering the large number of factors that modulate cell growth and differentiation, it might seem naive to assume that cell renewal is basically regulated according to a negative feedback principle, involving one tissue-specific chemical signal system in each organ/tissue. This concept is, nevertheless, about 60 years old. In 1937, Simms and Stillman found that some heat-labile factor, extracted from aorta, inhibited growth of tissue from chicken aorta. Twenty years later, Weiss and Kavanau (1957) formulated a broader hypothesis based on their own data, suggesting that differentiated cells in a tissue or organ generate an inhibitory signal that regulates the rate of cell division in the "proliferative pool" of that same tissue or organ, possibly also stimulating differentiation. Their work was continued by Iversen (1961) and Iversen and Evensen (1962), who analyzed epidermal cell kinetics during skin carcinogenesis, and Bullough (1962), who worked with epidermal regeneration after wounding. Bullough coined the name chalone, a name that was rapidly used for any growth-inhibitory activity that could be found in tissue/organ extracts, most of them more or less crude.

During the following decade, a large number of papers were published on "chalones", and several international meetings were held. On the basis of experimental data and a fair amount of speculation, a chalone was defined as a growth inhibitor that had the following properties: (1) total cell/tissue specificity, (2) species nonspecificity, (3) reversible effect, and (4) no toxicity. Theory and experimental data soon collided, and this fact, together with the general lack of success in purifying a "chalone", was disastrous. What happened then was recently summed up by Marshall and Lord (1996): "The rapid rise to fame of chalones in the 1960s was only paralleled by their speedy decline. The crude extracts produced were easy targets for criticism and much of the experimental work was viewed with considerable scepticism". The "cha-

[1] Institute of Pathology and [2] Pediatric Research Insitute, The National Hospital (Rikshospitalet), 0027 Oslo, Norway

Progress in Molecular and Subcellular Biology, Vol. 20
A. Macieira-Coelho (Ed.)

lone" research performed in the 1960s and 1970s was reviewed by Iversen in 1981.

In retrospect, many interesting observations were made during the two hectic "chalone" decades. Thus, the effect of the epidermal "chalone" seemed to be related to cyclic AMP and catecholamines (Bullough and Laurence 1964; Marks and Rebien 1972; Elgjo 1975), and the same inhibitory activity was found in extracts made from skin of different species (Bullough et al. 1967). Inhibition of DNA synthesis in Ehrlich ascites tumor cells was shown to be due to inhibition of DNA polymerase alpha and beta (Nakai 1976). Repeated applications of epidermal "chalone" (crude water extract) suppressed epidermal regeneration and hyperplasia after making wounds in bat wings (Iversen et al. 1974). Also, water extracts of mouse skin seemed to contain two different inhibitors with preferential activity at different cell-cycle phase transitions; one acting on cells in late G1 and another on cells in G2, as described below. However, the lack of "negative" controls, and the fact that none of the putative chalones had been purified and characterized 20 years after Bullough advanced the term chalone apparently made people lose interest in this type of research. Nevertheless, the chalone paradigm was not fruitless. Several laboratories continued their attempts to isolate and characterize the inhibitory factor(s) present in tissue/organ extracts. This, in turn, led to the discovery of the growth-inhibiting oligopeptides that are the subject of this chapter. As will be discussed later, the structure of these inhibiting oligopeptides can be the basis for designing peptidomimetics that should have a promising potential in clinical medicine.

2 Purification and Characterization of Endogenous Growth Inhibitors

2.1 Early Attempts at Purification

The first attempts at purification and biochemical characterisation of factors with the same properties as the more or less crude extracts indicated that they were fairly large molecules with an apparent molecular weight of 30–40 kDa (Hondius Boldingh and Laurence 1968). In the late 1970s, however, it became clear that the molecular weight was considerably less. In a review article in 1980 with the fitting name *The Incredible Shrinking Chalone*, Patt and Houck summed up the development in this field of research. In conclusion, they even suggested that the inhibitory activity could be represented by peptides. The following years of research fully corroborated this view, even though some larger inhibitors were still found. Thus, a lipophilic glycopeptide with molecular weight of about 10 kDa was found in skin extracts (Richter et al. 1988), but this substance has, to our knowledge, not been further characterized biochemically. Similarly, a 14.5-kDa polypeptide (mammary-derived growth inhibitor – MDGI) has been found in lactating bovine mammary tissue (Grosse et al. 1992). This growth-inhibiting polypeptide has 95% homology with bovine cardiac fatty acid-binding proteins (FABP), and some sequences are identical to some found in bovine growth hormone. It is not expressed in virgin mammary gland tissue and appears to be dependent on systemic hormones. For this reason, it is difficult to accept this factor as a locally acting growth inhibitor, regulating the rate of cell renewal in a normal or regenerating cell population.

Growth-inhibiting peptides were first identified in bone marrow cells and epidermis, the findings being made by three independent research groups. Here, the purification of the peptides found in bone marrow cells and leucocytes will be summarized, while the purification and identification of inhibitory peptides examined by our group will be described in more detail.

2.2 Structure and Biological Properties of Endogenous Bone Marrow Growth-Inhibiting Peptides

Paukovits and Laerum (1982) were the first to report on the characterization of an acidic pentapeptide (pyroGlu-Glu-Asp-Cys-LysOH) that inhibited proliferation of granulo/monocytopoietic cells at low concentrations (10^{-10} M). Later experiments demonstrated that the pentapeptide inhibited even CFU-S (Laerum and Paukovits 1984). No toxic effects were observed, even at high doses, and the effect was fully reversible. In the following years, it became evident that the pentapeptide was labile and easily formed a stimulating dimer (Laerum et al. 1988). Recently, a stable dimer has been constructed by replacing the -SH bridge with a methylene group (King et al. 1992). The significance of dimerization will be discussed below. Interestingly, in the bone marrow, the stimulatory effect of the dimer is mediated via stromal cells.

Another inhibitory oligopeptide has been isolated from fetal calf bone marrow (Lenfant et al. 1989). This tetrapeptide, N-acetyl-Ser-Asp-Lys-ProOH (AcSDKP), which is structurally different from the hemoregulatory pentapeptide found by Paukovits and Laerum (1982), is now trademarked as Gorolotide. It has some sequences in common with TNFα and thymosin-β4, and it is found in several organs (Guigon and Bonnet 1995). In the bone marrow, it inhibits CFU-S, CFU-A, HPP-CFC, and CFU-GM. This diverse activity has led Guigon and Bonnet (1995) to suggest that the action of AcSDKP is related to a mechanism rather than to a cell type. Apparently, it is not inhibitory in malignant cell populations.

2.3
Structure and Biological Properties of the Inhibitory Epidermal Pentapeptide

In 1984, a growth-inhibiting pentapeptide, pyroGlu-Glu-Asp-Ser-GlyOH, was isolated from water extracts of mouse skin (Elgjo and Reichelt 1984). Preliminary reports on its effects, including its biochemical properties, were published in the following years (Elgjo et al. 1986a,b; Reichelt et al. 1987). The purification procedures of N-substituted peptides from tissue/organ extracts will be described in more detail below.

The dose-response curve of the epidermal pentapeptide (EPP) is bell-shaped, with optimal effect when picomole doses are given to mice or nanomole concentrations are used in vitro. In addition to reversible inhibition of epidermal cell proliferation in vivo and in vitro, EPP enhances keratinization in vitro (Elgjo et al. 1986a). In accordance with this, the RNA profile is altered in cultured epidermal cells after treatment with EPP (Jensen et al. 1990)

After a single, intraperitoneal (ip) injection of picomole doses of EPP, both epidermal DNA synthesis and the mitotic rate exhibit oscillations lasting for at least 24h (Elgjo and Reichelt 1988). Such an oscillating pattern is to be expected in a cell population whose rate of cell renewal is regulated according to a negative feedback principle (Tustin 1948). Another relevant observation is that treatment with EPP is followed by a refractory period during which a second treatment with EPP has no effect.

Experiments with topical application of EPP have confirmed that both proliferation and differentiation are modulated so that differentiation and cell loss are in balance with the decreased rate of cell proliferation (Elgjo and Reichelt 1988, 1990). In vitro experiments have confirmed that a balance between cell differentiation and proliferation is maintained in both transformed mouse epidermal cells and in nontransformed rat tongue keratinocytes (Elgjo et al. 1991). This experiment showed, however, that the enhancing effect on differentiation, expressed as number of cornified envelopes, is stronger in nontransformed than in transformed keratinocytes. In these experiments, the effect on keratinization was observed at higher doses or concentrations of EPP than those that are optimal for mitosis inhibition. Also, Olsen and Elgjo (1989) showed that topical treatment with a 0.02% EPP cream inhibits UVB-induced epidermal hyperproliferation, while a 0.005% EPP cream only alters the cell kinetics pattern without suppressing proliferation.

Early experiments had indicated that the inhibitory effect of skin extracts on epidermal cell proliferation was in some way dependent on cyclic AMP (Bullough and Laurence 1964; Marks and Rebien 1972). This assumption was corroborated by in vivo experiments, first by using partially purified mouse skin extracts (Elgjo 1975), later by employing synthetic EPP (Elgjo and Reichelt 1994). In both experiments, beta-receptor blockade with propranolol abrogated the inhibitory effect. However, the experiment with synthetic EPP after beta-receptor blockade revealed that only cell flux at the G_2-M transition was affected, while cell flux at the G_1-S transition was not measurably influenced (Elgjo and Reichelt 1994).

EPP has an inhibitory effect not only on epidermis proper but even on some other ectodermal derivatives. Thus, it inhibits cell proliferation in hair follicles in vitro and in vivo (Paus et al. 1991), and regeneration of rat corneal epithelium following abrasion is delayed (Wang et al. 1996).

In 1991, Whitehead et al. demonstrated that EPP is rapidly broken down by a specific serum peptidase. This peptidase was recently purified and characterized by Bramucci et al. (1996). It is therefore likely that the minute amounts of EPP that are injected ip into mice, or added to a culture medium containing serum, are rapidly broken down. For that reason, the long-lasting effects of EPP seen in in vivo experiments must result from a strong effect of EPP on cells at certain points in the cell cycle. This could be related to findings by Bramucci et al. (1992), who showed that EPP is easily phosphorylated by protein kinase N II, and that the phosphorylated peptide binds to DNA. Also, phosphorylated EPP is more resistant to proteases. It is, however, not clear which phases of the cell cycle are susceptible to EPP. Early experiments showed that extracts of differentiated (keratinized) epidermal cells inhibited cells mainly at the G_1-S transition, while extracts made from separated epidermal basal layer cells inhibited cell flux chiefly at the G_2-M transition (Elgjo et al. 1971, 1972). It is still a moot question whether two different inhibitors exist, whether phosphorylation, for example, may alter cell-cycle phase specificity, or whether the concept of two different, cell-cycle-specific inhibitors could be related to different susceptibility to different concentrations of one and the same inhibitor. The fact that EPP acts at several cell-cycle transitions seems to point to the latter explanation. Some early in vivo experiments with partially purified extracts support this assumption (Elgjo et al. 1981). By combining several cell kinetic techniques, they could show that a single treatment was followed by an inhibition at several cell-cycle transitions.

As mentioned above, the hemoregulatory peptide easily forms a stimulatory dimer. In experiments that originally were designed to examine possible interaction between EPP and divalent cations, we found that EPP formed a dimer when mixed with equimolar concentrations of zinc. This dimer has a strong stimulatory effect on epidermal cell proliferation (Elgjo and Reichelt 1995), but here, too, the balance between cell proliferation and cell loss appears to be maintained. The dizinc-diEPP dimer is formed only when equimolar concentrations of the two substances are mixed. Excess of either of them results in inactive aggregates. The stimulating effect of a zinc-EPP dimer could be related to the old clinical observation that zinc-containing creams or ointments have a beneficent effect on epidermal wound healing. However, to our knowledge, this issue has not been pursued.

3 Other N-Substituted Inhibitory Oligopeptides

3.1 Inhibitory Oligopeptides in Liver, Colon, and Melanocytes

In 1987, a mitosis-inhibiting pentapeptide was isolated from the liver (Paulsen et al. 1987), using principally the same purification procedures as those that were elaborated during the search for an inhibitor in epidermis extracts. These procedures will be described below. The liver pentapeptide, pyroGlu-Gln-Gly-Ser-AsnOH, was found both in amidated and in deamidated forms, all active (Reichelt et al. 1990). The following year, a mitosis-inhibiting tripeptide, pyroGlu-His-GlyOH, was found in mouse intestinal extracts (Skraastad et al. 1988). Both peptides have biological properties that are very similar to those found for EPP. They have a bell-shaped dose-response curve and are inhibitory at picomole concentrations. Like EPP, the colon tripeptide inhibits cell flux at both the G_1-S and G_2-M transitions (Skraastad and Reichelt 1989a). Likewise, treatment with the tripeptide is followed by a refractory period during which a second treatment has no effect. Furthermore, the same study demonstrated that the tripeptide has an inhibitory effect even on epidermal cell proliferation. A comparison of the various peptides showed, however, that the kinetic response is different and that a far lower dose is needed of the tripeptide than of the others to inhibit cell proliferation in colon crypt epithelium.

Recently, we have isolated and characterized a growth-inhibitory tripeptide, pyroGlu-Phe-GlyNH_2 from nontransformed melanocytes and from cultured melanoma cells (B16 cells) (Gembitsky et al. 1997). This peptide, too, has a bell-shaped dose-response curve and optimal efficiency at picomole doses.

Another low molecular weight inhibitor has been found in Ehrlich ascites tumor cells (Gembitsky et al. 1996). This inhibitor appears to belong to the same family of inhibitory peptides as the others mentioned in this section, but it has not yet been fully characterized. Finally, we have isolated and characterized an inhibitory peptide from neuroblastoma cells and from neuroblastoma metastases. Work is in progress to examine more closely the properties of this peptide, whose structure indicates that it is another member of the N-substituted inhibitory oligopeptides.

3.2 Inhibitory N-Substituted Oligopeptides Not Related to Specific Organs/Tissues

Gianfranceschi and associates have for several years worked with oligopeptides that are structurally very similar to the inhibitory peptides described above, being N-substituted and containing the same selection of amino acids. These peptides were first extracted from deproteinized DNA from prokaryotic and eukaryotic cells, later from wheat-germ chromatin, which yielded sufficient quantities for isolation and biochemical characterization (Mancinelli et al. 1992). First, they were found to inhibit RNA synthesis in cells and in cell-free systems (Mancinelli et al. 1992). It was soon revealed that they were easily phosphorylated by protein kinase NII and exhibited a regulatory activity on DNA in vitro transcription (Gianfranceschi et al. 1994). The same group has also demonstrated that synthetic and native acidic low molecular weight peptides can serve as a competitive substrate for the phosphorylation of proteins whose activity is dependent on phosphorylation (Angiolillo et al. 1993). It has also been suggested that this group of peptides could act as regulators of RNA polymerase II transcription, since they act as inhibitors after being phosphorylated by protein kinase NII (Castigli et al. 1993). As mentioned above, this group has shown that even EPP is easily phosphorylated by protein kinase NII and binds to DNA in this form (Bramucci et al. 1992). Circumstantial evidence therefore exists that these small acidic N-substituted and chromatin-bound peptides are closely related to the N-substituted growth-inhibitory oligopeptides found in organ/tissue extracts.

4 N-Substituted Inhibitory Oligopeptides in Neoplasia

4.1 Carcinogenesis

Mouse experiments with ip EPP treatment before giving a complete carcinogen showed that the number of skin tumors was higher than in the control group that received no EPP (Iversen et al. 1989). This effect was ascribed to a synchronizing effect of EPP, delaying cells in late G_1 or early S. Thus, the same increased tumor yield was found when epidermal DNA synthesis was inhibited by pretreatment with hydroxyurea. In another experiment, EPP dissolved in acetone was given topically prior to TPA in a two-stage mouse skin carcinogenesis experiment (Iversen et al. 1993). Here, too, pretreatment with EPP enhanced tumorigenesis, raising the interesting question whether the mechanisms of promotion and complete carcinogenesis are, in principle, of the same nature. None of the other inhibitory peptides described above has been tested for their possible ability to prevent or inhibit tumor formation in carcinogenesis experiments.

4.2 N-Substituted Inhibitory Oligopeptides and Tumors

The liver pentapeptides have been examined in several tumor experiments. Both the amidated and the deamidated forms inhibit growth of rat hepatoma cells in vitro (Paulsen et al. 1991) and in vivo (Paulsen et al. 1992). Skraastad and Reichelt (1989b) had demonstrated that the colon tripeptide inhibits human adenocarcinoma cells (HT29) in vitro. Later, Paulsen (1993) found that the colon tripeptide inhibits growth even of subcutaneously transplanted adenocarcinoma cells (HT29) in mice. The inhibitory effect of the tripeptide in these experiments was most pronounced in the earlier phases of tumor growth. In addition to these two peptides, preliminary experiments with the melanocyte tripeptide have indicated that repeated ip injections decrease the average tumor load when B16 melanoma cells are transplanted subcutaneously into mice (unpubl. data).

In order to examine indirectly whether treatment with EPP would modify the ability of malignant cells to attach and form metastases, Ushmorov et al. (1992) examined the effect of EPP on adhesiveness of squamous carcinoma cells to different matrices. Their experiments demonstrated a significant alteration of adhesiveness in several lines of tumor cells. Their findings are possibly related to those by Paulsen et al. (1992), who found that treatment with liver pentapeptide inhibited the number and size of lung metastases after intravenous injection of MH_1C_1 rat hepatoma cells into young rats.

Experiments with Ehrlich ascites tumor cells have demonstrated that peptide-like inhibitory factors can be extracted from malignant cells (Gembitsky et al. 1996). Likewise, the melanocyte and the neuroblastoma peptides are found in malignant cells and inhibit growth of the two neoplastic cell types in culture. In addition to the experiment with EPP and transformed mouse epidermal cells mentioned above (Elgjo et al. 1991), EPP has been shown to inhibit human squamous carcinoma cells in vitro (Watt et al. 1989). The subject of endogenous inhibitory peptides and neoplasia has been reviewed by Elgjo (1988, 1993) and by Marshall and Lord (1996).

A systematic study of several malignant cell types in vitro has revealed that tumor cells contain a smaller amount of low molecular weight peptides than their normal counterparts, while the medium in which the cells grow has a far larger concentration of such peptides, many of which are phosphorylated (Reichelt et al. 1994). Malignant cells thus seem to secrete, or leak, small peptides into their surroundings. In vivo, excess peptides are probably broken down by the high concentration of proteases found in the stroma (Buö et al. 1993). Since many types of tumor cells are still capable of reacting to inhibitory peptides, they have obviously kept their ability to respond to inhibitory signals.

5 Purification Procedures Used for Characterization of N-Substituted Peptides

5.1 Extraction and Chromatography Systems

As referred to in Section 2.1, inhibitory polypeptides with far higher molecular weights than the N-substituted oligopeptides have been isolated by others. The different results could well be related to different extraction and purification procedures. A short description of the techniques used by us, and in principle

Table 1. General purification procedures

Solvent system	Column	Comments
1. 0.5 M acetic acid	Sephadex G25 1.6 × 90 cm	Separation of acid-soluble macromolecules from low MW proteins and peptides
2. a) Isocratic elution in 20 mM TFA, pH 2 b) n-Propanol/TFA 20 mM (0–60%)	Supersil 9 MM 10/20 ODS (C_{18}, reverse phase)	Separation of amino acids, salts, and phosphorylated peptides from non-phosphorylated peptides
3. a) 0.001 M formic acid b) 4 M ammonia	Dowex 50 (H^+-form) 1 × 20 cm	a) Nonprotonated peptides are eluted with this buffer b) Protonated peptides are eluted in 4 M ammonia
4. a) Water b) 4 M formic acid	Dowex 1 (formate form) 1 × 20 cm, 100–200 mesh	a) Peptides more anionic than formate retained, and the neutral ones eluted b) Strongly anionic peptides eluted in 4 M formic acid
5. 1 M acetic acid/20 mM hydrochloric acid	Fractogel MG2000 1.6 × 90 cm	Gives approximate MW determination
6. 5 mM citrate pH 7/10 mM citric acid	DEAE anion exchange membranes	Gradient anion exchange elution of peptides
7. a) TFA/acetonitrile (0–60%) b) HFBA[a]/acetonitrile (0–60%) c) TFA/methanol (0–80%)	Supersil 9 MM 10/20 ODS C_{18} reverse-phase gradient chromatography	Gradient reverse phase purification

[a] HFBA hexafluorobutyric acid.

also by Paukovits and Laerum, in purifying the hemoregulatory pentapeptide therefore seems warranted.

Pieces of the organ or tissue to be extracted are first homogenized in an electric mill cooled with liquid nitrogen. The resulting fine powder is then extracted either in distilled water or in 1 mM acetic acid to prevent proteolysis. After centrifugation, the supernatant is freeze-dried. The subsequent fractionation procedures are listed in Table 1. The fractions obtained at the different steps of purification are tested for inhibitory activity in vitro and/or in vivo. Offline peptide quantification is then performed of the final, active peak(s). The fraction(s) is dried overnight at 150°C and hydrolyzed in 2M KOH for 2h in a boiling water bath. Ninhydrin color is then developed in an acetate-cyanide buffer that has equimolar absorption coefficient for amino acids (Reichelt and Kvamme 1973).

5.2 Biochemical Characterization of Peptides

Inhibitory pure fractions are characterized biochemically by the following procedures: (1) after alkaline hydrolysis amino acid analysis is performed on an amino acid analyzer, using the ninhydrin technique; (2) since all inhibitory peptides found so far have been N-substituted, we use C-terminal carboxypeptidase digestion in combination with amino acid analysis. Recently, we have also used the C-terminal variation of Edman degradation; (3) pyroglutamyl aminopeptidase treatment to remove N-terminal pyroglutamyl, followed by dansylation of the remaining peptide and thin layer chromatographic identification of the next terminal amino acid, and (4) mass spectrometry.

5.3 Criteria for Purity

We use the following criteria to ascertain that a peptide is pure: (1) constant amino acid composition in attempts at further purification; (2) integral relationship of amino acids on analysis; (3) one N- and one C-terminal amino acid; (4) one peak on HPLC with the same composition on upturn and downturn run of the peak, and (5) one peak on mass spectrometry.

5.4 Criteria for Identity Between Native and Synthetic Peptides

We use the following criteria to establish whether an endogenous and a synthetic peptide are identical: (1) HPLC cochromatography in several systems; (2) same bioactivity in vitro and/or in vivo; and (3) identical mass on mass spectrometry.

6 "Chalones" and Inhibitory N-Substituted Oligopeptides: Similarities and Differences

In retrospect, it is not possible to tell whether the inhibitory "activities" found in a large number of organ/tissue extracts and called chalones are identical with the mitosis-inhibiting N-substituted oligopeptides described in this chapter. Also, it is difficult to tell whether the larger growth-inhibiting molecules found in, e.g., epidermis (Richter al. 1988) and lactating mammary tissue (Grosse et al. 1992) represent large entities of which inhibitory peptides are an active factor. However, in many ways, the N-substituted inhibitory peptides and the larger inhibiting molecules behave in the same way. Thus, the lipophilic glycoprotein found in skin extracts (Richter et al. 1988) is soluble in both water and organic solvents, indicating that it could contain some N-substituted peptide. Higher concentrations (molarity) are needed to inhibit epidermal cell proliferation but this could, however, be taken as an indication that each molecule holds only one active small molecular weight inhibitory peptide. The inhibitory protein found in lactating mammary gland tissue also bears some similarity to N-substituted inhibitory peptides. Thus, its activity is abrogated by serum and insulin (Grosse et al. 1992). The same phenomenon was found for the colon tripeptide (Skraastad and Reichelt 1989b), and later even for the peptide isolated from melanocytes (unpubl. data).

7 Possible Strategies for Clinical Use of Endogenous Growth Inhibitors

7.1 Possible Use of Endogenous Inhibitors in Medicine

Some animal experiments have indicated that endogenous inhibitory peptides can be used to suppress tumor growth, either in their native form or as synthetic peptides (Rytömaa and Kiviniemi 1968, 1970; Rytömaa et al. 1976; Paulsen et al. 1991, 1992; Paulsen 1993). Recent studies have shown that even the growth of murine melanoma cells (B16) injected subcutaneously into mice is reduced by the melanocyte peptide (unpubl. results). In addition, several of the experiments described in the previous sections have demonstrated that endogenous inhibitory oligopeptides can be found in malignant tumors, and that tumor cells in vitro are partially inhibited by the particular peptide found in these malignant cells and in their normal counterparts. The reason why this type of inhibitory peptides can be, at least theoretically, of clinical use is that relatively high doses are needed to inhibit malignant cells, while the corresponding normal cells react only to low doses.

Because malignant cells and their normal counterparts respond to different levels of peptide concentration it should be possible to protect normal cells during irradiation or treatment with cytostatics by using a low peptide dose that will only delay normal cell proliferation. Some experiments of this type have been performed. Both the synthetic tetrapeptide AcSDKP (Bogden et al. 1991; Grillon et al. 1994) and the synthetic pentapeptide pEEDCK (Paukovits et al. 1993) have been shown to protect bone marrow cells in connection with treatment with cytostatics. A similar use of the colon tripeptide may be possible, but has not been investigated.

A third and relevant aspect of studies on endogenous growth-inhibiting peptides in neoplasia is that further research could possibly teach us more about how neoplastic cells are regulated, alternatively, which properties they have lost so that they do not respond normally to growth-modulating signals.

7.2 Possible Use of Stimulating Dimers and Peptidomimetics

The fact that the two pentapeptides EPP and the hemoregulatory peptide (pEEDCK) form growth-stimulating dimers is interesting from both a biological and a practical point of view. It is quite conceivable that they exist both as monomers and dimers in an organ or tissue, and that the ratio between the two forms could depend on the microenvironment and its requirement for cell renewal. This obviously needs further study.

The practical aspect of the ability to form stimulating dimers has already been exploited as regards the hemoregulatory peptide; stable, stimulatory peptidomimetics have been made on the basis of its structure (Pelus et al. 1993). In short, the reducible S-S bond in the pEEDCK dimer has been replaced with an isosteric ethylene spacer resulting in a novel peptide SK&F 107647: (pyroGlu-Glu-Asp)$_2$-**Sub**-(Lys)$_2$, where Sub = (2*S*,7*S*)-2,7-diaminosuberic acid. Structure-activity relationship (SAR) studies of SK&F 107647 (Bhatnagar et al. 1996a) have revealed several stringent structural requirements that are necessary for biological activity. Also, SAR studies of the spacer connecting the two halves of the SK&F 107647 molecule (Bhatnagar et al. 1996a,b) have shown several important conformational features.

Based on SAR studies of SK&F 107647, Bhatnagar and coworkers (1996b) introduced a pharmacophore model of this peptide. This pharmacophore model was applied successfully to design a series of dramatically simpler peptidomimetic analogs, e.g., N,N′-Bis(pyroglutamyl-β-alanyl)-1,4-diaminobenzene, that display similar hematopoietic activity (Callahan et al. 1997).

The synthetic hemoregulatory peptide SK&F 107647 and its analogs have several properties that could be used in clinical medicine. In addition to accelerating recovery of the bone marrow in murine bone marrow transplantation studies (Veiby and Olsen 1995), they enhance survival in experimental infection models (DeMarsh et al. 1994, 1996; Bhatnagar et al. 1996b). Currently, SK&F 107647 is in phase II clinical trials for prevention of infectious complications associated with neutropenia in oncology patients (King et al. 1995; Brocks et al. 1996).

A similar development of growth-stimulating peptidomimetics based on the structure of the dizinc-diEPP dimer should be possible, but has so far not been launched.

8
Concluding Remarks

The concept of growth regulation by inhibiting factors is now accepted by most researchers in the field. Most likely, tissue homeostasis (steady-state) is the result of the growth-modulating effects of both stimulating and inhibiting factors. The fact that lost tissue is replaced by exactly the same amount of identical tissue in regenerating organs must, however, be based on a negative feedback mechanism (Tustin 1948; Weiss and Kavanau 1957). Inhibiting effects are, however, more difficult to investigate than stimulating ones, since so many unspecific stimuli can suppress cell proliferation, either directly or via stress hormones. Also, it is difficult to examine the effect of one single factor on a normal cell population or organ, since a change in the concentration of one factor will be met by adaptive adjustments in the concentration of others. Biologically active factors will a bell-shaped dose-response curve are particularly difficult to investigate, since the optimal concentration is rarely known in advance.

The N-substituted growth-inhibiting oligopeptides examined so far have no species specificity but a tissue preference at low concentrations. This fact, together with their structural similarity to several peptide hormones and some neurotransmitters, could indicate that they belong to an evolutionary old signal system. At the present time, we do not know whether they can be put to use in clinical medicine. Most likely, they would not be used directly but be the basis of synthetic peptidomimetics. At any rate, a better knowledge of their mode of action, their receptor system, and their possible interaction with other growth-modifying factors and their receptors would greatly increase our understanding of normal growth regulation and, hence, our understanding of defects in the regulation of growth and differentiation in malignant cells.

References

Angiolillo A, Bramucci M, Marsili V, Panara F, Miano A, Amici D, Gianfranceschi GL (1993) Phosphorylation of synthetic acidic peptides by casein kinase II. Evidence for competition with phosphorylation of proteins involved in transcription. Mol Cell Biochem 125:65–72

Bhatnagar PK, Agner EK, Alberts D, Arbo BE, Callahan JF, Cuthbertson AS, Engelsen SJ, Fjerdingstad H, Hartmann M, Heerding D, Hiebl J, Huffman WF, Hysben M, King AG, Kremminger P, Kwon C, LoCastro S, Lovhaug D, Pelus LM, Petteway S, Takata JS (1996a) Structure-activity relationships of novel hematoregulatory peptides. J Med Chem 39:3814–3819

Bhatnagar PK, Alberts D, Callahan JF, Heerding D, Huffman WF, King AG, LoCastro S, Pelus LM, Takata JS (1996b) Development of a pharmacophore model for novel hematoregulatory peptide. J Am Chem Soc 118:12862–12863

Bogden AE, Carde P, Deschamps de Paillete E, Moreau JP, Tubiana M, Frindel E (1991) Amelioration of chemotherapy-induced toxicity by co-treatment with AcSDKP, a tetrapeptide inhibitor of haemoatopoietic stem cell proliferation. Ann NY Acad Sci 628:126–139

Bramucci M, Miano A, Amici D (1992) Epidermal inhibitory pentapeptide phosphorylated in vitro by calf thymus protein kinase II is protected from serum enzyme hydrolysis. Biochem Biophys Res Commun 183:474–480

Bramucci M, Miano A, Quassinti L, Maccari E, Canofeni S, Amici D (1996) Purification and characterisation of swine serum proteinase which hydrolyses epidermal inhibitory pentapeptide. Biochim Biophys Acta 1290:184–190

Brocks DR, Freed MI, Martin DE, Sellers TS, Mehdi N, Citerone DR, Boppana V, Levitt B, Davies BE, Nemunaitis J, Jorkasky DK (1996) Interspecies pharmacokinetics of a novel hematoregulatory peptide (SK&F 107647) in rats, dogs, and oncological patients. Pharm Res 13:794–797

Bullough WS (1962) The control of mitotic activity in adult mammalian tissues. Biol Rev 37:307–342

Bullough WS, Laurence EB (1964) Mitotic control by internal secretion: the role of the chalone-adrenalin complex. Exp Cell Res 33:176–194

Bullough WS, Laurence EB, Iversen OH, Elgjo K (1967) The vertebrate epidermal chalone. Nature 214:578–580

Buö L, Lyberg T, Jörgensen L, Johansen HT, Aasen AO (1993) Location of plasminogen activator (PA) and PA inhibitor in human colorectal adenocarcinomas. APMIS 101:235–241

Callahan JF, Alberts D, Burgess JL, Chen L, Heerding D, Huffman WF, King AG, LoCastro S, Pelus LM, Bhatnagar PK (1997) SAR of a series of related peptidomimetics derived from the novel hematoregulatory peptide, SK&F 107647. 15th Am Pept Symp Abstr 73, Nashville, Tennessee, 1997

Castigli E, Mancinelli M, Mariggio A, Gianfranceschi GL (1993) Possible specific activation of RNA synthesis in PC-12 cell isolated nuclei by small acidic peptides. Am J Physiol 265:C1220–C1223

DeMarsh PL, Sucoloski SK, Frey CL, Koltin Y, Actor P, Bhatnagar PK, Petteway SR (1994) Efficacy of the hematoregulatory peptide SK&F 107647 in experimental systemic *Candida albicans* infections in normal and immunosuppressed mice. Immunopharmacology 27:199–206

DeMarsh PL, Wells GL, Lewandowski TF, Frey CL, Bhatnagar PK, Ostovic EJR (1996) Treatment of experimental Gram-negative and Gram-positive bacterial sepsis with the hematoregulatory peptide SK&F 107647. J Infect Dis 173:203–211

Elgjo K (1975) Epidermal chalone and cyclic AMP: an in vitro study. J Invest Dermatol 64:14–18

Elgjo K (1988) Inhibitory endogenous growth factors and cancer. In: Iversen OH (ed) Theories of carcinogenesis. Hemisphere Publishing Corporation, Cambridge, pp 93–96

Elgjo K (1993) Pentapeptide growth inhibitors: carcinogenesis and cancer. In: Iversen OH (ed) New frontiers in cancer causation. Taylor & Francis, Washington, pp 125–135

Elgjo K, Reichelt KL (1984) Purification and characterization of a mitosis inhibiting epidermal peptide. Cell Biol Int Rep 8:379–382

Elgjo K, Reichelt KL (1988) Structure and function of growth inhibitory epidermal pentapeptide. Ann NY Acad Sci 548:197–203

Elgjo K, Reichelt KL (1990) Mouse epidermal cell renewal after topical treatment with different concentrations of the epidermal peptide pyroGlu-Glu-Ser-GlyOH. Virchows Archiv B Cell Pathol 59:89–93

Elgjo K, Reichelt KL (1994) Beta-receptor blockade by propranolol modifies the effect of the inhibitory endogenous epidermal pentapeptide on epidermal cell flux at the G_2-M transition but not at the G_1-S transition. Epith Cell Biol 3:32–37

Elgjo K, Reichelt KL (1995) Zinc and the mitosis-inhibitory epidermal pentapeptide (EPP) form a stimulatory dimer. Arch Dermatol Res 287:735–739

Elgjo K, Laerum OD, Edgehill W (1971) Growth regulation in mouse epidermis I. G2-inhibitor present in the basal cell layer. Virchows Arch B Cell Pathol 8:277–283

Elgjo K, Laerum OD, Edgehill W (1972) Growth regulation in mouse epidermis II. G1-inhibitor present in the differentiating cell layer. Virchows Arch B Cell Pathol 10:228–236

Elgjo K, Clausen OPF, Thorud E (1981) Epidermis extracts (chalone) inhibit cell flux at the G1-S, S-G2, and G2-M transitions in mouse epidermis. Cell Tissue Kinet 14:21–29

Elgjo K, Reichelt KL, Hennings H, Michael D, Yuspa SH (1986a) Purified epidermal pentapeptide inhibits proliferation and enhances terminal differentiation in cultured mouse epidermal cells. J Invest Dermatol 87:555–558

Elgjo K, Reichelt KL, Edminson P, Moen E (1986b) Endogenous peptides in epidermal growth control. In: Baserga R, Foa P, Metcalf D, Polli EE (eds) Biological regulation of cell proliferation. Raven Press, New York, pp 259–265

Elgjo K, Reichelt KL, Clausen OPF, de Angelis P (1991) Inhibitory epidermal pentapeptide modulates proliferation and differentiation of transformed mouse epidermal cells in vitro. Virchows Arch B Cell Pathol 60:161–164

Gembitsky DS, Popova NY, Antokhin AI, Podusov YN, Karpets LZ, Adrianov NV, Romanov YA (1996) The Ehrlich ascites carcinoma (EAC) contains 300–500 Da inhibitor(s) suppressing DNA synthesis in cultured EAC cells and EAC cell division in tumor-bearing mice. Oncol Rep 3:333–338

Gembitsky DS, Reichelt KL, Haakonsen P, Paulsen JE, Elgjo K (1998) Identification of a melanocyte growth-inhibiting tripeptide and determination of its structure. J Peptide Res 51:80–84

Gianfranceschi GL, Czerwinski A, Angiolillo A, Marsili V, Castigli E, Mancinelli L, Miano A, Bramucci M, Amici D (1994) Molecular models of small phosphorylated chromatin peptides. Structure-function relationship and regulatory activity on in vitro transcription and on cell growth and differentiation. Peptides 15:7–13

Grillon C, Bonnet D, Mary J-Y, Lenfant M, Najman A, Guigon M (1994) The tetrapeptide AcSerAspLysPro (Seraspenide), a haemotopoietic inhibitor, may reduce the in vitro toxicity of 3′azido-3′deoxythymidine to human haematopoietic progenitors. Stem Cells 11:455–464

Grosse R, Böhmer F-D, Binas B, Kurtz A, Spitzer E, Müller T, Zschiesche W (1992) Mammary-derived growth inhibitor (MDGI). Cancer Treat Res 61:69–96

Guigon M, Bonnet D (1995) Inhibitory peptides in hematopoiesis. Exp Hematol 23:477–481

Hondius Boldingh W, Laurence EB (1968) Extraction, purification and preliminary characterisation of the epidermal chalone. Eur J Biochem 5:191–198

Iversen OH (1961) The regulation of cell numbers in epidermis. A cybernetic point of view. Acta Pathol Microbiol Scand Suppl 148:91–96

Iversen OH (1981) The chalones. In: Baserga R (ed) Tissue growth factors. Hand Exp Pharm, vol 57. Springer, Berlin Heidelberg New York, pp 491–550

Iversen OH, Evensen A (1962) Experimenal skin carcinogenesis in mice. Acta Pathol Microbiol Scand Suppl 156

Iversen OH, Bhangoo KS, Hansen K (1974) Control of epidermal cell renewal in the bat web. Virchows Arch B Cell Pathol 16:157–179

Iversen OH, Elgjo K, Reichelt KL (1989) Enhancement of methylnitrosourea-induced tumorigenesis and carcinogenesis in hairless mice by pretreatment with a mitosis-inhibiting epidermal pentapeptide. Carcinogenesis 10:241–244

Iversen OH, Elgjo K, Paulsen JE, Reichelt KL (1993) Moderate enhancement of the promotion phase of skin tumorigenesis in hairless mice by topical treatment with a mitosis-inhibiting epidermal pentapeptide. Carcinogenesis 14:2537–2542

Jensen PKA, Elgjo K, Laerum OD, Bolund L (1990) Synthetic epidermal pentapeptide and related growth regulatory peptides inhibit proliferation and enhance differentiation in primary and regenerating cultures of human epidermal keratinocytes. J Cell Sci 97:51–58

King AG, Talmadge JE, Badger Am, Pelus LM (1992) Regulation of colony-stimulating activity production from bone marrow stromal cells by the hematoregulatory peptide HP5b. Exp Hematol (Copenhagen) 20:223–228

King AG, Scott R, Wu DW, Strickler J, McNulty D, Scott M, Johanson K, McDevitt P, Trulli S, Jonak Z, Bhatnagar P, Balcarec J, Pelus L (1995) Characterization and purification of a stromal cell-derived hematopoietic synergistic factor induced by a novel hematoregulatory compound, SK&F 107647. Blood 86, Suppl 1:310a

Laerum OK, Paukovits WR (1984) Inhibitory effects of a synthetic pentapeptide on haemopoietic stem cells in vitro and in vivo. Exp Hematol (Copenhagen) 12:7–17

Laerum OD, Sletvold O, Bjerknes R, Eriksen JA, Johansen JH, Schanche JS, Teveraas T, Paukovits WR (1988) The dimer of hemoregulatory peptide (HP5b) stimulates human and mouse myelopoiesis in vitro. Exp Hematol (Copenhagen) 16:274–280

Lenfant M, Wdzieczak-Bakala J, Guittet E, Promé JC, Sotty D, Frindel E (1989) Sequence determination of an inhibitor of hemopoietic pluripotent stem cell proliferation. Proc Natl Acad Sci USA 36:779–782

Mancinelli L, Castigli E, Qualadrucci P, Gianfranceschi GL, Bramucci M, Miano A, Amici D (1992) Small acidic peptides from wheat-germ chromatin. 1. Isolation and biochemical characterization. Physiol Chem Phys Med NMR 24:97–107

Marks F, Rebien W (1972) The second messenger system of mouse epidermis I. Properties and β-adrenergic activation of adenylate cyclase in vitro. Biochim Biophys Acta 284:556–567

Marshall E, Lord BI (1996) Feedback inhibitors in normal and tumor tissues. Int Rev Cytol 167:185–201

Nakai GS (1976) Ehrlich ascites tumor (EAT) chalone effects on nascent DNA synthesis and DNA polymerase alpha and beta. Cell Tissue Kinet 9:553–563

Olsen WM, Elgjo K (1989) UVB-induced epidermal hyperproliferation is partially inhibited by a single treatment with an endogenous epidermal pentapeptide. J Invest Dermatol 94:101–106

Patt LM, Houck JC (1980) The incredible shrinking chalone. FEBS Lett 120:163–170

Paukovits WR, Laerum OD (1982) Isolation and synthesis of a hemoregulatory peptide. Z Naturforsch 37c:1297–1300

Paukovits MR, Moser MH, Paukovits JB (1993) Pre-CFU-S quiescence and stem cell exhaustion after cytostatic drug treatment. Protective effects of the inhibitory peptide pGlu-Glu-Asp-Cys-Lys (pEEDCK). Blood 81:1755–1761

Paulsen JE (1993) The synthetic colon peptide pyroGlu-His-GlyOH inhibits growth of human colon carcinoma cells (HT-29) transplanted subcutaneously into athymic mice. Carcinogenesis 14:1719–1721

Paulsen JE, Reichelt KL, Petersen A (1987) Purification and characterization of a growth inhibitory hepatic peptide. Virchows Arch Cell Pathol 54:152–154

Paulsen JE, Sundly Hall K, Endresen L, Rugstad HE, Reichelt KL, Elgjo K (1991) The peptide pyroGlu-Gln-Gly-Ser-Asn, isolated from mouse liver, inhibits growth of rat hepatoma cells in vitro. Carcinogenesis 12:207–210

Paulsen JE, Sundby Hall K, Rugstad HE, Reichelt KL, Elgjo K (1992) The synthetic hepatic petides pyroglutamylglutamylglycylserylasparagine and puroglutamylglutamylglycylserylaspartic acid inhibit growth of MH_1C_1 rat hepatoma cells transplanted into buffalo rats or athymic mice. Cancer Res 52:1218–1221

Paus P, Stenn KL, Elgjo K (1991) The epidermal pentapeptide pyroGlu-Glu-Asp-Ser-GlyOH inhibits murine hair growth in vivo and in vitro. Dermatologica 183:173–178

Pelus LM, King AG, Broxmeyer HE, DeMarsh PL, Petteway SR, Bhatnagar PK (1993) In vivo modulation of hematopoiesis by a novel hematoregulatory peptide. Exp Hematol 22:239–247

Reichelt KL, Kvamme E (1973) Histamine-dependent formation of N-acetyl-aspartyl peptides in mouse brain. J Neurochem 21:849–859

Reichelt KL, Elgjo K, Edminson P (1987) Isolation and structure of an epidermal mitosis-inhibiting pentapeptide. Biochem Biophys Res Commun 146:1493–1501

Reichelt KL, Paulsen JE, Elgjo K (1990) Isolation of a growth and mitosis inhibitory peptide from mouse liver. Virchows Arch B Cell Pathol 59:137–142

Reichelt KL, Haakonsen P, Paulsen JE, Kornstad R, Elgjo K (1994) Difference in the distribution of low molecular weight peptides and amino acids in normal and malignant cells. Oncol Rep 1:1107–1112

Richter KH, Clauss M, Höfle W, Schnapke R, Marks F (1988) The epidermal G1-chalone: an endogenous tissue-specific inhibitor of epidermal cell proliferation. Ann NY Acad Sci 548:204–210

Rytömaa T, Kiviniemi K (1968) Chloroma regression induced by the granulocytic chalone. Nature 222:995–996

Rytömaa T, Kiviniemi K (1970) Regression of generalized leukemia in rat induced by the granulocyre chalone. Eur J Cancer 6:401–410

Rytömaa T, Vilpo JA, Levanto A, Jones WA (1976) Effect of granulocyte chalone on acute and chronic granulocytic leukaemia in man. Report of seven cases. Scand J Haematol Suppl 27:1–28

Simms HS, Stillman NP (1937) Substances affecting adult tissue is vitro II. A growth inhibitor in adult tissue. J Gen Physiol 20:621–629

Skraastad Ö, Reichelt KL (1989a) Further studies on the biological characteristics of an endogenous colon mitosis inhibitor: comparison with some structurally related peptides. Virchows Archiv B Cell Pathol 56:321–325

Skraastad Ö, Reichelt KL (1989b) An endogenous colon mitosis inhibitor reduces proliferation of colon carcinoma cells (HT 29) in serum-restricted medium. Virchows Arch Cell Pathol 56: 393–396

Skraastad Ö, Fossli T, Edminson PD, Reichelt KL (1988) Purification and characterization of a mitosis inhibitory tripeptide from mouse intestinal extracts. Epithelia 1:107–119

Tustin A (1948) Automatic controls. Simon & Shuster, New York, pp 10–31

Ushmorov AA, Danilov AO, Zubova SG, Elgjo K, Okulov VB (1992) Modulation of KB and A431 cell adhesion by epidermal growth inhibitory pentapeptide. Virchows Arch B Cell Pathol 62:353–356

Veiby OP, Olsen WM (1995) Accelerated hematopoietic recovery with the hemoregulatory peptide dimer SK&F 107647 in bone marrow transplantation. Bone Marrow Transplant 15:305–311

Wang Xiao Ling, Elgjo K, Haaskjold E (1996) Regeneration of rat corneal epithelium is delayed by the inhibitory epidermal pentapeptide (EPP). Acta Opththalmol Scand 74:361–363

Watt FM, Reichelt KL, Elgjo K (1989) Pentapetide inhibitor of epidermal mitosis: production and responsiveness in cultures of normal, transformed, and neoplastic human keratinocytes. Carcinogenesis 10:2249–2253

Weiss P, Kavanau JL (1957) A model of growth and growth control in mathematical terms. J Gen Physiol 41:1–47

Whitehead PA, Robinson PA, Hume WJ, Keen JN (1991) Identification and partial characterisation of a serum enzyme which hydrolyses epidermal inhibitory pentapeptide. Biochim Biophys Res Commun 175:978–985

Endogenous Angiogenesis Inhibitors: Angiostatin, Endostatin, and Other Proteolytic Fragments

Yihai Cao[1]

1 Introduction

In mammals, the vasculature develops by vasculogenesis and angiogenesis. Vasculogenesis is the de novo blood vessel formation from endothelial cells which are born from precursor cell types including hemangioblasts and angioblasts (Risau 1995). Vasculogenesis usually occurs during early embryonic development, including the formation of heart and aorta (Risau and Flamme 1995). Angiogenesis represents the process of growth of new capillaries from preexisting blood vessels (Folkman and Shing 1992). In adults, new blood vessels are generated virtually via angiogenesis. The vasculature remains quiescent in the adult mammal, except for transient processes of neovascularization in the female reproductive system. In response to an appropriate growth stimulus, endothelial cells can degrade the basement membrane locally. Simultaneously, the quiescent endothelial cells change their morphology, proliferate, migrate, invade into the surrounding stroma tissue, form microtubes, and sprout new capillaries. This complex process of angiogenesis implies the presence of multiple controls of the system, which can be switched on and off within a short period.

Angiogenesis has been implicated in the development and progression of pathogenic processes of a variety of disorders, including diabetic retinopathy, psoriasis, rheumatoid arthritis, cardiovascular diseases, and tumor growth (Folkman 1995a). For the past three decades, a large body of work by a number of laboratories has provided both direct and indirect evidence that tumor growth and its metastasis are accompanied by the growth of new blood vessels. At the prevascular stage, a solid tumor rarely grows bigger than 2–3 mm^3 and may contain a few million cells (Folkman 1995a). Cells in prevascular tumors may proliferate as rapidly as those in vascularized expanding tumors. However, the growth rate of cells in avascular tumors reaches an equilibrium with their death rate.

[1] Laboratory of Angiogenesis Research, Microbiology and Tumor Biology Center, Karolinska Institute, 171-77, Stockholm, Sweden

Progress in Molecular and Subcellular Biology, Vol. 20
A. Macieira-Coelho (Ed.)

The switch to the angiogenic phenotype requires a local change of balance between angiogenic and angiogenic inhibitors. Among angiogenic factors, families of fibroblast growth factor (FGF) and vascular endothelial growth factor (VEGF)/vascular permeability factor (VPF) are most commonly expressed in tumors. While FGFs display their biological effects on a variety of cell types, VEGF/VPF appears to be the most selective growth factor acting on endothelial cells. These two families of angiogenic factors can promote angiogenesis in a synergistic manner (Pepper et al. 1992). Levels of expression of VEGF mRNA and protein are markedly upregulated in the majority of human tumors including bladder, breast, lung, gastrointestinal, ovary, prostate, glioblastoma, hemangioma, and retinoblastoma (Ferrara and Davis-Smyth 1997). High concentrations of bFGF were found in both serum and urine of patients with various cancers (Nguyen et al. 1994). Although upregulation of angiogenic factors is necessary for a tumor to switch on its angiogenic phenotype, production of angiogenesis inhibitors has to be simultaneously downregulated. Increasing evidence demonstrates that downregulation of angiogenesis inhibitors is equally important as upregulation of angiogenic factors in the switch of angiogenic phenotype of a tumor (Folkman 1995b). For example, an angiogenesis inhibitor, thrombospondin-1, has been found to be downregulated in a number of tumors (Dameron et al. 1994; Grossfeld et al. 1997).

While biological functions and molecular mechanisms of angiogenic factors including FGF and VEGF families have been extensively characterized, little is known about the molecular aspects of angiogenesis inhibitors. The discovery of specific endothelial inhibitors such as angiostatin and endostatin not only increases our understanding of physiologic and pathologic functions of these negative regulators, but also provides an important therapeutic strategy for cancer treatment. This chapter will discuss our current understanding of functional properties of these endogenous angiogenesis inhibitors.

2
Angiostatin

2.1
Discovery

Inhibition of metastatic tumor growth by primary tumor mass has been observed in some clinical malignancies. The removal of certain primary tumors in patients such as breast and colon carcinomas can be followed by a rapid growth of distant metastases (Southam and Brunschwig 1961; Warren et al. 1977; Woodruff. 1980; Fidler and Balch 1987; Clark et al. 1989). A number of animal experiments also demonstrate that some primary tumors can inhibit the growth of their metastases (Gorelik et al. 1978, 1980; Bonfil et al. 1988).

Metastatic growth can also suppress the growth of a primary tumor (Yuhas and Pazmino 1974). In addition, partial removal of a primary tumor can accelerate the growth rate of a residual tumor (Lange et al. 1980; Fisher et al. 1989). Although several hypotheses, such as mechanical release of tumor cells during surgical procedures, concomitant immunity, and depletion of available nutrients by the primary tumor, have emerged to explain the phenomenon (Gorelik 1983; Prehn 1991, 1993), none of these has provided the molecular mechanisms by which tumor growth is inhibited by tumor mass. A recent study on tumor angiogenesis has provided the most compelling explanation for this phenomenon (O'Reilly et al. 1994). An angiogenesis inhibitor named angiostatin was isolated from both serum and urine of mice bearing a transplantable murine Lewis lung carcinoma (3LL) in syngeneic C57B16/J mice (O'Reilly et al. 1994).

The endothelial inhibitory activity was purified by an in vitro assay of inhibition of basic fibroblast growth factor (bFGF)-stimulated bovine capillary endothelial cell proliferation. Angiostatin is a circulating angiogenesis inhibitor produced in association with the primary Lewis lung tumor growth. This 38-kDa inhibitor accumulates in the circulation in the presence of growing primary tumor and disappears from the circulation after removal of the primary tumor. Thus, resection of the primary Lewis lung carcinoma results in depletion of circulating angiostatin and promotes neovascularization and growth of lung micrometastases.

2.2
Structure and Generation

Microsequence analysis of angiostatin purified from both urine and serum samples from tumor (3LL)-bearing animals revealed greater than 98% identity to an internal fragment of mouse plasminogen with an N-terminus at amino acid valine 98 (O'Reilly et al. 1994). Based on its molecular weight, mouse angiostatin is predicted to contain a C-terminus at amino acid approximately 440 of plasminogen. Thus, angiostatin should contain the first four of five triple loop disulfide-linked structures, known as kringle structures of plasminogen. Proteolytic fragments generated by in-vitro limited elastase proteolysis of human plasminogen, which are compatible to the murine angiostatin, contain the inhibitory activity on endothelial cell proliferation (O'Reilly et al. 1994). Purified human angiostatin in the initial studies contains three molecular weight species of apparent molecular masses of 40, 42, and 45 kDa; each of the three bands comparably inhibited endothelial cell proliferation (O'Reilly et al. 1994). Microsequence analysis of each of the three bands revealed an identical N-terminus at amino acid 97 or 99, suggesting that the C-terminal length of these fragments is variable. Although the intact plasminogen molecule contains the same kringle structures, it is inactive in inhibition of endothelial cell growth, in vivo neovascularization, and metastatic tumor growth (O'Reilly et al. 1994, 1996). These data imply that angiostatin may contain a different dimensional structure than the kringle structure of plasminogen. It is also speculated that the endothelial inhibitory activity is hidden in the intact plasminogen molecule.

Although angiostatin is detected in vivo in association with 3LL Lewis lung tumor growth, it is unlikely that tumor cells produce angiostatin directly because they lack a detectable amount of mRNA for angiostatin and plasminogen (Y. Cao et al., unpubl. observ.). Angiostatin can be generated in vitro by limited elastase proteolysis of plasminogen (O'Reilly et al. 1994; Cao et al. 1996). A recent study has uncovered the mechanism responsible for the in vivo production of angiostatin in syngeneic C57B16/J mice bearing Lewis lung carcinomas (3LL) (Dong et al. 1997). The mediator of angiostatin production in 3LL-Lewis lung carcinoma is found to be the tumor-infiltrating macrophage (Dong et al. 1997). Immunohistochemical staining of subcutaneous tumor sections discovers the infiltration of macrophages, and high levels of expression of metalloelastase (MME) mRNA are detected in tumor tissues (Dong et al. 1997). Successive passages of cultures established from tumors demonstrate that the generation of angiostatin is directly correlated with the macrophage-derived metalloelastinolytic activity. The incubation of plasminogen with 3LL cells cultured in vitro does not produce angiostatin, whereas cocultures of macrophages with 3LL cells convert plasminogen to angiostatin. Further studies indicate that the metalloelastinolytic activity in macrophages can be upregulated by the cytokine, GM-CSF, secreted by 3LL tumor cells. Thus, 3LL tumor cells communicate with macrophages through the secretion of GM-CSF, which increases the production of elastase activity and generation of angiostatin (Fig. 1).

It appears that MME released from tumor-infiltrating macrophages is not the only source responsible for angiostatin production in tumors. Three human prostate carcinoma cell lines (PC-3, DU-145, and LN-CaP) have been found to produce proteolytic activity that generates angiostatin from plasminogen or plasmin (Gately et al. 1997). The enzymatic activity released by these human cell lines is independent from that of tumor-infiltrating macrophages (Fig. 1). Studies with protease inhibitors demonstrate that a serine proteinase is essential for angiostatin generation. Thus, at least two known sources of proteases in association with tumors can convert plasminogen to angiostatin in vivo, MME released from tumor-infiltrating macrophages, and serine proteases produced directly from tumor cells (Fig. 1).

2.3
Antiendothelial, Antiangiogenic and Antitumor Effects

Purified angiostatin specifically and reversibly inhibits proliferation of endothelial lineages, including bovine capillary endothelial (BCE), bovine aorta en-

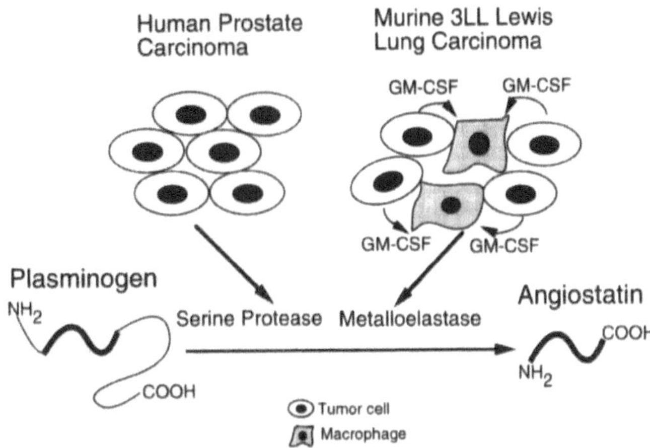

Fig. 1. Generation of angiostatin by tumors. Angiostatin is generated by proteolytic degradation of plasminogen in association with tumor growth. Serine-like proteases released by human prostate carcinoma cells and a metalloelastase secreted by tumor-infiltrating macrophages in response to a cytokine (*GM-CSF*) produced by the Lewis lung tumor cells can convert plasminogen into angiostatin

dothelial (BAE), human umbilical vein endothelial (HUVE), and malignant mouse hemangioendothelioma (EOMA) cells in a dose-dependent manner (O'Reilly et al. 1994). In contrast, concentrations of angiostatin for maximal inhibition of endothelial cell proliferation are not inhibitory on a variety of normal and neoplastic nonendothelial cell lines, including 3T3 fibroblasts, bovine aorta smooth muscle cells, bovine retinal pigment epithelial cells, human fetal fibroblasts, and 3LL murine Lewis lung carcinoma cells (O'Reilly et al. 1994). Thus, angiostatin specifically inhibits endothelial cell proliferation.

Studies of smaller fragments of human angiostatin on inhibition of endothelial proliferation demonstrate that a functional difference is present among individual kringle structures (Cao et al. 1996). For example, appropriately folded recombinant kringle 1 exhibits potent inhibitory activity on BCE cells (Fig. 2). Recombinant kringle 2 and kringle 3 also display endothelial inhibition. In contrast, kringle 4 exerts ineffective inhibition on endothelial cell proliferation (Cao et al. 1996). These data indicate that the endothelial inhibitory activity of angiostatin is shared by kringle 1, kringle 2, and kringle 3, but probably not by kringle 4. Among the tandem kringle arrays, the recombinant kringle 2–3 fragment exerts inhibitory activity similar to kringle 2 alone. However, relative to kringle 2–3, a marked enhancement of inhibition is observed when individual kringle 2 and kringle 3 are added together to endothelial cells (Cao et al. 1996). This implies that it is necessary to open the interkringle disulfide bridge between kringle 2 and kringle 3 in order to obtain the maximal inhibitory effect of kringle 2–3. In view of variable lysine-binding affinity of

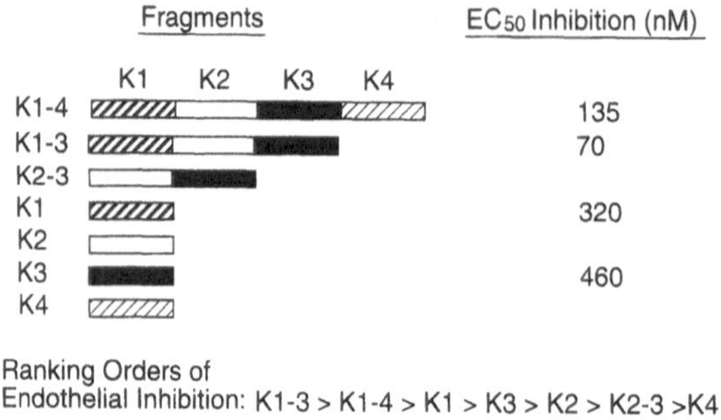

Fragments (K1 K2 K3 K4)	EC50 Inhibition (nM)
K1-4	135
K1-3	70
K2-3	
K1	320
K2	
K3	460
K4	

Ranking Orders of
Endothelial Inhibition: K1-3 > K1-4 > K1 > K3 > K2 > K2-3 >K4

Fig. 2. Antiendothelial growth activity of kringle structures of angiostatin. Angiostatin consists of the first four kringle structures (*K1-4*) of plasminogen. Smaller fragments derived from angiostatin including kringle 1 (*K1*), kringle 2 (*K2*), and kringle 3 (*K3*) inhibit capillary endothelial cell growth in vitro. Kringle 4 (*K4*) does not exhibit inhibitory activity on endothelial cells. *Numbers* represent protein concentrations to reach 50% of maximal inhibition

homologous domains, it would appear that lysine-binding capability does not correlate with relative inhibitory effects of the kringle-containing fragments. Folding studies indicate that the anti-proliferative activity of angiostatin is largely abolished after reduction/alkylation (Cao et al. 1996). Thus, appropriate folding of kringle structures as tandem domains held together by intrachain and interchain disulfide bonds is essential for angiostatin to maintain its full antiendothelial activity.

In the chick embryo chorioallantoic membrane (CAM), purified human angiostatin induces avascular zones over a concentration range of 0.1–100 μg/embryo. The dose-dependent inhibition reaches saturation at approximately 100 μg/embryo. Systemic administration of human angiostatin in mice implanted with basic fibroblast growth factor in corneal micropockets (80–100 ng bFGF/cornea) significantly inhibits corneal neovascularization induced by bFGF. At the concentration of 50 mg/kg every 12 h, angiostatin inhibits new vessel growth by 85% compared to controls.

Systemic administration of human angiostatin potently inhibits the growth of transplanted human and murine primary tumors in mice (O'Reilly et al. 1996; Sim et al. 1997). The growth of three aggressive primary murine tumors (Lewis lung carcinoma, T241 fibrosarcoma, and reticulum cell sarcoma) is inhibited by an average of 84% at doses of 100 mg/kg/day. These tumors produce poor response to other therapies (O'Reilly et al. 1996). Inhibition of primary tumor growth becomes apparent at 10 mg/kg/day and increasing doses of angiostatin correlate with increased antitumor efficacy. Systemic treatment of human tumors growing in immunodeficient mice produces an almost complete suppression of primary tumor growth. Angiostatin inhibits the growth of human breast carcinoma by 95%, colon carcinoma by 97%, and prostate carcinoma by almost 100% (O'Reilly et al. 1996). In the colon and

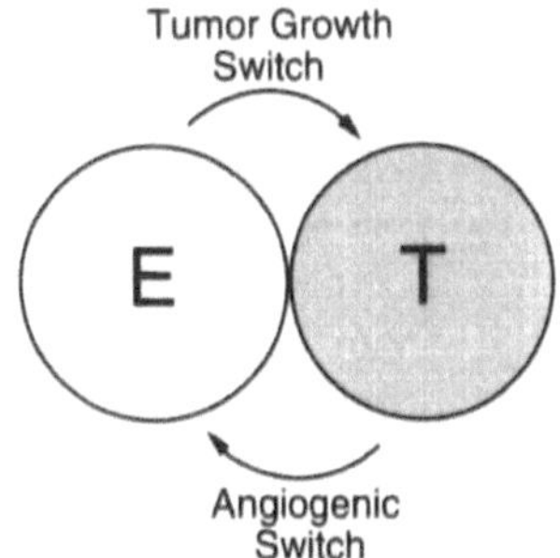

Fig. 3. Interrelationship between endothelial compartment (*E*) and tumor compartment (*T*). Tumor cells (*T compartment*) produce angiogenic factors such as VEGF/VPF and FGFs and angiogenesis inhibitors including angiostatin and endostatin. The switch of angiogenic phenotype of the tumor requires upregulation of angiogenic factors and down-regulation of angiogenesis inhibitors. Once the tumor becomes angiogenic, tumor-infiltrating endothelial cells (*E compartment*) produce growth stimulators to support tumor growth. These paracrine factors can be survival factors and growth factors, which can turn on tumor growth

breast carcinoma-bearing mice, tumors regrow within 2 weeks after withdrawal of angiostatin treatment. Angiostatin treatment does not result in weight loss or other toxicity in mice, including those that receive 100 mg/kg/day and in immunocompromised mice for as long as 60 days. The lack of toxicity suggests that the effective antitumor therapy by angiostatin can function directly against the endothelial compartment of a tumor, which demonstrates the importance of the endothelial cell compartment in controlling tumor growth (Fig. 3).

Histological studies reveal that indexes of tumor cell proliferation and apoptosis reach a net balance in angiostatin-treated human dormant tumors (O'Reilly et al. 1994, 1996; Holmgren et al. 1995). In the angiostatin-treated mice, the apoptotic index of tumor cells can increase to five times than that of control mice, whereas tumor cell proliferative rate remains at the same level before and after exposure to angiostatin. Thus, angiostatin can cause human primary carcinomas to regress to a dormant state as defined by a balance of tumor-cell proliferation and apoptosis. In addition, the angiostatin-induced dormant tumors lack neovascularization, as detected by the von Willebrand factor (Holmgren et al. 1995).

Similarly, primary tumor-produced and systemically administered angiostatin can induce dormant lung metastases in Lewis lung carcinoma-bearing mice (O'Reilly et al. 1994; Holmgren et al. 1995). Metastases are dependent on angiogenesis in at least two steps of the metastatic events. First, metastatic tumor cells must exit from a primary tumor which has been vascularized. Second, upon arrival at their target organ, metastatic tumor cells must undergo neovascularization in order to grow to a clinically detectable size. In the presence of angiogenesis inhibition, metastatic tumor cells form microscopic perivascular cuffs around the preexisting microvessel from which they probably left the circulation. The colonies of dormant lung micrometastases rarely expand beyond 0.3 mm in diameter; they proliferate as rapidly as fast-growing metastases, they undergo a high apoptotic rate, and they lack neovascularization. The mechanism by which angiostatin therapy leads to an increased death rate of tumor cells is not known. It is speculated that complete inhibition of tumor angiogenesis may result in a loss of survival factors essential for tumor cells. These tumor cell survival factors can be either from the circulation as endocrine factors and/or from the endothelial cell as paracrine factors. Alternatively, a decreased tumor-cell number in a dormant tumor may also limit the production of survival factors for tumor cells as autocrine factors.

2.4 Kringle 5

Amino-acid sequence alignment of individual kringle domains of human plasminogen shows that kringle 5 displays remarkable sequence identity with kringle 1 (57.5%), kringle 2 (46.25%), kringle 3 (48.75%), and kringle 4 (52.5%), respectively. Based on the primary structure similarities with other kringle domains, especially kringle 1 (highest sequence identity), kringle 5 of plasminogen has recently been demonstrated to be a potent endothelial-specific inhibitor (Cao et al. 1997). Kringle 5 obtained as a proteolytic fragment of human plasminogen displays potent inhibitory effect on bovine capillary endothelial cells with a half-maximal concentration of approximately 50 nM. Thus, kringle 5 would apprear more potent than angiostatin on inhibition of bFGF-stimulated capillary endothelial cell growth. Appropriately folded recombinant mouse kringle 5 protein, expressed in *E. coli*, exhibits a comparable inhibitory activity as the proteolytic kringle 5 fragment (Cao et al. 1997). Kringle 5 is thus far the most potent inhibitory fragment derived from plasminogen. The identification of kringle 5 as a potent endothelial cell-specific inhibitor increases our knowledge in understanding the role of plasminogen kringles in inhibiting endothelial cell proliferation.

3 Endostatin

3.1 Structure

Many human and murine malignant tumors produce inhibitors of angiogenesis (Chen et al. 1995). Using a similar strategy of isolation of angiostatin, O'Reilly et al. (1997) have identified and purified another potent inhibitor specific to endothelial cells, endostatin, produced by a murine hemangioendothelioma (EOMA). Endostatin is purified from the conditioned medium of EOMA cells that specifically inhibits endothelial cell proliferation. This inhibitor migrates to a molecular mass of 20kDa under reducing conditions by SDS-PAGE (O'Reilly et al. 1997).

Amino-acid sequence analysis reveals that endostatin is a C-terminal fragment of collagen XVIII. N-terminal sequence analysis confirms that the purified endostatin is contained in the C-terminal noncollagen (NC1) portion of collagen XVIII (Oh et al. 1994; Rehn and Pihlajaniemi 1994). Collagen XVIII is a new member of the collagen-like protein family, which comprises of an N-terminal region with at least three spliced variants, collagen-like repeats, and the NC1 domain (Muragaki et al. 1995; Rehn and Pihlajaniemi 1995).

3.2 Antiangiogenic and Antitumor Activities

Similarly to angiostatin, endostatin specifically inhibits capillary endothelial cell proliferation, but not other cells, including bovine aorta smooth muscle cells, bovine retinal pigment epithelial cells, 3T3 fibroblasts, mink lung epithelial cells, EOMA hemangioendothelioma cells, and Lewis lung carcinoma cells (O'Reilly et al. 1997). In the chick chorioallantoic membrane, purified recombinant endostatin displays a potent effect on inhibition of new blood vessel growth. In the murine Lewis lung carcinoma model system, i.e., rapid growth of lung metastases after removal of the primary tumor, the growth of lung metastases is almost completely suppressed by the systemic administration of endostatin at a dose 0.3mg/kg/day given subcutanously. Systemic administration of precipitated and nonrefolded recombinant endostatin expressed in *E. coli* allows sustained release of the protein from a subcutaneous pellet (O'Reilly et al. 1997). The growth of Lewis lung primary tumors is potently suppressed by endostatin therapy. At 10mg/kg/day, tumor growth is inhibited by 97%. Almost complete suppression of primary tumor growth is observed at the dose of 20mg/kg/day. Immunohistochemical studies reveal that the residual microscopic tumors lack neovascularization. The growth of several other mouse primary tumors, including B16F10, melanoma, T241 fibrosarcoma, and EOMA hemangioendothelioma, is also regressed to a dormant state for as long as endostatin is administered for treatment. There is no weight loss or any evidence of toxicity in any of the mice treated with endostatin.

Table 1. Endogenous angiogenesis inhibitors

Inhibitors	Reference
Angiostatin	O'Reilly et al. (1994)
Chemokine gro-β	Cao et al. (1995)
Endostatin	O'Reilly et al. (1997)
Inerferon-α	Ribatti et al. (1996)
Interleukin-12 (IL-12)[a]	Voest et al. (1995)
Chemokine IP-10	Angiolillo et al. (1995); Strieter et al. (1995)
Platelet factor-4 (PF-4)	Maione et al. (1990)
Prolactin-RP	Jackson et al. (1994)
Soluble FGF-R	Hanneken et al. (1994)
Thrombospondin-1 (TSP-1)	Good et al. (1990)
TIMPs (TIMP-1,2,3)	Moses et al. (1990)
16-kDa prolactin	Clapp et al. (1993)

[a] IL-12 does not inhibit endothelial cell growth in vitro.

4 Other Angiogenesis Inhibitors

4.1 Endogenous Inhibitors

In addition to angiostatin and endostatin, several other endogenous inhibitors of angiogenesis have been recently identified although not all associated with tumor growth (Table 1). Under physiological conditions, these inhibitors prevent blood-vessel growth from angiogenic stimuli. They include a number of proteins that have several functions in addition to inhibitory effect on angiogenesis. For example, chemokines of IP-10 and gro-beta are potent chemoattractants for leukocytes (Baggiolini et al. 1994), metalloproteinase inhibitors (TIMPs) are potent inhibitors for proteases (Knox et al. 1997), and soluble FGF-receptors do not only block the FGF-mediated angiogenic effects but also inhibit other functions mediated by FGFs. The only known angiogenesis inhibitors that specifically inhibit endothelial cell growth are angiostatin and endostatin.

4.2 Proteolytic Fragments

Of the 12 known angiogenesis inhibitors (Table 1), several are proteolytic fragments (Table 2). It appears that the generation of endogenous inhibitors in vivo from large precursor proteins with distinct functions is a recurrent theme in the inhibition of angiogenesis. In addition to angiostatin and endostatin, the 16-kDa N-terminal fragment of prolactin has been characterized as an

Table 2. Proteolytic fragments that inhibit angiogenesis

Source	M/W (kDa)	Reference
Fibronectin	29	Homanberg et al. (1985)
Prolactin	16	Clapp et al. (1993)
Angiostatin	38	O'Reilly et al. (1994)
Platelet factor-4[a]	7.8	Gupta et al. (1995)
Endostatin	20	O'Reilly et al. (1997)

[a] An N-terminally truncated fragment of PF-4 molecule inhibits endothelial cell growth and angiogenesis.

antiangiogenic domain (Clapp et al. 1993; D'Angelo et al. 1995). Similarly to angiostatin and endostatin, the intact parental molecule of prolactin lacks the inhibitory activity on endothelial cells, nor is it an angiogenesis inhibitor. Platelet factor-4 is a relatively weak endothelial cell inhibitor (Maione et al. 1990). However, a proteolytic fragment with the N-terminally truncated sequence of PF-4 increases its inhibitory activity by greater than 50-fold (Gupta et al. 1995). A fibronectin fragment derived from plasmin-digestion also inhibits endothelial cell growth (Homandberg et al. 1985).

Thus, proteolytic processing plays critical dual roles in control of angiogenesis. When the process of angiogenesis begins, proteolytic degradation of the basement membrane surrounding quiescent endothelial cells is a prerequisite for endothelial cell growth in vivo. Once new blood vessels have been formed, they may require proteolytic fragments to control the over-neovascularization. The molecular mechanisms underlying how protease activity is regulated in control of angiogenesis is not known, nor do we know the substrate specificity of these proteases. However, comparison of amino-acid sequences of these antiangiogenic fragments does not reveal a common cleavage site, suggesting that more than one proteases participate in this process. Furthermore, one angiostatic fragment can be generated by two or more proteases. Indeed, angiostatin produced in association with tumor growth can be generated by both serine-like proteases and a metalloelastase (Dong et al. 1997; Gately et al. 1997).

4.3 Small Peptides

Small synthetic peptides with a length of between 4 and 20 amino acids derived from a number of proteins have been found to inhibit endothelial cell growth and/or angiogenesis (Table 3). Comparison of amino-acid sequences of these peptides reveals no common motif, nor have significant homologies been detected among these peptides. Little is known as to whether similar small

Table 3. Antiendothelial peptides[a]

Peptide sequence	Length[b]	Source	Reference
PLYKKIIKKLLES	13	Platelet factor-4	Maione et al. (1990)
CDPGYIGSR	9	Laminin	Sakamato et al. (1991)
FCYWKVCW or FCFWKTCT	8	Somatostatin	Woltering et al. (1991)
GRGD	4	Fibronectin, others	Eijian et al. (1991)
NGVQYRN	7	Procollagen, TSP	Tolsma et al. (1993)
NIPPITCVQNGLRY	14	Collagen I	Tolsma et al. (1993)
SPWSSCSVTCGD-GVITRIR	19	Thrombospondin-1	Tolsma et al. (1993)
cyclo-RGDfV	4	Integrin $\alpha_v\beta_3$, others	Brooks et al. (1994)
VIGYSGDRC	9	EGF	Nelson et al. (1995)
LRAPLIPMEH CTTAFFETCD	20	SPARC	Sage et al. (1995)

[a] Many of these synthetic peptides also inhibit angiogenesis in vivo.
[b] Amino acid numbers.

peptides are also naturally present in vivo. Further work should be carried out to study their in vivo stability and their potency in suppression of neovascularization.

5 Clinical Applications

In conclusion, the discovery of angiogenesis inhibitors such as angiostatin and endostatin not only facilitates our understanding of the regulation of the angiogenesis process under physiological and pathological conditions, but also allows us to develop novel therapeutic strategies to prevent blood-vessel growth in diseases. This approach is already beginning. In addition to the above-discussed angiostatin and endostatin, monoclonal antibodies that neutralize the actions of VEGF completely block tumor growth in mice (Kim et al. 1993). Similarly, a soluble receptor for VEGF also blocks tumor growth in vivo (Millauer et al. 1994). Abrogation of the functions of $\alpha_v\beta_3$ or $\alpha_v\beta_5$ specifically expressed in growing endothelial cells dramatically retards tumor growth (Brooks et al. 1994; Friedlander et al. 1995). -C-X-C- chemokines including PF-4, gro-β, and IP-10 have been shown to impair tumor growth (Maione et al. 1991; Luster and Leder 1993; Cao et al. 1995). Some of these endogenous angiogenesis inhibitors have already given rise to a promising treatment in cancer patients. For example, Interferon alpa-2a has been used successfully in treatment life-threatening hemangioma in children (Ezekowitz et al. 1994, 1995).

Thus, angiogenesis inhibitors seem likely to become one of several important therapeutic strategies in cancer and other vascularized diseases. Preclinical and clinical studies have provided several important clues for development of antiangiogenic therapy. Because several of the angiogenesis inhibitors specifically target the proliferating endothelial cell compartment, it is less likely that they will cause immune suppression, bone marrow suppression, and gastrointestinal symptoms. Resistance to angiogenesis inhibitors has not been observed in animal studies. Tumors which are resistant to chemotherapy can be suppressed by angiogenesis inhibitors. Because endogenous angiogenesis inhibitors are normal constituents in the body, they less likely cause immune reactions. A combination of antiangiogenic therapy and cytotoxic therapy or immune therapy may become more effective because the combined treatment is directed at different compartments.

For all these applications, more potent and specific angiogenesis inhibitors need to be identified and more knowledge should be gained in the characterization of molecular and cellular mechanisms.

Acknowledgments. The research in the author's laboratory is supported by the Swedish Medical Council, K97-12P-11819-02B and the Swedish Cancer Foundation, 96 1607; 3811-B96-01XBA. The author also would like to thank Jacob Farnebo for computer assistance during the preparation of this manuscript.

References

Angiolillo AL, Sgadari C, Taub DD, Liao F, Farber JM, Meaheshwari S, Kleinman HK, Reaman GH, Tosato G (1995) Human interferon-inducible protein 10 is a potent inhibitor of angiogenesis in vivo. J Exp Med 182:155–162

Baggiolini M, Dewald B, Moser B (1994) Interleukine-8 and related chemotactic cytokines, - CXC and CC chemokines. Adv Immunol 55:97–197

Bonfil RD, Ruggiero RA, Bustuoabad OD, Meiss RP, Pasqualini CD (1988) Role of concomitant resistance in the development of murine lung metastases. Int J Cancer 41:415–422

Brooks PC, Montgomery AM, Rosenfeld M, Reisfeld RA, Hu T, Klier G, Cheresh DA (1994) Integrin $\alpha_v\beta_3$ antagonists promote tumor regression by inducing apoptosis of angiogenic blood vessels. Cell 79:1157–1164

Cao Y, Chen C, Weatherbee JA, Tsang M, Folkman J (1995) gro-beta, a -C-X-C- chemokine, is an angiogenesis inhibitor that suppresses the growth of Lewis lung carcinoma in mice. J Exp Med 182:2069–2077

Cao Y, Ji R-W, Davidson D, Schaller J, Marti D, Sohndel S, McCance SG, O'Reilly MS, Llinas M, Folkman J (1996) Kringle domains of human plasminogen: characterization of the antiproliferative activity of endothelial cells. J Biol Chem 271:29461–29467

Cao Y, Chen A, An SSA, Ji R-W, Davidson D, Cao Y, Llinas M (1997) Kringle 5 of plasminogen: a novel inhibitor of endothelial cell growth. J Biol Chem 272:22924–22928

Chen C, Parangi S, Tolentino M, Folkman J (1995) A strategy to discover circulating angiogenesis inhibitors generated by human tumors. Cancer Res 55:4230–4233

Clapp C, Martial JA, Guzman RC, Rentierdelrue F, Weiner RI (1993) The 16-kilodalton N-terminal fragment of human prolactin is a potent inhibitor of angiogenesis. Endocrinology 133:1292–1299

Clark WH, Elder DE, Guerry DIV, Braitman LE, Trock BJ, Schultz D, Synnestevdt M, Halpern AC (1989) Model predicting survival in stage I melanoma based on tumor progression. J Natl Cancer Inst 81:1893–1904

Dameron KM, Volpert OV, Tainsky MA, Bouck N (1994) Control of angiogenesis in fibroblasts by p53 regulation of thrombospondin-1. Science 265:1582–1584

D'Angelo G, Struman I, Martial J, Weiner RI (1995) Activation of mitogenactivated protein kinases by vascular endothelial growth factor and basic fibroblast growth factor in capillary endothelial cells is inhibited by the antiangiogenic factor 16-kDa N-terminal fragment of prolactin. Proc Natl Acad Sci USA 92:6374–6378

Dong Z, Kumar R, Yang X, Fidler IJ (1997) Macrophage-derived metalloelastase is responsible for the generation of angiostatin in Lewis lung carcinoma. Cell 88:801–810

Eijan AM, Davel L, Oisgold-Data S, de Lustig ES (1991) Modulation of tumor-induced angiogenesis system. Mol Biother 3:38–40

Ezekowitz RAB, Mulliken J, Folkman J (1994) Interferon alfa-2a therapy for life-threatening hemangiomas of infancy. N Engl J Med 330:300

Ezekowitz RAB, Mulliken J, Folkman J (1995) Interferon alfa-2a therapy for life-threatening hemangiomas of infancy. N Engl J Med 333:595–596

Ferrara N, Davis-Smyth T (1997) The biology of vascular endothelial growth factor. Endocr Rev 18:4–25

Fidler IJ, Balch CM (1987) The biology of cancer metastasis and implications for therapy. Curr Probl Surg 24:137–209

Fisher B, Gunduz N, Coyle J, Rudock C, Saffer E (1989) Presence of growth-stimulating factor in serum following primary tumor removal in mice. Cancer Res 49:1996–2001

Folkman J (1995a) Angiogenesis in cancer, vascular, rheumatoid and other disease. Nat Med 1:27–31

Folkman J (1995b) Clinical applications of research on angiogenesis. N Engl J Med 333:1757–1763

Folkman J, Shing Y (1992) Angiogenesis. J Biol Chem 267:10931–10934

Friedlander M, Brooks PC, Shaffer RW, Kincaid CM, Varner JA, Cheresh DA (1995) Definition of two angiogenic pathways by distinct α_v integrins. Science 270:1500–1502

Gately S, Twardowski P, Stack MS, Patrick M, Boggio L, Cundiff DL, Schnaper HW, Madison L, Volpert O, Bouck N, Enghild J, Kwaan HC, Soff GA (1997). Human prostate carcinoma cells express enzymatic activity that converts human plasminogen to the angiogenesis inhibitor, angiostatin. Cancer Res 56:4887–4890

Good DJ, Polverini PJ, Rastinejad F, LeBeau MM, Lemons RS, Frazier WA, Bouck NP (1990) A tumor suppressor-dependent inhibitor of angiogenesis is immunologically and functionally indistinguishable from a fragment of thrombospondin. Proc Natl Acad Sci USA 87:6624–6628

Gorelik E (1983) Concomitant tumor immunity and the resistance to a second tumor challenge. Adv Cancer Res 39:71–120

Gorelik E, Segal S, Feldman M (1978) Growth of a local tumor exerts a specific inhibitory effect on progression of lung metastases. Int J Cancer 21:617–625

Gorelik E, Segal S, Fredman M (1980) Control of lung metastasis progression in mice: role of growth kinetics of 3LL Lewis lung carcinoma and host immune reactivity. J Natl Cancer Inst 65:1257–1264

Grossfeld GD A GD, Stein JP, Bochner BH, Esrig D, Groshen S, Dunn M, Nichols PW, Taylor CR, Skinner DG, Cote RJ (1997) Thrombospondin-1 expression in bladder cancer: association with p53 alterations, tumor angiogenesis, and tumor progression. J Natl Cancer Inst 89:219–227

Gupta SK, Hassel T, Singh JP (1995) A potent inhibitor of endothelial cell proliferation is generated by proteolytic cleavage of the chemokine platelet factor 4. Proc Natl Acad Sci 92:7799–7803

Hanneken A, Ying W, Ling N, Baird A (1994) Identification of soluble forms of the fibroblast growth factor receptor in blood. Proc Natl Acad Sci USA 91:9170–9174

Homandberg GA, Williams JE, Grant DBS, Eisenstein R (1985) Heparin-binding fragments of fibronectin are potent inhibitors of endothelial cell growth. J Am Pathol 120:327–332

Holmgren L, O'Reilly MS, Folkman J (1995) Dormancy of micrometastases: balanced proliferation and apoptosis in the presence of angiogenesis suppression. Nat Med 1:149–153

Jackson D, Volpert OV, Bouck N, Linzer DIH (1994) Stimulation and inhibition of angiogenesis by placental proliferin and proliferin related protein. Science 266:1581–1584

Kim KJ, Li B, Winer J, Armanini M, Gillett N, Phillips HS, Ferrara N (1993) Inhibition of vascular endothelial growth factor-induced angiogenesis suppresses tumour growth in vivo. Nature 362:841–844

Knox JB, Sukhova GK, Whittemore AD, Libby P (1997). Evidence for altered balance between matrix metalloproteinases and their inhibitors in human aortic diseases. Circulation 95:205–212

Lange P, Hekmat K, Bosl G, Kennedy BJ, Fraley EE (1980) Accelerated growth of testicular cancer after cytoreductive surgery. Cancer 45:1498–1506

Luster AD, Leder P (1993) IP-10, a -C-X-C- chemokine, elicits a potent thymus-dependent antitumor response in vivo. J Exp Med 178:1057–1065

Maione TE, Gray GS, Hunt AJ, Donner AL, Sharpe RJ (1991) Inhibition of tumor growth in mice by an analogue of platelet factor 4 that lacks affinity for heparin and retains potent angiostatic activity. Cancer Res 51:2077–2083

Millauer B, Shawver LK, Plate KH, Risau W, Ullrich A (1994) Glioblastoma growth inhibited in vivo by a dominant-negative Flk-1 mutant. Nature 367:576–579

Moses MA, Sudhalter J, Langer R (1990) Identification of an inhibitor of neovascularization from cartilage. Science 248:1408–1410

Muragaki Y, Timmons S, Griffith CM, Oh SP, Fadel B, Quertermous T, Olsen BR (1995) Mouse col18a1 is expressed in a tissue-specific manner as three alternative variants and is localized in basement membrane zones. Proc Natl Acad Sci USA 92:8763–8767

Nelson J, Allen WE, Scott WN, Bailie JR, Walker B, McFerran NV (1995) Murine epidermal growth factor (EGF) fragment (33–42) inhibits both EGF and laminin-dependent endothelial cell motility and angiogenesis. Cancer Res 55:3772–3776

Nguyen M, Watanabe H, Budson AE, Richie JP, Hayes DF, Folkman J (1994) Elevated levels of an angiogenic peptide, basic fibroblast growth factor, in the urine of patients with a wide spectrum of cancers. J Natl Cancer Inst 86:356–361

O'Reilly MS, Holmgren L, Shing Y, Chen C, Rosenthal RA, Moses M, Lane WS, Cao Y, Sage EH, Folkman J (1994) Angiostatin: a novel angiogenesis inhibitor that mediates the suppression of metastases by a Lewis lung carcinoma. Cell 79:315–328

O'Reilly MS, Holmgren L, Chen C, Folkman J (1996) Angiostatin induces and sustains dormancy of human primary tumors in mice. Nat Med 2:689–692

O'Reilly MS, Boehm T, Shing Y, Fuhai N, Vasios G, Lane WS, Flynn E, Birkhead JR, Olsen B, Folkman J (1997) Endostatin: an endogenous inhibitor of angiogenesis and tumor growth. Cell 88:1–20

Oh SK, Kamagata Y, Muragaki Y, Timmons S, Ooshima A, Olsen BR (1994) Isolation and sequencing of cDNAs for proteins with multiple domains of Gly-Xaa repeats identify a distinct family of collagenous proteins. Proc Natl Acad Sci USA 91:4229–4233

Pepper MS, Ferrara N, Orci L, Montesano R (1992) Potent synergism between vascular endothelial growth factor and basic fibroblast growth factor in the induction of angiogenesis in vitro. Biochem Biophys Res Commun 189:824–831

Prehn RT (1991) The inhibition of tumor growth by tumor mass. Cancer Res 51:2–4

Prehn RT (1993) Two competing influences that may explain concomitant tumor resistance. Cancer Res 53:3266–3269

Rehn M, Pihlajaniemi T (1994) a 1 (XVIII), a collagen chain with frequent interruptions in the collagenous sequence, a distinct tissue distribution, and homology with type XV collagen. Proc Natl Acad Sci USA 91:4234–4238

Rehn M, Pihlajaniemi T (1995) Identification of three N-terminal ends of type XVIII collagen chains and tissue-specific differences in the expression of the corresponding transcripts. J Biol Chem 270:4705–4711

Ribatti D, Vacca A, Iurlaro M, Ria R, Roncali L, Dammacco F (1996) Human recombinant interferon alpha-2a inhibits angiogenesis of chick area vasculosa in shell-less culture. Int J Microcir Clin Exp 16:165–169

Risau W (1995) Differentiation of endothelium. FASEB J 9:926–933

Risau W, Flamme I (1995) Vasculogenesis. Annu Rev Cell Dev Biol 11:73–91

Sage EH, Bassuk JA, Vost JC, Folkman MJ, Lane TF (1995) Inhibition of endothelial cell proliferation by SPARC is mediated through a Ca (2+)-binding EF-hand sequence. J Cell Biochem 57:127–140

Sakamato N, Iwahana M, Tanaka NG, Osaka Y (1991) Inhibition of angiogenesis and tumor growth by a synthetic laminin peptide, CDPGYIGSR-H2. Cancer Res 51:903–906

Sim BKL, Oreilly MS, Liang H, Fortier AH, He WX, Madsen JW, Lapcevich R, Nacy CA (1997) A recombinant human angiostatin protein inhibits experimental primary and metastatic cancer. Cancer Res 57:1329–1334

Southam CM, Brunschwig A (1961) Quantitative studies of autotransplantation of human cancer. Cancer 14:971–978

Strieter RM, Polverini PJ, Arenberg DA, Kunkel SL (1995) The role of CXC chemokines as regulators of angiogenesis. Shock 4:155–160

Tolsma SS, Volpert OV, Good DJ, Frazier WA, Polverini PJ, Bouck N (1993) Peptides derived from two separate domains of the matrix protein thrombospondin-1 have antiangiogenic activity. J Cell Biol 122:497–511

Voest EE, Kenyon BM, O'Reilly MS, Truitt G, D'Amato RJ, Folkman J (1995) Inhibition of angiogenesis in vivo by interleukin 12. Natl Cancer Inst 87:581–586

Warren BA, Chauvin WJ, Philips J (1977) Blood-Bone tumor emboli and their adherence to vessel walls. Prog Cancer Res Ther:185–197

Woltering EA, Barrie R, O'Dorisio TM, Arce D, Ure T, Cramer A, Holms D, Robertson J, Fassler J (1991) Somatostain analogues inhibit angiogenesis in the chick chorioallantoic membrane. J Surg Res 50:245–251

Woodruff M (1980) The interactions of cancer and host. Grune and Stratton, New York

Yuhas JM, Pazmino NH (1974) Inhibition of subcutaneously growing line 1 carcinomas due to metastatic spead. Cancer Res 34:2005–2010

Inhibitors of Preadipocyte Replication: Opportunities for the Treatment of Obesity

James L. Kirkland[1] and Charles H. Hollenberg[2]

1 Introduction

The prevalence of obesity is increasing rapidly in Western countries, and associations between obesity and atherosclerosis, diabetes, and other serious diseases are becoming increasingly apparent (Rosenbaum et al. 1997). Frustratingly little progress has been made in developing effective treatments for obesity. Appetite-suppressant drugs, dietary, and behavioral approaches so far result in either loss of only a small percent of fat mass or in transient weight loss with rebound weight gain, often within a matter of months. Indeed, around 97% of obese subjects who lose weight through dieting regain the lost weight within 1 year. The potential side effects of centrally acting appetite suppressants are a concern. Furthermore, although the amount of fat tissue can increase or decrease more rapidly and to a greater extent than most other tissues, appetite suppression, dietary restriction, or surgical methods for reducing caloric intake can result in loss not only of fat mass, but also of muscle and other components of lean body mass. Hence, approaches specifically targeted to fat tissue may prove to be an attractive strategy for treating obesity and its complications.

Development of obesity results in part from fat cell enlargement due to increased stores of triglyceride in each cell, but also from increases in fat cell number (Hirsch 1971; Salans et al. 1993). During fat tissue gain, fat cell volume increases initially and is then followed by appearance of new fat cells (Björntorp 1991). Obesity caused by prolonged high fat feeding and massive obesity are particularly likely to be associated with hyperplasia as well as fat cell hypertrophy. Hyperplastic obesity can occur even in adult life and is not, as was once thought, restricted to the maturational phase of the life span. Fat cell hyperplasia may even contribute to the difficulty in achieving sustained reversal of obesity through dieting since adipose tissue is an active storage site

[1] Boston Medical Center, 88 East Newton Street, Room F435, Boston, MA 02118, USA
[2] President, Cancer Care Ontario, 620 University Avenue, 16th Floor, Toronto, Ontario, M5G 2L7, Canada

Progress in Molecular and Subcellular Biology, Vol. 20
A. Macieira-Coelho (Ed.)

which regulates both food intake and caloric expenditure by releasing leptin and through other mechanisms.

Over the past few years, advances in the understanding of fat tissue cell dynamics have indicated that there may be opportunities for preventing and treating obesity and its complications by altering fat cell number and size. Among other approaches, development of specific inhibitors of the replication of preadipocytes, the precursor cells in adipose tissue which can differentiate into fat cells, may lead to interventions useful in obesity.

2 Preadipocytes

The existence of precursor cells in fat tissue capable of replicating or of differentiating into fat cells was initially suspected from isotopic labeling studies (Hollenberg and Vost 1968; Frolich et al. 1972; Klyde and Hirsch 1979). In tritiated thymidine pulse-chase studies in rats, thymidine uptake occurred first in the denser stromal-vascular fraction of fat tissue collagenase digests and later in the lighter fat cell fraction. This implied the presence of cells which contain little lipid and which are capable of replicating in the stromal-vascular fraction, and that these cells subsequently differentiated into less dense fat cells. These studies also indicated that fat cell precursors are present in fat depots even of adult animals. Studies demonstrating increases in fat cell number in adult experimental animals fed high fat diets and fat tissue regeneration in adult animals following partial lipectomy further supported this conclusion (Lemmonier 1972; Faust et al. 1978). The culture of cells from fat tissue explants and the stromal-vascular fraction of fat tissue digests which are capable of replicating and differentiating into fat cells confirmed the existence of preadipocytes (Fig. 1; Ng et al. 1971; Smith 1972; Poznanski et al. 1973). It turns out that around a third of cells in fat depots are fat cells, a third are preadipocytes, and the remainder are vascular and other cell types (Kirkland et al. 1994; Kirkland and Dobson 1997). In their undifferentiated state, cultured preadipocytes resemble fibroblasts. After addition of agents which induce differentiation, these cells assume a rounded shape associated with cytoskeletal changes and loss of mobility. They then accumulate lipid as multiple small vesicles which later coalesce into a single, large lipid droplet that displaces the nucleus laterally (Figs 1, 2).

Agents that initiate or promote differentiation of cultured preadipocytes vary considerably in effectiveness among species, but share in common effects on several signaling pathways: (1) various tyrosine kinase pathways (e.g., IGF-1), (2) the adenylyl cyclase/phosphodiesterase system, (3) nuclear receptor pathways (steroid/thyroid/peroxisome proliferator activator/retinoid receptors), and (4) protein kinase C systems (Wiederer and Löffler 1987; Hauner et al. 1989; Xu and Björntorp 1990; Cornelius et al. 1994; Safonova et al. 1994;

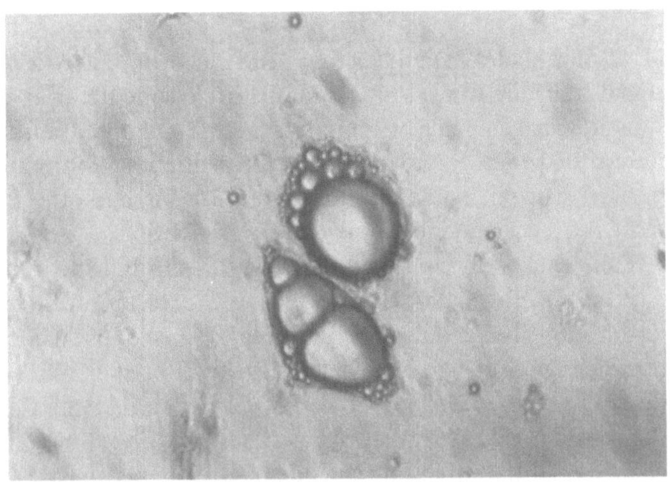

Fig. 1. Differentiated preadipocytes. Preadipocytes were isolated from the perirenal depot of a 3-month-old male Fischer 344 rat, cloned by plate dilution, and exposed to a differentiation-inducing enriched medium. (Kirkland et al. 1990)

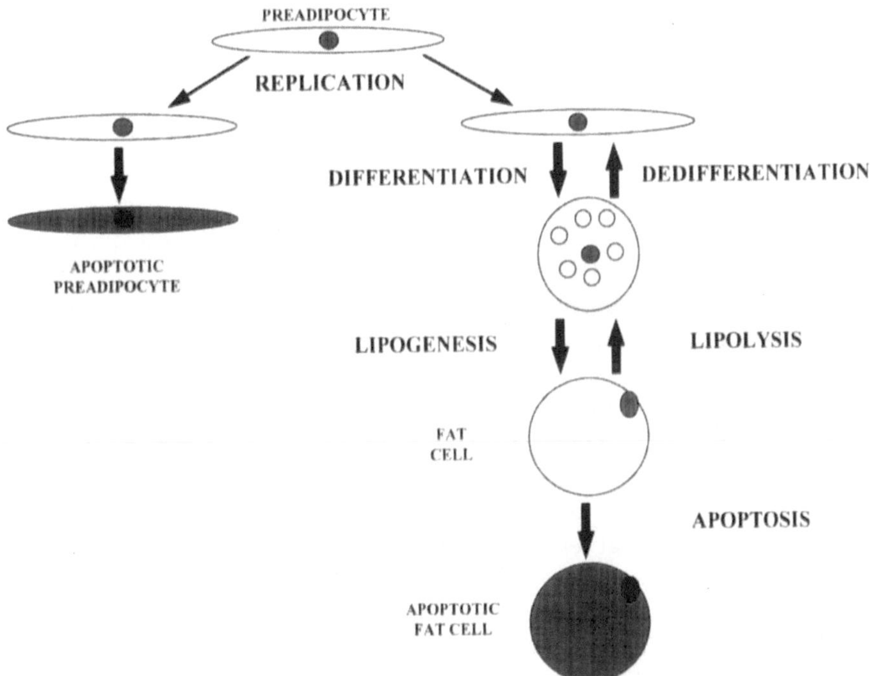

Fig. 2. Preadipocyte replication and differentiation. Undifferentiated preadipocytes can replicate, differentiate into fat cells, or undergo apoptosis. During differentiation, multiple lipid droplets (*white circles*) accumulate and coalesce into a single droplet which laterally displaces the nucleus (*dark circle*). As preadipocytes differentiate, their replicative capacity declines. Under appropriate conditions, fat cells undergo apoptosis (*dark cell*) or lipolysis with loss of lipid and dedifferentiation to reacquire the properties of preadipocytes, including the capacity to replicate. (Kirkland and Dobson 1997 by copyright permission of the Journal of the American Geriatrics Society)

MacDougald and Lane 1995; Smas and Sul 1995). There is redundancy in these signal transduction systems: modulators of one pathway can, in some cases, substitute for agents which modulate other separate transduction pathways to induce or promote preadipocyte differentiation (Smas and Sul 1995). Once initiated, insulin, substrates for lipid synthesis, and other conditions can promote and accelerate preadipocyte differentiation.

Following induction of differentiation through these signal transduction pathways, coordinated changes in the expression of over 600 genes occurs, leading to the acquisition and maintenance of the fat cell phenotype (Spiegelman and Farmer 1982; Spiegelman et al. 1983; Bernlohr et al. 1985; Cook et al. 1985; Dobson et al. 1987; Dani et al. 1989; Cornelius et al. 1994; Campfield et al. 1995; MacDougald and Lane 1995; Smas and Sul 1995). These changes in differentiation-dependent gene expression are orchestrated by several transcription factors including CCAAT-enhancer binding proteins (C/EBPα, β, and δ), peroxisome proliferator activator receptor γ (PPARγ), and others (reviewed in MacDougald and Lane 1995; Smas and Sul 1995; Kirkland and Dobson 1997). Overexpression of some of these transcription factors, including C/EBPα and PPARγ, is sufficient to induce the differentiation of preadipocytes (Lin and Lane 1994; Hu et al. 1995; Wu et al. 1995; Yeh et al. 1995).

Genes controlling production of cytoskeletal elements, including β-actin, α-tubulin, and vimentin, are among the earliest genes whose expression changes during preadipocyte differentiation (Spiegelman and Farmer 1982; Spiegelman and Ginty 1983; Bernlohr et al. 1985; Cook et al. 1985; Teichert-Kuliszewska et al. 1996). These changes are associated with the acquisition of a rounded shape by preadipocytes. Next, changes occur in expression of lipoprotein lipase and collagen isoforms, and subsequently of the adipocyte fatty acyl binding protein and glycerol-3-phosphate dehydrogenase (a key lipogenic enzyme) (Dani et al. 1989; MacDougald and Lane 1995). These changes in lipogenic enzyme activities contribute to lipid accumulation. Fat cell size ultimately depends on the balance between rates of lipogenesis and lipolysis. Late in the differentiation process, increased expression of adipsin (complement factor D) and tumor necrosis factor α (TNFα) occur.

Preadipocytes and fat cells are not only active in lipid metabolism, they also produce or process several hormones, paracrine factors, and other substances, some of which influence fat depot growth. Among these are leptin (a centrally acting hormone which induces satiety), sex steroids (which are activated by fat cell aromatase), angiotensinogen (which may regulate blood flow to adipose tissue), lipoprotein lipase, type I plasminogen activator inhibitor (which may contribute to the enhanced risk of thrombosis in obesity), apolipoprotein E, IGF-1 binding proteins, adipsin, protein C3, factor B, a stimulator of preadipocyte differentiation produced by adipocytes, monobutyrin (an angiogenic factor), bFGF, 32- and 66-kDa proteins with bFGF-like domains,

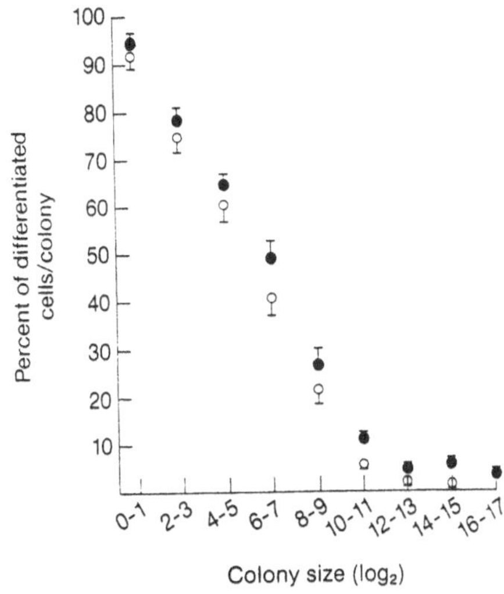

Fig. 3. Relation between preadipocyte capacities for replication and differentiation. Perirenal (●) and epididymal (○) preadipocytes were cloned from 3-month-old Fischer 344 rats and exposed for 2 weeks to a differentiation-inducing medium which also permits preadipocyte replication. Percent of differentiated cells within each resulting colony is shown. Capacity of cells for replication (log_2 colony number to reflect number of population doublings in 2 weeks) was inversely related to capacity for differentiation. At all colony sizes, perirenal clones contained a higher percent of differentiated cells than epididymal clones ($p < 0.0001$ by ANOVA). Means ±1 SEM of ten experiments are shown. (Kirkland et al. 1990 by copyright permission of the American Physiological Society)

TNFα, interleukin-1-β-converting enzyme, and a low molecular weight inhibitor of preadipocyte replication (Dobson et al. 1987; Gaskins et al. 1990; Frederich et al. 1992; Kirkland et al. 1993b; Zhang et al. 1994; Lau et al. 1996; Shimomura et al. 1996). Some of these substances appear to be autocrine regulators of preadipocyte replication, including bFGF (the production of which is greatest by preadipocytes and declines during differentiation), the 32- and 66-kDa proteins with bFGF-like domains (which are produced more extensively by preadipocytes from obese than lean subjects), TNFα, and the low molecular weight inhibitor.

As preadipocytes differentiate, they lose the capacity to replicate (Ailhaud et al. 1989; Entenmann and Hauner 1996). Preadipocytes with the greatest replicative potential have the lowest capacity for differentiation and are presumably at an earlier stage of adipocyte development than cells with lower replicative potential (Fig. 3). A round of replication may be necessary for the preadipocytes at the earliest stage of development to acquire the ability to differentiate (Entenmann and Hauner 1996; Yu et al. 1997). Unlike preadipocyte cell lines, both rat and human partially differentiated, lipid-containing preadipocytes can still undergo limited replication (Prins and O'Rahilly 1997). Under appropriate conditions (at least in vitro), fat cells can dedifferentiate with loss of lipid stores, reversion to the morphologic appearance of preadipocytes, and reacquisition of the ability to replicate (Sugihara et al. 1986; Cheng et al. 1994). Also, fat cells and possibly preadipocytes can undergo apoptosis (Chawla and Lazar 1994; Prins et al. 1994a,b; Prins and O'Rahilly 1997). Hence, preadipocyte number depends on the formation of new preadipocytes through preadipocyte replication or possibly fat cell dedifferentiation, and preadipocyte removal through differentiation into fat cells or cell death. Fat cell number depends on new fat cell formation through preadipocyte differentiation and fat cell loss through cell death or possibly dedifferentiation into preadipocytes. Preadipocyte and fat cell turnover continue throughout the life span, as demonstrated in tritiated thymidine pulse-chase studies (Hollenberg and Voss 1968; Frolich et al. 1972; Klyde and Hirsch 1979), but the rate of fat cell turnover in adult humans or experimental animals is not known with certainty, nor are the relative rates of the various cell dynamic processes (replication, interconversion of preadipocyte subtypes, differentiation, dedifferentiation, and cell death). Isotope labeling studies have indicated that the turnover of fat cells is in excess of 4 months in adult rats (Klyde and Hirsch 1979). However, certain conditions, such as severe food restriction or uncontrolled diabetes mellitus, can result in a rapid decline in fat cell number (Miller et al. 1983; Geloen et al. 1989).

3 Relationship Between Host Characteristics and Adipose Cell Dynamics

Changes in preadipocyte number, replicative capacity, and capacity for differentiation appear to be linked to fat depot size. Clonal composition with respect to replicative capacity can be determined by culturing individual preadipocytes and counting the number of cells in resultant colonies after 2 or 3 weeks. Even when derived from the same fat depot of the same individual, the replicative capacities of individual preadipocytes vary considerably (Djian et al. 1983, 1985; Wang et al. 1989; Kirkland et al. 1990). Indeed, in humans and rats there appear to be at least two preadipocyte subtypes, the one type of preadipocyte probably being at an earlier stage along the pathway to becoming a fat cell than the other, which has a lower proliferative capacity and a higher capacity for differentiation (Fig. 4; Kirkland et al. 1993a; Yu et al. 1997). Nevertheless, when the frequency of clones of varying replicative capacities (clonal composition) is examined, populations of preadipocytes exhibit reproducible characteristics. For example, the clonal composition of the right and left epididymal fat depots from the same individual is identical bilaterally (Table 1). Replicative potential varies among different fat depots as do many other characteristics of preadipocytes from different depots (Björntorp et al. 1980; Djian et al. 1983; Wang et al. 1989; Kirkland et al. 1990, 1996). In rats, the

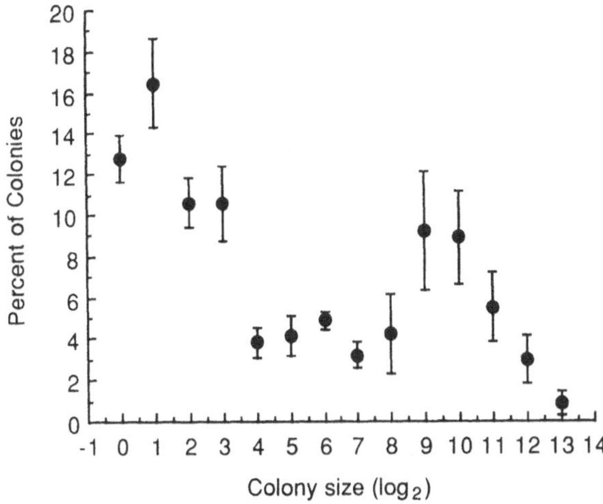

Fig. 4. Frequency distribution of colony sizes of human preadipocytes. Human omental preadipocytes were cloned by plate dilution in eight separate experiments. The mean (±1 SEM) percent of colonies at each population doubling level is presented as a function of the number of population doublings achieved after 6 weeks in culture in a medium which does not induce differentiation. There is considerable variation among individual preadipocytes from the same depot in capacities for replication. Furthermore, there appear to be two preadipocyte subtypes with respect to replicative potential. (Kirkland et al. 1993a by copyright permission from Obesity Research)

Table 1. Colony size distributions of preadipocytes from right and left epididymal and perirenal depots

Colony size ($\log_2$ cell number)	Percent distribution of colony size			
	Perirenal		Epididymal	
	Right	Left	Right	Left
0–9	40	41	60	64
10–12	36	40	31	29
13–16	24	20	9	7

Fat depots were removed from a single 143-g rat and colonies, each originating from a single preadipocyte, were cultured for 3 weeks, when cell numbers were determined in each colony. Despite considerable variation in replicative potential among animals, the percentages of preadipocytes which underwent 0–9, 10–12, and 13–16 divisions were similar bilaterally. (Wang et al. 1989 by copyright permission of the American Society for Clinical Investigation)

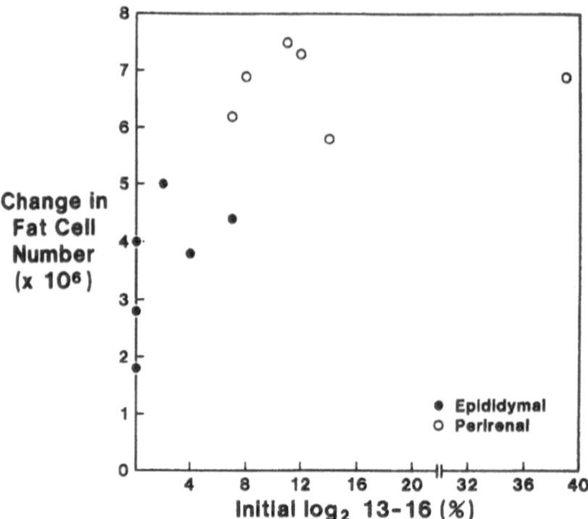

Fig. 5. Relation between preadipocyte replicative capacity and increase in fat cell number with growth. Right or left perirenal or epididymal fat depots were removed from individual 160–170-g Sprague-Dawley rats at 0 time and the rats were allowed to recover. After 70 days, the contralateral pad was removed from the same animals. Preadipocytes and fat cells were isolated from the fat pads collected at day 0, and fat cells were isolated from the opposite fat pads collected at day 70. Preadipocyte replicative potential was analyzed in fat pads collected at day 0 by cloning preadipocytes, culturing the cloned cells for 3 weeks, and expressing the size of resulting colonies as $\log_2$ to give the number of population doublings which occurred during the 3 weeks in culture. The percent of colonies arising from cells with the greatest replicative potential (which had achieved 13 or more population doublings within 3 weeks) was determined. Fat cell number was determined at day 0 and in the opposite pad at day 70. The increase in fat cell number over the 70 days was related to the percent of preadipocytes capable of extensive replication observed at day 0. *Each point* represents a single rat. (Wang et al. 1989 by copyright permission of the American Society for Clinical Investigation)

replicative potential of preadipocytes from the perirenal depot is greater than that of epididymal preadipocytes (Fig. 3). While there are interanimal differences in clonal composition, variation among animals is always in the same direction when epididymal and perirenal depots are compared (Wang et al. 1989).

Clonal composition with respect to replicative potential and mature fat cell number are related. Both fat cell number and preadipocyte clonal composition are identical bilaterally in paired fat depots. Fat cell number increases more extensively during early development in perirenal tissue in which preadipocytes have a high replicative potential than in epididymal tissue in which preadipocyte replicative potential is lower (Wang et al. 1989). Furthermore, there is a relationship between the percentage of preadipocytes with high replicative potential in the fat depots of an animal and subsequent increases in fat cell number during growth of the same animal (Fig. 5). Uncontrolled diabetes mellitus results in rapid loss of fat mass, fat cells, and preadipocytes, and we have observed reduced replicative capacity of preadipocytes cultured from diabetic rats (J.L. Kirkland, C.H. Hollenberg,

S. Kindler, W.S. Gillon, unpubl. observ.). Hypophysectomy results in a rapid decrease in fat depot size and the replicative potential of preadipocytes from hypophysectomized rats is lower than that of sham-operated animals (Kirkland et al. 1992). Obesity in some human subjects is associated with increases in both preadipocyte replicative potential and fat cell number (Roncari et al. 1981). Interestingly, dedifferentiated fat cells from some massively obese humans also have a greater capacity for replication than cells from lean subjects (Roncari et al. 1986). When rats are fed a high saturated fat diet, perirenal fat cell number increases and preadipocytes isolated from their perirenal depots exhibit a higher replication rate than cells from control animals (Shillabeer and Lau 1994). Unlike replication, preadipocyte differentiation is not affected by this dietary manipulation (Wang et al. 1989). Hence, there are links among preadipocyte replicative potential, fat cell number, and depot size, at least under some conditions.

Capacity of preadipocytes to undergo differentiation is also linked to fat depot size. Preadipocyte capacity for differentiation, fat cell size, and fat depot size all decline during senescence (Kirkland et al. 1990). Preadipocyte replicative potential also declines with age. However, both preadipocyte and fat cell number increase during senescence, and the decrease in fat depot size is caused by a decrease in fat cell size (Kirkland et al. 1994). The increased preadipocyte and fat cell numbers could result from changes in capacities for apoptosis and dedifferentiation with age (Kirkland and Dobson 1997). Hence, the relation between preadipocyte replicative potential and fat depot size is affected by several aspects of fat cell dynamics in addition to replicative potential, including capacities for differentiation and possibly dedifferentiation and apoptosis. Development of approaches to prevent or treat the hyperplastic component of obesity using inhibitors of preadipocyte replication will need to take account of this. However, the observations that preadipocytes from some massively obese human subjects have increased replicative potential, as well as the other evidence linking replicative potential to fat depot mass reviewed above, support the exploration of methods for specifically interfering with preadipocyte replication to treat obesity. Furthermore, induction of differentiation in the early-stage preadipocyte subtype (with high replicative potential and low capacity for differentiation) appears to require a round of replication (Yu et al. 1997). Hence, inhibiting preadipocyte replication could reduce fat depot enlargement and development of obesity by preventing differentiation as well as by blunting increases in preadipocyte number.

4 Control of Preadipocyte Replication

The control of preadipocyte replication has received less attention than that of preadipocyte differentiation, and even less is known about fat cell dedifferentiation and fat cell and preadipocyte apoptosis. Most of the knowledge about preadipocyte replication concerns agents which stimulate replication. Much less is known about inhibitors and very little is known about potential differences between the underlying mechanisms that regulate preadipocyte replication and those that regulate replication of other cell types.

Preadipocyte replication can be affected by influences external to preadipocytes, including hormones, paracrine factors, blood and nutrient supply, and anatomic constraints, as well as influences intrinsic to preadipocytes, including autocrine factors and prior replicative histories of individual cells in fat depots. The intrinsic capacity of preadipocytes for replication varies among fat depots, and this may contribute to differing patterns of fat distribution among individuals. Furthermore, fat cells in different depots vary in hormone sensitivity (Kissebah and Krakower 1994). For example, castration enhances epididymal preadipocyte replicative capacity and reduces capacity for differentiation of epididymal but not perirenal preadipocytes (Lacasa et al. 1997). These effects appear to be related to an increase in mitogen-activated protein kinase activity in preadipocytes following castration and are partially reversed by testosterone administration to the animals. Conceivably, then, agents that inhibit preadipocyte replication could be developed that would alter fat tissue regional distribution.

Certain influences which promote preadipocyte differentiation also cause decreased replicative capacity (Fig. 3). Some of the mechanisms responsible for the inverse relationship between differentiation and replication involve C/EBPα, a key transcription factor in the control of preadipocyte differentiation. Overexpression of C/EBPα in nonadipose cell types inhibits replication (Timchenko et al. 1996). This antiproliferative effect of C/EBPα appears to be mediated by p21. Overexpression of C/EBPα increases p21 levels but does not affect levels of CDK2, CDK4, proliferating cell nuclear antigen, p53, c-fos, p16, or p27. Antisense p21 prevents C/EBPα-induced proliferative arrest. C/EBPα overexpression increases p21 levels partly by enhancing p21 transcription and partly by stabilizing p21 protein. Whether these effects of C/EBPα on p21 expression in nonadipose cell types also occur in preadipocytes is not yet clear, nor is it clear whether p21 affects C/EBPα expression. C/EBPα induces expression of growth-arrest-associated genes (gadd) in preadipocytes (Constance et al. 1996). Additionally, retinoblastoma protein, which modulates replication by negatively regulating E2F-1, physically interacts with C/EBPα, β, and γ in murine lung bud fibroblasts induced to differentiate into adipocyte-like cells (Chen et al. 1996). This interaction with retinoblastoma protein promotes binding of C/EBPs to cognate DNA sequences in vitro and enhances the transactivation of a C/EBP-responsive promoter in cells. Hence, there are close links between cell-cycle machinery and processes responsible for preadipocyte differentiation, and C/EBPα figures prominently in these links.

Hormones, paracrine factors, and other agents which affect rodent, human, and aneuploid cell-line preadipocyte replication directly or indirectly through

Table 2. Hormones, paracrine factors, and other agents which affect replication of human, rodent, or aneuploid cell line preadipocytes

Agent	Effect on replication	Effect on differentiation
bFGF	↑ or –	↑ or ↓
32-kDa protein with bFGF-like domain	↑	
66-kDa protein with bFGF-like domain	↑	
Platelet-derived growth factor	↑	
Epidermal growth factor	↑	↑ or ↓
Estrogens (17β-estradiol)	↑	
Endothelial cell conditioned medium	↑	
Fat derived factor	↑	
Pituitary peptides	↑	
Phorbol esters	↑	↑ or ↓
Retinoids (high concentration)	↑	↓
α_2-Adrenergic agonists	↑	
Insulin*	↑ or ↓	↑
IGF-1*	↑ or ↓	↑
Glucocorticoids*	↓	↑
Growth hormone	↑ or ↓	↑ or ↓
63 kD serum protein*	↓	↑
PPARγ ligands* (thiazolidinediones, prostaglandin J, indomethacin)	↓	↑
Pancreatic polypeptide	↓	
Tumour necrosis factor α	↓	↓

Agents that induce or promote differentiation with associated decline in replication are indicated (*). References: Dixon-Shanies et al. (1975); Roncari and Van (1978); Roncari (1981); Kuri-Harcuch and Marsch-Moreno (1983); Torti et al. (1985); Lau et al. (1987); Serrero (1987); Cooper and Roncari (1989); Navre and Ringold (1989); Torti et al. (1989); Lau et al. (1990); Teichert-Kuliszewska et al. (1992); Kirkland et al. (1993b); Bouloumié et al. (1994); Chawla and Lazar (1994); Safonova et al. (1994); Teichert-Kuliszewska et al. (1994); Forman et al. (1995); Kliewer et al. (1995); Lehman et al. (1995); MacDougald and Lane (1995); Tang et al. (1995); Wright and Hausman (1995); Lau et al. (1996); Wabitsch et al. (1996).

effects on differentiation are listed in Table 2. It should be noted that effects of these agents vary considerably among species and between euploid primary culture cells and aneuploid cell lines (Hauner et al. 1995). Insulin and IGF-1 can stimulate preadipocyte replication, but can also result in reduced proliferation as a result of promoting differentiation (Dixon-Shanies et al. 1975; Wright and Hausman 1995). Insulin may also have a direct or indirect effect on apoptosis (see above). Undifferentiated preadipocytes have few insulin receptors but have IGF-1 receptors (Smith et al. 1988). Therefore, the high concentrations of insulin which affect undifferentiated preadipocytes may act by cross-reacting with IGF-1 receptors. α_2-Adrenergic agonists stimulate proliferation of rat preadipocytes through the mitogen-activated protein kinase pathway (Bouloumié et al. 1994). Retinoids enhance preadipocyte replication at high concentrations, either stimulate differentiation at low concentrations or inhibit differentiation at high concentrations, and induce apoptosis in preadipocyte-like cell lines (Chawla and Lazar 1994; Safonova et al. 1994). 17β-Estradiol stimulates replication of cultured human preadipocytes by inducing production of mitogens (Roncari and Van 1978; Cooper and Roncari 1989). The relevance of these in vitro findings has been demonstrated in vivo: ovariectomy blunts the peripubertal increase in preadipocyte numbers in some depots in female rats (Krakower et al. 1988). Epidermal, platelet-derived, and basic fibroblast growth factors (bFGF) stimulate preadipocyte replication (Roncari 1981; Navre and Ringold 1989; Lau et al. 1990, 1996; MacDougald and Lane 1995). Human preadipocytes (particularly from obese subjects) produce factors which stimulate rat preadipocyte replication and which are 32- and 66-kDa proteins with domains homologous to bFGF (Lau et al. 1987; Teichert-Kuliszewska et al. 1992). Endothelial cells also produce preadipocyte mitogens related to bFGF (Lau et al. 1996). Rat fat cells produce a 20-kDa protein which is mitogenic in preadipocyte cell lines (Aoki et al. 1990). High fat diets enhance production of this protein (Aoki et al. 1993). More of this protein is produced by perirenal than epididymal fat cells, which is interesting, since cultured perirenal preadipocytes have greater replicative potential than epididymal cells (Djian et al. 1983; Wang et al. 1989; Kirkland et al. 1990). Rat preadipocytes contain bFGF mRNA. We have found that levels of bFGF mRNA decrease during differentiation and that this decrease is blunted with increasing donor age, indicating the existence of an autocrine control mechanism (J.L. Kirkland, C.H. Hollenberg, S. Kindler, W.S. Gillon, unpubl. observations).

Several agents inhibit preadipocyte replication. A few are known which do so without promoting differentiation (e.g., pancreatic polypeptide, TNFα, and a factor which prevents stimulation of proliferation by endothelial cell-conditioned medium), while many of the known agents which can result in reduced proliferation (including insulin, IGF-1, growth hormone, glucocorticoids, and a 63-kDa serum protein) do so as a result of the replicative decline which accompanies differentiation. Of course, nutrient availability, space, pH, temperature, and other in vivo or in vitro conditions can also affect preadipocyte replication.

The first group of inhibitors, those that decrease preadipocyte replicative capacity by inducing or promoting differentiation, have variable effects on fat depot size. This is, in part, a result of opposite effects on replication and differentiation, and also because of direct effects of these agents on fat metabolism. For example, growth hormone can either stimulate preadipocyte differentiation through promoting IGF-1 production, or inhibit lipogenic enzyme expression and enhance lipolysis through direct effects on fat cells (Hauner 1994; Wabitsch et al. 1996). Growth hormone can also either enhance or inhibit replication. Effects of these differentiation inducer-replication inhibitors can vary among depots. For example, glucocorticoids can cause subcutaneous fat loss and visceral fat enlargement, perhaps because of variation in fat cell glucocorticoid receptor numbers among depots (Prins and O'Rahilly 1997).

TNFα belongs to the second group of preadipocyte replication inhibitors which do not promote differentiation. TNFα, which is produced by preadipocytes as well as other cell types, can cause decreased fat mass (Torti et al. 1985). However, TNFα also inhibits differentiation, causes dedifferentiation of fat cells, may cause apoptosis, results in insulin resistance, and induces lipolysis (Roncari 1981; Torti et al. 1989; Hotamisligil and Spiegelman 1994; Hauner et al. 1995; Prins and O'Rahilly 1997). It is not clear whether the decrease in fat mass caused by TNFα results principally from its inhibition of replication or from its many other effects on fat cells and preadipocytes. Enhanced fat tissue TNFα expression in obesity could be a mechanism to prevent further fat cell hyperplasia (Hotamisligil and Spiegelman 1994). Studies are not available of effects on fat mass of the other inhibitors which act without enhancing differentiation (pancreatic polypeptide and an antagonist of effects of endothelial cell conditioned medium), but pancreatic polypeptide does act centrally to cause satiety (Berntson et al. 1993).

A low molecular weight, specific inhibitor of preadipocyte replication has been found in conditioned medium (CM) prepared from preadipocytes cultured from the epididymal or perirenal fat depots of old (24 months) Fischer 344 rats. Antiproliferative activity of the CM was determined by diluting the CM with fresh culture medium, applying this for 48h to 3-month-old rat preadipocytes (test cells) in their exponential growth phase, and then by adding ^{3}H thymidine for another 24h. Comparisons were made using CM prepared from preadipocytes cultured from young (3 months) rats (young CM), which had little inhibitory activity, and CM prepared from preadipocytes of old (24 months) rats (old CM). The availability of young CM with little inhibitory activity as a negative control facilitates study of the inhibitor from old animals and has allowed the avoidance of many of the problems associated with work on replication inhibitors.

As little as 0.5% old CM resulted in a significant reduction of ^{3}H thymidine incorporation by test cells compared to effects of young CM and 40% old CM caused a 75% inhibition of replication. Increasing CM concentrations resulted in a log-linear decrease in test cell DNA synthesis. Increasing age of animals from which CM was prepared was associated with a significant increase in the potency of CM. This finding was replicated in coculture experiments using transwell inserts. Preadipocyte differentiation resulted in a decrease in CM antiproliferative potency. The factor(s) in CM responsible for inhibiting DNA synthesis is a small molecule (<3500D) that is pronase-sensitive and heat-stable. Its effect is not blocked by serum, it is not cytotoxic, and its effects reversed after removal from test cells. It did not affect the replication of a number of other cell types. This inhibitory activity may contribute to the declining replicative capacity of preadipocytes with increasing donor age

(Kirkland et al. 1990). Potential relationships between this factor and an age-related fibroblast inhibitory peptide need to be elucidated (Macieira-Coelho and Soderberg 1993; Macieira-Coelho 1996). Further work is underway to determine the identity of the preadipocyte inhibitory activity.

5 Potential Therapeutic Strategies

Approaches for treating obesity directed at fat tissue itself are much more likely to provide means for targeting specific depots than are approaches directed at controlling appetite or nutrient absorption. The development of specific inhibitors of preadipocyte replication could potentially be very important in preventing and treating obesity, hyperlipidemias, diabetes, and other disorders. Such inhibitors would need to be specific to adipose tissue, free of enhancing effects on differentiation or other cell dynamic processes which could increase fat mass, and must meet the many other criteria necessary for therapeutic agents (including solubility, stability, lack of side effects, and ease of administration). Agents which block preadipocyte mitogens specifically without effects on other cell types would also be useful. Agents targeted to specific fat depots would be particularly valuable. While no such agents are on the horizon, the framework for beginning to develop such agents now exists. There has been an exponential increase in understanding about the basic of biology of preadipocytes, albeit mainly resulting from studies of preadipocyte differentiation rather than replication. It is now possible to prepare primary cultures of rodent and human preadipocytes which are virtually free of other potentially contaminating cell types (Wang et al. 1989; Kirkland et al. 1990). Serum-free approaches for culturing these cells have been developed and are being improved (Broad and Ham 1983; Serrero and Mills 1987; Hauner et al. 1989; Wabitsch et al. 1996). Development of replication inhibitors would be best conducted using human preadipocytes in primary culture rather than the available aneuploid rodent cell lines, since the latter have unlimited replicative potential and other characteristics distinct from those of preadipocytes in fat depots or in primary culture. Another advantage of using euploid preadipocytes is that effects of inhibitors on preadipocytes from different fat depots can be determined, facilitating the development of agents with site-specific effects. Finally, several hormones, paracrine factors, and their signaling pathways are known to affect the replication of preadipocytes. This provides clues for the selection of candidate compounds for further investigation.

A great deal of work needs to be done to verify if this approach will be feasible. Although there are strong indications that reduced preadipocyte replicative capacity is associated with decreased fat depot growth, effects of altering preadipocyte replicative potential directly on subsequent development of obesity need to be elucidated fully. Effects on blood lipid profiles of reducing the availability of adipocytes in which to store ingested calories need to be determined. More needs to be understood about mechanisms controlling preadipocyte replication, particularly the hormones, paracrine factors, and signaling pathways involved, and much more needs to be understood about the mechanisms of action and effects of currently available inhibitors of preadipocyte replication. Using recently developed, improved human preadipocyte culture techniques, high throughput screens for preadipocyte replication inhibitors need to be developed.

There are tremendous potential benefits from inhibitors of preadipocyte replication, most of the necessary technology to develop them is available, and the time has come to pursue vigorously this promising approach for tackling a major public health problem.

Acknowledgments. This work was supported by the Adipocyte Core of the Boston Obesity/Nutrition Research Center (NIH DK46200), grant NIH AG/DK13925, and the Evans Foundation of Boston University. The authors are grateful for the assistance of Ms. J. Armstrong and Ms. A. Varkas.

References

Ailhaud G, Dani C, Amri E (1989) Coupling growth arrest and adipocyte differentiation. Environ Health Perspect 80:17–23

Aoki N, Kawada T, Umeyama T, Sugimoto E (1990) Protein factor obtained from rat adipose tissue specifically permits the proliferation of the 3T3-L1 and Ob1771 cell lines. Biochem Biophys Res Commun 171:905–912

Aoki N, Kawada T, Sugimoto E (1993) Level of preadipocyte growth factor in rat adipose tissue which specifically permits the proliferation of preadipocytes is affected by restricted energy intake. Obesity Res 1:126–131

Bernlohr DA, Bolanowski MA, Kelly TJ, Lane MD (1985) Evidence for an increase in transcription of specific mRNA's during differentiation of 3T3-L1 preadipocytes. J Biol Chem 260:5563–5567

Berntson GG, Zipf WB, O'Dorisio TM, Hoffman JA, Chance RE (1993) Pancreatic polypeptide infusions reduce food intake in Prader-Willi syndrome. Peptides 14:497–503

Björntorp P (1991) Adipose tissue distribution and function. Int J Obesity 15:67–81

Björntorp P, Karlsson M, Petterson P, Sypniewska G (1980) Differentiation and function of rat adipocyte precursor cells in primary culture. J Lipid Res 21:714–723

Bouloumié A, Planat V, Devedjian JC, Valet P, Saulnier-Blache JS, Record M, Lafontan M (1994) α_2-Adrenergic stimulation promotes preadipocyte proliferation. J Biol Chem 269:30254–30259

Broad TE, Ham RG (1983) Growth and adipose differentiation of sheep preadipocyte fibroblasts in serum-free medium. Eur J Biochem 135:33–39

Campfield LA, Smith FJ, Guise Y, Devos R, Burn P (1995) Recombinant mouse OB protein: evidence for a peripheral signal linking adiposity and central neural networks. Science 269:546–549

Chawla A, Lazar MA (1994) Peroxisome proliferator and retinoid signalling pathways coregulate preadipocyte phenotype and survival. Proc Natl Acad Sci USA 91:1786–1790

Chen PL, Riley DJ, Chen Y, Lee WH (1996) Retinoblastoma protein positively regulates terminal differentiation through direct interaction with C/EBPs. Genes Dev 10:2794–2804

Cheng A, Deitel M, Roncari D (1994) The biochemistry and molecular biology of human adipocyte reversion. Int J Obesity 18(Suppl 2):112

Constance CM, Morgan IV JI, Umek RM (1996) C/EBPα regulation of the growth-arrest-associated gene *gadd*45. Mol Cell Biol 16:3878–3883

Cook KS, Hunt CR, Spiegelman BM (1985) Developmentally regulated mRNAs in 3T3-adipocytes: analysis of transcriptional control. J Cell Biol 100:514–520

Cooper SC, Roncari DAK (1989) 17-beta-estradiol increases mitogenic activity of medium from cultured preadipocytes of massively obese persons. J Clin Invest 83:1925–1929

Cornelius P, MacDougald OA, Lane MD (1994) Regulation of adipocyte development. Annu Rev Nutr 14:99–129

Dani C, Bertrand B, Bardon S, Doglio A, Amri E, Grimaldi P (1989) Regulation of gene expression by insulin in adipose cells: opposite effects on adipsin and glycerophosphate dehydrogenase genes. Mol Cell Endocrinol 63:199–208

Dixon-Shanies D, Rudick J, Knittle JL (1975) Observations on the growth and metabolic functions of cultured cells derived from human adipose tissue. Proc Soc Exp Med Biol 149:541–545

Djian P, Roncari DAK, Hollenberg CH (1983) Influence of anatomic site and age on the replication and differentiation of rat adipocyte precursors in culture. J Clin Invest 72:1200–1208

Djian P, Roncari DAK, Hollenberg CH (1985) Adipocyte precursor clones vary in capacity for differentiation. Metabolism 34:880–883

Dobson DE, Groves DL, Spiegelman BM (1987) Nucleotide sequence and hormonal regulation of glycerophosphate dehydrogenase mRNA during adipocyte and muscle differentiation. J Biol Chem 262:1804–1809

Entenmann G, Hauner H (1996) Relationship between replication and differentiation in cultured human adipocyte precursor cells. Am J Physiol 270:C1011–C1016

Faust IM, Johnson PR, Stren JS, Hirsch J (1978) Diet-induced adipocyte number increase in adult rats: a new model of obesity. Am J Physiol 235:E279–E289

Forman B, Tontonoz P, Chen J, Brun R, Spiegelman BM, Evans RM (1995) 15-deoxy-delta12,14-prostaglandin J2 is a ligand for the adipocyte determination factor PPAR γ. Cell 83:803–812

Frederich RC, Kahn BB, Peach MJ, Flier JS (1992) Tissue-specific nutritional regulation of angiotensinogen in adipose tissue. Hypertension 19:339–344

Frolich J, Vost A, Hollenberg CH (1972) Organ culture of rat white adipose tissue. Biochim Biophys Acta 280:579–587

Gaskins HR, Kim JW, Wright JT, Rund LA, Hausman GJ (1990) Regulation of insulin-like growth factor-1 ribonucleic acid expression, polypeptide secretion, binding protein activity by growth hormone in porcine preadipocyte cultures. Endocrinology 126:622–630

Geloen A, Roy PE, Bukowiecki LJ (1989) Regression of white adipose tissue in diabetic rats. Am J Physiol 257:E547–E553

Hauner H (1994) Prevention of adipose tissue growth. Int J Obesity 20(Suppl 2):147

Hauner H, Entenmann G, Wabitsch M, Gaillard D, Ailhaud G (1989) Promoting effect of glucocorticoids on the differentiation of human adipocyte precursor cells cultured in a chemically defined medium. J Clin Invest 84:1663–1670

Hauner H, Röhrig K, Petruschke T (1995) Effects of epidermal growth factor (EGF), platelet-derived growth factor (PDGF) and fibroblast growth factor (FGF) on human adipocyte development and function. Eur J Clin Invest 25:90–96

Hirsch J (1971) Adipose cellularity in relation to human obesity. In: Stollerman GH (ed) Advances in internal medicine, 17. Year Book Medical Publishers, Chicago, pp 289–300

Hollenberg CH, Vost A (1968) Regulation of DNA synthesis in fat cells and stromal elements from rat adipose tissue. J Clin Invest 47:2485–2498

Hotamisligil GS, Spiegelman BM (1994) Tumor necrosis factor alpha: a key component of the obesity-diabetes link. Diabetes 43:1271–1278

Hu E, Tontonoz P, Spiegelman BM (1995) Transdifferentiation of myoblasts by the adipogenic transcription factors PPARγ and c/EBPα. Proc Natl Acad Sci USA 92:9856–9860

Kirkland JL, Dobson DE (1997) Preadipocyte function and aging: links between age-related changes in cell dynamics and altered fat cell function. J Am Geriatr Soc 45:959–967

Kirkland JL, Hollenberg CH, Gillon WS (1990) Age, anatomic site, and the replication and differentiation of adipocyte precursors. Am J Physiol 258:C206–C210

Kirkland JL, Hollenberg CH, Gillon W, Kindler S (1992) Effect of hypophysectomy on rat preadipocyte replication and differentiation. Endocrinology 131:2769–2773

Kirkland JL, Hollenberg CH, Gillon WS (1993a) Two preadipocyte subtypes cloned from human omental fat. Obesity Res 1:87–91

Kirkland JL, Hollenberg CH, Kindler S (1993b) Ageing results in increased elaboration of autocrine inhibitors of replication by rat preadipocytes. Clin Invest Med 16:B56

Kirkland JL, Hollenberg CH, Kindler S, Gillon WS (1994) Effects of age and anatomic site on preadipocyte number in rat fat depots. J Gerontol 49:B31–B35

Kirkland JL, Hollenberg CH, Gillon WS (1996) Effects of fat depot site on differentiation-dependent gene expression in rat preadipocytes. Int J Obesity 20:S102–S107

Kissebah AH, Krakower GR (1994) Regional adiposity and morbidity. Physiol Rev 74:761–811

Kliewer SA, Lenhard JM, Wilson TM, Patel I, Morris DC, Lehman JM (1995) A prostaglandin J2 metabolite binds peroxisome proliferator-activated receptor γ and promotes adipocyte differentiation. Cell 83:813–819

Klyde BJ, Hirsch J (1979) Isotopic labelling of DNA in rat adipose tissue: evidence for proliferating cells associated with mature adipocytes. J Lipid Res 20:691–704

Krakower GR, James RG, Arnaud C, Etienne J, Keller RH, Kissebah AH (1988) Regional adipocyte precursors in the female rat: influence of ovarian factors. J Clin Invest 81:641–648

Kuri-Harcuch W, Marsch-Moreno M (1983) DNA synthesis and cell division related to adipose differentiation of 3T3 cells. J Cell Physiol 114:39–44

Lacasa D, Garcia E, Henriot D, Agli B, Giudicelli Y (1997) Site-related specificities of the control by androgenic status of adipogenesis and mitogen-activated protein kinase cascade/c-*fos* signalling pathways in rat preadipocytes. Endocrinology 138:3181–3186

Lau DCW, Roncari DAK, Hollenberg CH (1987) Release of mitogenic factors by cultured preadipocytes from massively obese human subjects. J Clin Invest 79:632–636

Lau DCW, Shillabeer G, Wong KL, Tough SC, Russell JC (1990) Influence of paracrine factors on preadipocyte replication and differentiation. Int J Obesity 14:193–201

Lau DCW, Shillabeer G, Li ZH, Wong KL, Varzaneh FE, Tough SC (1996) Paracrine interactions in adipose tissue development and growth. Int J Obesity 20(Suppl 3):S16–S25

Lehman JM, Moore LB, Smith-Oliver TA, Wilkison WO, Wilson TM, Kliewer SA (1995) An antidiabetic thiazolidinedione is a high-affinity ligand for peroxisome proliferator-activated receptor γ (PPAR γ). J Biol Chem 270:12953–12956

Lemmonier D (1972) Effect of age, sex, and site on the cellularity of adipose tissue in mice and rats rendered obese by a high fat diet. J Clin Invest 51:2907–2915

Lin FT, Lane MD (1994) CCAAT/enhancer binding protein α is sufficient to initiate the 3T3-L1 adipocyte differentiation program. Proc Natl Acad Sci USA 91:8758–8761

MacDougald OA, Lane MD (1995) Transcriptional regulation of gene expression during adipocyte differentiation. Annu Rev Biochem 64:345–373

Macieira-Coelho A (1996) A growth inhibitor implicated in the growth arrest of human fibroblasts. FEBS Lett 378:61–63

Macieira-Coelho A, Soderberg A (1993) Growth inhibitory activity in extracts from human fibroblasts. J Cell Physiol 154:92–100

Miller WHJ, Faust IM, Goldberger AC, Hirsch J (1983) Effects of severe long-term food deprivation and refeeding on adipose tissue cells in the rat. Am J Physiol 245:E74–E80

Navre M, Ringold GM (1989) Differential effects of fibroblast growth factor and tumor promoters on the initiation and maintainance of adipocyte differentiation. J Cell Biol 109:1857–1863

Ng CW, Poznanski WJ, Borowieki M, Reimer G (1971) Differences in growth in vitro of adipose cells from normal and obese patients. Nature 231:445

Poznanski WJ, Waheed I, Van RLR (1973) Human fat cell precursors. Morphologic and metabolic differentiation in culture. Lab Invest 29:570–576

Prins JB, O'Rahilly S (1997) Regulation of adipose cell number in man. Clin Sci 92:3–11

Prins JB, Walker NI, Winterford CM (1994a) Human adipocyte apoptosis occurs in malignancy. Biochem Biophys Res Commun 205:625–630

Prins JB, Walker NI, Winterford CM, Cameron DP (1994b) Apoptosis of human adipocytes in vitro. Biochem Biophys Res Commun 201:500–507

Roncari DAK (1981) Hormonal influences on the replication and maturation of adipocyte precursors. Int J Obesity 5:547–552

Roncari DAK, Van RLR (1978) Promotion of human adipocyte replication by 17β-estradiol in culture. J Clin Invest 62:503–508

Roncari DAK, Lau DCW, Kindler S (1981) Exaggerated replication in culture of adipocyte precursors from massively obese persons. Metabolism 30:425–427

Roncari DAK, Kindler S, Hollenberg CH (1986) Excessive proliferation in culture of reverted adipocytes from massively obese subjects. Metabolism 35:1–4

Rosenbaum M, Leibel RL, Hirsch J (1997) Obesity. N Engl J Med 337:396–407

Safonova I, Darimont C, Amri EZ, Grimaldi P, Ailhaud G, Reichert U, Shroot B (1994) Retinoids are positive effectors of adipose cell differentiation. Mol Cell Endocrinol 104:201–211

Salans L, Cushman SW, Weissman RE (1993) Studies of human adipose tissue: adipose cell size in nonobese and obese patients. J Clin Invest 92:1543–1547

Serrero G (1987) EGF inhibits the differentiation of adipocyte precursors in primary culture. Biochem Biophys Res Commun 146:194–202

Serrero G, Mills D (1987) Differentiation of newborn rat adipocyte precursors in defined serum-free medium. In Vitro 23:63–66

Shillabeer G, Lau DCW (1994) Regulation of new fat cell formation in rats: the role of dietary fats. J Lipid Res 35:592–600

Shimomura I, Funahashi T, Takahashi M, Maeda K, Kotani K, Nakamura T, Yamashita S, Miura M, Fukuda Y, Takemura K, Tokunaga K, Matsuzawa Y (1996) Enhanced expression of *PAI-1* in visceral fat: possible contributor to vascular disease in obesity. Nat Med 2:800–803

Smas CM, Sul HS (1995) Control of adipocyte differentiation. Biochem J 309:697–710

Smith PJ, Wise LS, Berkowitz R, Wan C, Rubin CS (1988) Insulin-like growth factor-1 is an essential regulator of the differentiation of 3T3-L1 adipocytes. J Biol Chem 263:9402–9408

Smith U (1972) Studies of human adipose cells in culture. Anat Rec 172:597–602

Spiegelman BM, Farmer SR (1982) Decreases in tubulin and actin gene expression prior to morphological differentiation of 3T3 adipocytes. Cell 29:53–60

Spiegelman BM, Ginty CA (1983) Fibronectin modulation of cell shape and lipogenic gene expression in 3T3-adipocytes. Genes Dev 35:657–666

Spiegelman BM, Frank M, Green H (1983) Molecular cloning of mRNA from 3T3 adipocytes. J Biol Chem 258:10083–10089

Sugihara H, Yonemitsu N, Miyabara S, Yum K (1986) Primary cultures of unilocular fat cells: characteristics of growth in vitro and changes in differentiation properties. Differentiation 31:42–49

Tang B, Jeoung DI, Sonenberg M (1995) Effect of human growth hormone and insulin on [^{3}H]thymidine incorporation, cell cycle progression, and cyclin D expression in 3T3-F442A preadipose cells. Endocrinology 136:3062–3069

Teichert-Kuliszewska K, Hamilton BS, Deitel M, Roncari DAK (1992) Augmented production of heparin-binding mitogenic proteins by preadipocytes from massively obese persons. J Clin Invest 90:1226–1231

Teichert-Kuliszewska K, Hamilton BS, Deitel M, Roncari DAK (1994) Decreasing expression of a gene encoding a protein related to basic fibroblast growth factor during differentiation of human preadipocytes. Biochem Cell Biol 72:54–57

Teichert-Kuliszewska K, Hamilton BS, Roncari DAK, Kirkland JL, Gillon WS, Deitel M, Hollenberg CH (1996) Increasing vimentin expression associated with differentiation of human and rat preadipocytes. Int J Obesity 20(Suppl 2):S108–S113

Timchenko NA, Wilde M, Nakanishi M, Smith JR, Darlington GJ (1996) CCAAT/enhancer-binding protein α (C/EBPα) inhibits cell proliferation through the p21 (WAF-1/CIP-1/SDI-1) protein. Genes Devel 10:804–815

Torti FM, Dieckmann B, Beutler B, Cerami A, Ringold GM (1985) A macrophage factor inhibits adipocyte gene expression: an in vitro model of cachexia. Science 229:867–869

Torti FM, Torti SV, Larrick JW, Ringold GM (1989) Modulation of adipocyte differentiation by tumor necrosis factor and transforming growth factor beta. J Cell Biol 108:1105–1113

Wabitsch M, Heinze E, Hauner H, Shymko RM, Teller WM, De Meyts P, Ilondo MM (1996) Biological effects of human growth hormone in rat adipocyte precursor cells and newly differentiated adipocytes in primary culture. Metabolism 45:34–42

Wang H, Kirkland JL, Hollenberg CH (1989) Varying capacities for replication of rat adipocyte precursor clones and adipose tissue growth. J Clin Invest 83:1741–1746

Wiederer O, Löffler G (1987) Hormonal regulation of the differentiation of rat adipocyte precursor cells in primary culture. J Lipid Res 28:649–658

Wright JT, Hausman GJ (1995) Insulinlike growth factor-1 (IGF-1)-induced stimulation of porcine preadipocyte replication. In Vitro 31:404–408

Wu Z, Xie Y, Bucher NLR, Farmer SR (1995) Conditional ectopic expression of C/EBPβ in NIH-3T3 cells induces PPARγ and stimulates adipogenesis. Genes Dev 9:2350–2363

Xu X, Björntorp P (1990) Effects of dexamethasone on multiplication and differentiation of rat adipose precursor cells. Exp Cell Res 189:247–252

Yeh WC, Cao M, Classon M, McKnight SL (1995) Cascade regulation of terminal adipocyte differentiation by three members of the C/EBP family of leucine zipper proteins. Genes Dev 9:168–181

Yu ZK, Wright JT, Hausman GJ (1997) Preadipocyte recruitment in stromal vascular cultures after depletion of committed preadipocytes by immunotoxicity. Obesity Res 5:9–15

Zhang Y, Proenca R, Maffei M, Barone M, Leopold L, Friedman J (1994) Positional cloning of the mouse obese gene and its human homologue. Nature 372:425–432

Growth Inhibitors for Mammary Epithelial Cells

Ralf Brandt[1] and Andreas D. Ebert[2]

1
Introduction

In general, we know more about what stimulates the growth of organs than about what regulates their growth. This is particularly true for the mammary gland, in which the systemic mitogenes, the mammogenic hormones, which drive growth have been examined by generations of endocrinologists. However, we do not know which factors or hormones define the final size of a mammary gland even in continued presence of powerful mitogenes. Nor do we know which factors define the typical pattern of ducts and alveoli during the complex process of morphogenesis of the mammary gland during pregnancy and lactation. The mammary gland is a special object for examining processes such as growth and differentiation for several reasons. (1) Breast cancer is one of the leading illnesses during the lifetime of a woman and therefore it is of interest to know more about growth regulation of the mammary epithelia. (2) Branching morphogenesis of the mammary gland is reminiscent of the developmental pattern in other structures such as the lung, kidney, exocrine organs, and various organs of the reproductive tract. (3) Most mammary growth occurs after birth and can be easily studied in mouse, which is large enough to work on conveniently and in which the glands are in an easily accessible location. (4) Finally, cell culture systems either for monolayer growing immortalized, transformed or primary mammary epithelial cells (MEC) and whole-mount culture systems for the mammary gland are well developed.

In contrast to the steadily increasing number of growth factors capable of stimulating cells to synthesize DNA and to divide, very little is known about their potential antagonists, which are supposed to contribute balanced growth and development by inhibiting cellular proliferation. This becomes clear when we take into account that all factors so far proposed as pure growth inhibitors may activate other functions. Stopping cell growth is not essentially the start of

[1] Novartis Pharma Inc., Oncology N K-681.5.42, 4002 Basel, Switzerland
[2] Freie Universität Berlin, Universitätsklinikum Benjamin Franklin, Frauenklinik und Poliklinik, Hindenburgdamm 30, 12200 Berlin, Germany

Progress in Molecular and Subcellular Biology, Vol. 20
A. Macieira-Coelho (Ed.)

cell death. It is known especially for the mammary gland and other organs with cycling development of the epithelium that functional differentiation or programmed cell death terminates cell growth (proliferation). Cells usually fulfill their function, e.g., milk protein synthesis and milk secretion, for a certain time and then die by programmed cell death (Atwood et al. 1995). It is therefore not surprising that some of the classic growth inhibitors for the mammary epithelium have been more or less redefined as differentiation factors. They act as growth inhibitors depending on the stage of mammary gland development and/or the expression level of their antagonists (growth factors), such as EGF, TGF-alpha, NDF, IGF-1, and others. On the other hand, it has been shown that the mammary-derived growth inhibitor (MDGI) is a member of the fatty acid-binding protein gene family, tempting the question of the role of fatty acids in controlling growth of the mammary epithelium. Furthermore, for TGF-β it has been shown that, depending on the developmental stage of the mammary gland, it can not only inhibit growth, but also modulate differentiation of the mammary epithelium. Proof of the existence of pure growth inhibitors acting in an autocrine or paracrine pathway in normal mammary gland physiology would help to explain how this periodically regenerating epithelium proceeds through different cycles of ductal growth and lobuloalveolar differentiation.

Loss of corresponding signals may also be related to the mechanism of escape of transformed cells from normal control. What is the evidence for the participation of growth inhibitors in the normal development of the mammary gland? In fact, ductal growth, morphogenesis, and functional differentiation can be driven in vitro by a combination of steroid hormones and prolactin (Imagawa et al. 1990). Neither stimulatory nor inhibitory growth factors need to be added to achieve glandular development. On the other hand, ablation of endocrine glands causes the inhibition of mammary epithelial cell growth in vivo. These findings lead to the question of whether locally acting factors, the synthesis of which is under hormonal control, could be involved. It is in our interest to search for these peptides, in particular, for growth inhibitors. The present chapter describes in detail the hitherto best-investigated growth inhibitors for mammary epithelial cells, the family of beta-type transforming growth factors (TGF βs), and the mammary-derived growth inhibitor (MDGI), a 14.5-kDa polypeptide that we believe is involved in the local control of proliferation and differentiation of the mammary gland. Another important growth inhibitor for the mammary epithelium is the tumor necrosis factor (TNF), which is expressed rather in the stroma than in the mammary epithelium, acting most likely indirectly on mammary epithelial cells. Mammastain, another growth inhibitor, is not discussed in this chapter due to the restricted number of publications (Ervin et al. 1989). Since TGF-β is discussed in a later chapter of this Volume, we omit discussing its general biology and TGF-β receptors unless necessary for understanding the topic.

2
The Role of Transforming Growth Factor-β (TGF-β) in Mammary Gland Development

Members of the transforming growth factor-β superfamily have emerged as key regulators of many aspects of cellular physiology, with potent effects on cell growth and differentiation. In particular, TGF-βs are strong growth inhibitors for many epithelial cell types, including the mammary epithelial cell, while being growth-stimulatory for many cells of mesenchymal origin (reviewed in Moses et al. 1990; Sporn and Roberts 1990a,b). In vitro experiments suggest that TGF-βs are highly multifunctional and are probably key components in cellular signaling paths regulating and integrating cell growth and differentiation in the mammary gland. As is emerging for other growth factors, the effects of TGF-β depend critically on the nature of the mammary epithelial cell, on its current state of differentiation, on environmental influences, such as the composition of the extracellular matrix, and the spectrum of other growth factors acting on the cell at the same time in its microenvironment (Sporn and Roberts 1988; Atwood et al. 1995). Growth inhibition by TGF-β is generally reversible, unless coupled to terminal differentiation (Masui et al. 1986). However, TGF-β may be associated with or may drive apoptotic cell death in some systems (Martikainen et al. 1990; Kyprianou et al. 1991). The change of state induced in the mammary epithelial cell by TGF-β is an integrated function of all these parameters and is related to its multifunctionality. Thus results obtained in vitro should be extrapolated with extreme caution to the complex in vivo situation, and preferably should be confirmed by suitable in vivo experiments. Whole-mount organ culture (Binas et al. 1992; Brandt et al. 1997), implanted pellets (Daniel and Robinson 1992) releasing TGF-β alone or in combination with other growth factors, or transgenic mice overexpressing TGF-β_1 in the mammary gland (Pierce et al. 1993; Robinson et al. 1993a,b) are more suitable to study TGF-βs regulatory role for mammary gland development than trivial monolayer culture of mammary epithelial cells, in which they usually cannot undergo differentiation. In this chapter we do not discuss the main biology of TGF-β in detail or its structural features and receptors unless it is essential to understand its function for mammary epithelial cell growth and differentiation. To obtain information about these topics refer to the chapter by Herrera (The Growth Inhibitory Effect of TGF-β) later in this Volume.

2.1
TGF-β Is a Multifunctional Factor of Normal Mammary Gland Development

TGF-β develops different functionality for the development of the normal mammary gland, depending mainly on the physiological state of the gland, as well as on the expression of hormones and growth factors. TGF β's biological activity may be different for mammary epthelial cells in combination with other growth factors also expressed from the mammary gland than its application alone, which underlines the multifunctionality of TGF-β (Sporn and Roberts 1985). It has been observed in the mouse that even during periods of peak growth and development of the mammary gland >95% of the glandular epithelium is quiescent, despite the presence of high levels of mitogens, and the fact that the quiescent epithelium retains the capacity to proliferate and form new ducts when transplanted onto cleared mammary fat pads (Kyprianou et al. 1991). These observations implicate an endogenous growth inhibitor which must normally prevent epithelial proliferation except in the region of the developing endbuds (Kyprianou et al. 1991). In this way, ductal hypertrophy is prevented and clear spaces are maintained between mature ducts for the subsequent development of secretory alveoli. The potent inhibitory effects of TGF-β in combination with other growth inhibitors such as MDGI, antiestrogens or progestins make this protein family a plausible candidate for endogenous inhibitors in the mammary gland.

2.1.1 TGF-β Can Act Exogenously on Mammary Development in Vivo

TGF-β_1 implanted in developing mouse mammary glands in slow-release plastic pellets had a potent inhibitory effect on the growth and morphogenesis of the mammary ductal tree (Silberstein and Daniel 1987b). Endbuds regressed, and epithelial DNA synthesis and ductal development were inhibited. The effect of TGF-β_1 was reversible and no cytotoxicity or dysplasia were observed over extended periods of treatment. The morphology of TGF-β_1-inhibited ducts was similar to normal quiescent ducts in untreated glands, indicating that TGF-β_1 had many of the properties predicted for the endogenous natural growth regulator. Similar data have been obtained on pregnant or hormone-treated mice and TGF-β_1-overexpressing transgenic mice, showing that the growth-inhibitory effect was highly cell-specific (Daniel et al. 1989; Jhappan et al. 1993). There was no inhibition of development of the lobular-alveolar secretory structures in pregnant or hormone-primed animals. Inhibition was only observed in the rapidly proliferating epithelium of the mammary endbuds, and not in the surrounding stromal tissues, nor was there inhibition of the maintenance of DNA synthesis in the lumenal epithelium of the ducts. Further analysis indicated that exogenous TGF-β stimulated the stromal cells surrounding affected endbuds to deposit glycosaminoglycans and collagen, suggesting an important role for endogenous TGF-β in the organization of periductal stroma (Silberstein et al. 1990). Matrix deposition by the stromal cells was dependent on the presence of endbud epithelium, since no stimulation of periductal matrix synthesis was observed, nor was there increased synthesis in stromal cells lying near the TGF-β implant or in cells adjacent to the endbud. This indicates the highly cell-type-specific nature of the effects of TGF-β in vivo and the importance of interactions between distinct cell types in determining the response. However, testing the effect of TGF-β in vitro revealed another biological role of TGF-β which might play a role in mammary gland development (Miettinen et al. 1994). Interestingly, TGF-βs are highly expressed during embryogenesis in tissues where branching morphogenesis is occurring, such as lung and salivary gland (Millan et al. 1991). Localization and dynamics of extracellular matrix deposition in the postpartum development of the mammary ductal system and in the development of these embryonic structures are similar (Silberstein et al. 1990). Thus, TGF-β may play critical regulatory roles in organizing this type of development in a number of organ systems. It has been demonstrated that TGF-β, while transdifferentiating mammary epithelia cells, modulates the expression and intracellular distribution of desmoplakins, E-cadherin, vinculin, and ZO-1. The resulting pattern is similar to that found in fibroblasts (Miettinen et al. 1994). Further studies have shown that the differentiation process is kinase-dependent and transduced by the Tsk7L receptor (Miettinen et al. 1994).

2.1.2 Evidence for Endogenous TGF-β in the Mammary Gland

The most comprehensive analyses of TGF-β expression in the mammary gland have been performed in the mouse (Robinson et al. 1991). Northern blot analysis indicated expression of TGF-β_1 mRNA at high levels during all developmental stages except lactation. In situ studies in 5-week-old animals localized this mRNA to the epithelium of ducts and endbuds, and to adjacent stromal cells. Complementary immunohistochemical studies showed intense staining for TGF-β_1 in the extracellular matrix surrounding growth-inhibited ducts, but not around ductal endbuds or small lateral branches. It is thought that the extracellular localization of TGF-β represents a kind of store for release on demand for stopping cell proliferation in certain areas of the mammary gland. Interestingly, a nonhomogenous staining pattern has been shown to be consistent with the local loss of extracellular TGF-β, being important in allowing endbud development and lateral branching (Silberstein et al. 1992). In situ studies suggested that the extracellular TGF-β protein could be of either stromal or epithelial origin. Interestingly, in midpregnancy animals, the fibrous matrix around the mature ducts stained strongly for TGF-β_1, whereas the matrix surrounding the secretory epithelium of alveoli stained lightly or not at all, suggesting that the loss of extracellular TGF-β may also be necessary to allow secretory differentiation to occur. By contrast, TGF-β was shown to be strongly represented in the cytoplasm of rapidly dividing epithelial cells in the endbuds in 5-week-old animals, and there was scattered expression in nondividing ductal cells. It has been proposed that this is important in maintaining the stem cells in a pluripotent, undifferentiated state (Silberstein et al. 1992). Human studies have been limited to the nonpregnant, nonlactating adult breast, but broadly similar expression patterns for TGF-β were seen. Immunohistochemical studies have demonstrated the presence of all three mammalian TGF-β subtypes in the normal breast (McCune et al. 1992). TGF-β_{1-3} were detected intracellularly in the epithelium of lobules and ducts. In human, no TGF-β protein could be demonstrated in stromal fibroblasts, myoepithelium, and fat (Silberstein et al. 1992). There is evidence that in the mouse the different subtypes of TGF-β appear to be differentially expressed during the various stages of mammary gland development, which could be indicative of different biological activities (Dublin et al. 1993; Atwood et al. 1995).

2.1.3
In Vitro Studies for TGF-β Activity

Normal human mammary epithelial cells (HMEC) derived from reduction mammoplasties or from morphological normal areas of mastectomy tissue respond to TGF-β by assuming a flattened elongated morphology and showing extensive growth inhibition when cultured on plastic surfaces under serum-free conditions (Hosobuchi and Stampfer 1989; Valverius et al. 1989; Takahashi et al. 1990). There appears to be some individual variation in the degree of sensitivity of the response, but generally inhibition was seen at relatively low TGF-β doses (ED_{50} 1–20 pM), and 80% of the cell population was inhibited at maximal doses (Hosobuchi and Stampfer 1989; Valverius et al. 1989). Blocking the TGF-β pathway by neutralizing anti-TGF-β antibody stimulated growth in rat mammary epithelial cells, supporting the hypothesis of an autocrine loop involving endogenous TGF-β (Ethier and Van de Velde 1990). The biological activity of TGF-β and its expression depends on the cellular environment. Human mammary epithelial cells cultured in a type I collagen gel matrix, with the matrix either adhering to the plastic surface or floating in the medium, did not respond to growth inhibition by TGF-β (Takahashi et al. 1990; Streuli et al. 1993). Controversially, TGF-β_1 is locally expressed in vivo, but in most mammary epithelial cells in vitro (Heine et al. 1987; Thompson et al. 1989; Gatherer et al. 1990). To this end, experiments have demonstrated that the level of TGF-β_1 expression is higher on plastic, but is strongly downregulated when cells are cultured on a reconstituted basement membrane matrix (ECM). Additionally, TGF-β_1 expression is not downregulated when cells were cultured on attached collagen I gels, whereas cells growing on floating collagen I gels synthesize a basement membrane inhibiting TGF-β_1 expression (Streuli et al. 1993) allowing contraction of cells and development of three-dimensional structures. The mRNA level is tenfold and the protein level is three to fivefold higher in cells cultured on plastic compared to cells cultured on ECM. These substratum-dependent expression data could not be demonstrated for TGF-β_2 (Streuli et al. 1993). This suggests that, in vivo, TGF-β_1 may play a major role in regulating ECM synthesis, cell-ECM interaction, and functional differentiation, whereas TGF-β_2 may be more important in morphogenic processes. Interestingly, compared to control, TGF-β-treated cells were prevented from forming radial extensions from floating cell mass, while having no effect on overall levels of DNA synthesis. The inhibition of formation of radial extensions in floating gels can be understood as inhibition of lateral branching in vivo. Atwood et al. (Atwood et al. 1995) overcame this problem by using a culture model close to the in vivo situation. Using a whole-mount mammary organ culture (WMO), the role of lactogenic hormones to control TGF-βs expression pattern and its linkage to modulate involution of the mammary epithelium by apoptosis has been studied. Changes in expression patterns of TGF-βs using the WMO were not distinguishable from involution in vivo (Atwood et al. 1995). TGF-β_1 expression was increased threefold during involution of the mouse mammary gland, whereas no change in TGF-β_2 expression was observed to day 9 postweaning. The expression of the TGF-β_3 gene increased sixfold by day 3 postweaning and remained at this level for additional 6 days. Both TGF-β_1 and β_2 increased in vitro after withdrawal of lactogenic hormones from the culture medium and remained to day 15 (Atwood et al. 1995). TGF-β_2 expression was reduced by 50% during in vitro involution. In parallel, modulation of DNA fragmentation by TGF-β was demonstrated, which underlined the role of TGF-βs in apoptosis (Atwood et al. 1995; Yamamoto et al. 1996). Other studies demonstrated that TGF-β_1 expression is already increased 2.5-fold during midpregnancy (Yamamoto et al. 1994). It is thought that TGF-β_1 downregulates milk protein expression during pregnancy, which was demonstrated by in vitro organ culture studies (Yamamoto et al. 1994). Additionally, a binding study using ^{125}I-TGF-β_1 as ligand has shown that pregnant mouse mammary epithelial cells possessed a single class of high affinity TGF-β_1-binding sites. These results suggest the presence of a TGF-β autocrine mechanism in pregnant mouse mammary epithelial cells (Yamamoto et al. 1994). As already outlined, TGF-β has transdifferentiation capabilities. Treatment of NmuMG normal immortalized mammary epithelial cells with TGF-β_1 induced phenotypic changes which made cells elongated and spindle-shaped, i.e., characteristics of fibroblasts. It has been shown that this reversible effect of mesenchymal transdifferentiation is pronounced at concentrations lower than 1 ng/ml. There are as yet no reports describing these phenomena in vivo or in whole-mount organ cultures. It might be possible that this transdifferentiation of an epithelial cell into a mesenchymal stromal cell triggers a not yet understood interaction of terminal differentiated mammary tissue and surrounding paracrine active stroma.

Taken together, these results in vitro suggest that mammary epithelial cells are potentially sensitive to the growth-inhibitory effects of TGF-β, but that growth inhibition is only observed in the context of a particular cellular architecture or microenvironment. To this end, it is thought that exogenous TGF-β in vivo is only growth inhibitory for the epithelial cells of the ductal endbuds and not for epithelia in other areas of the mammary gland at a different physiological stage, since these are invested in a different extracellular matrix and are in communication with different cell types.

2.2 TGF-β Is a Growth Inhibitor Modulating Differentiation of the Mammary Gland

Messenger RNA levels for TGF-$\beta_{1\text{-}3}$ decreased in the mouse mammary gland during lactation, and no TGF-β_1 protein was observed in secretory structures, suggesting that a decrease in TGF-β levels might be necessary to allow secretory differentiation to occur (Robinson et al. 1991). Indeed, direct effects of TGF-β_1 on the differentiation of isolated mammary epithelial cells have been demonstrated in vitro. TGF-β treatment of HC11 mouse mammary epithelial cells antagonized the induction of β-casein synthesis by dexamethasone and prolactin (Mieth et al. 1990). The effect was selective, since there was no change in overall protein synthesis, and it was not simply a consequence of the TGF-β-induced growth inhibition. Furthermore, TGF-β decreased synthesis of β-casein in cells that had already been induced to differentiate. This is consistent with roles for TGF-β in preventing secretory differentiation or in switching off lactation. Conversely, TGF-β upregulated the expression of milk-fat globule antigen/epithelial membrane antigen in HMECs, indicating that its effects on differentiation markers are complex and may depend on the preexisting state of differentiation or the specialized nature of the target epithelium (Hubbs et al. 1989). Although there is no experimental evidence to date, it is likely that TGF-β may also have indirect effects on mammary epithelial differentiation, through effects on the underlying stroma, possibly involving changes in matrix production. To determine the effects of TGF-β in vivo implants releasing TGF-$\beta_{1\text{-}3}$ have been introduced directly in front of endbuds into the mammary glands of subadult virgin mice (Robinson et al. 1991; Silberstein and Daniel 1987a,b). The implants locally and reversibly inhibited ductal growth, suggesting that TGF-β may help to regulate ductal penetration of the fatty stromal tissue in juvenile mice. In contrast, introduction of similar TGF-β-releasing implants into hormonally or pregnancy-induced mammary glands failed to overtly affect lobuloalveolar development (Daniel and Robinson 1992). Another approach to determine the role of TGF-β in vivo was the construction of transgenic mice expressing TGF-β_1 either under the control of MMTV-LTR or the whey acidic protein (WAP) promoter (Pierce et al. 1993, 1995). Both promoters are mammary gland-specific, whereas the MMTV-LTR shows also some promoter activity in other organs as well as in the mammary gland between pregnancies (Bouchard et al. 1989; Matsui et al. 1990; Dardick et al. 1992). In female mice carrying the TGF-β transgene driven by MMTV-LTR, a reduction of the ductal tree has been observed already at the age of 7 weeks soon after estrus begins and was most apparent at the age of 13 weeks, as ductal growth in the normal gland declines. However, during pregnancy, alveolar outgrowth developed from the hypoplastic ductal tree and lactation occurred (Pierce et al. 1993). In mice carrying the TGF-β_1 transgene driven by the WAP promoter, which is specifically activated during late pregnancy, the mammary epithelium developed mostly normally until first pregnancy. Between day 15 and 19 of pregnancy, coinciding with increased expression of the TGF-β_1 transgene, alveolar development of the transgenic mammary gland has been shown to be greatly inhibited compared to the robust growth demonstrated by the nontransgenic control glands (Jhappan et al. 1993). The overall ductal development was shown to be normal, but endbuds with numerous mitotic figures persisted in transgenic females even at 19 days of pregnancy, resulting in milk deficiency. Similar results were shown reconstituting the mammary gland by transplanting transgenic mammary epithelial cells into a cleared fat pad of pregnant, nontransgenic host animals. Interestingly, staining with an anti-TGF-β_1 antibody was mostly cytoplasmic and luminal. In addition, strong immunostaining was shown to be associated with periductal extracellular matrix of ductal structures (Jhappan et al. 1993). Accordingly, the ectopic expression of WAP-TGF-β not only impairs lobular progenitors, but also promotes an early senescence of the regenerative capacity of the mammary ductal epithelium. TGF-α transgene overexpression under the control of the MMTV-LTR promoter/enhancer induced massive development of adenocarcinomas during the first 8 months in the mammary gland (Matsui et al. 1990; Sandgren et al. 1990), which could not be demonstrated for TGF-β transgenic animals (Jhappan et al. 1993). Cross-breeding of both types of transgenic mice markedly suppressed hyperplasia and the number of mammary adenocarcinomas was reduced (Pierce et al. 1993), although widely scattered hyperplastic alveolar outgrowth remained, indicating that expression of the TGF-β_1 transgene did not completely suppress the effects of the TGF-alpha transgene but rescued totally from malignancy. The expression pattern of both the EGFR and its ligand TGF-alpha did not change in double transgenic mice (Pierce et al. 1993). Interestingly, in TGF-β_1 transgenic mice, no malignant tumors could be induced by DMBA treatment (Pierce et al. 1993). In both the MMTV-LTR and WAP-promoter-driven transgenic models, no adenocarcinomas have been developed, underlining the TGF-β's role in growth inhibition of normal mammary epithelia. Nevertheless, the morphology of the mammary gland in both models is different, which could be explored by transient activity of the WAP promoter during pregnancy and the less restricted action of the MMTV-LTR. TGF-β knockout mice are lethal as homozygotes, so that the loss of TGF-β function so far could not be studied in vivo (Dickson et al. 1995). Conditional knockout techniques will be the way to overcome this problem.

In summary, all three mammalian TGF-β isoforms are present in the normal mammary gland, and experiments both in vitro and in vivo point to roles for TGF-β in regulating growth, differentiation, and function. TGF-βs probably have locally different effects, depending on the target cells and the stage of development or maturation. However, the combined evidence suggests an important role for TGF-β_1 in the periductal fibrous stroma in inhibiting local proliferation of epithelial cells, thereby suppressing lateral branching and ductal hyperplasia, inhibiting functional differentiation of the mammary epithelium. Other, less understood roles in maintaining ductal stem cell populations and regulating secretory differentiation are also likely. Due to TGF-βs expression in both epithelial and mesenchymal cells TGF-β could act in a paracrine and/or autocrine manner in both cell systems to play a major role in regulating growth and differentiation of the mammary gland.

2.2.1 TGF-β's Role in Mammary Gland Tumorigenesis

Experimental evidence increasingly suggest important autocrine or paracrine roles for TGF-β in inhibiting the proliferation of mammary epithelial cells in the normal breast. Thus, it is reasonable to ask whether lesions in TGF-β-inhibitory paths might contribute to mammary carcinogenesis.

2.2.2 Alterations in TGF-β Production

A detailed immunohistochemical study, using a large battery of anti-TGF-β antisera, indicated little difference in TGF-β expression patterns in the epithelia of normal human breast when compared with fibrocystic change, epithelial hyperplasia, sclerosing adenosis, fibroadenoma, cystosarcoma phylloides, and several breast carcinoma variants (McCune et al. 1992). TGF-$\beta_{1\text{-}3}$ were present intracellularly in most mammary epithelia, including carcinomas (Dublin et al. 1993). The only exception among the carcinomas was the epithelia within mucin pools in mucinous carcinoma, which do not show any expression. As with the normal human breast, little or no TGF-β is observed intracellularly in stromal cells. However, particularly strong staining for TGF-$\beta_{1/3}$ has been observed in the extracellular matrix of the desmoplastic stroma, suggesting that TGF-β may play a role in the induction or maintenance of the desmoplastic response. TGF-β staining is associated with a positive epithelial component. With the exception of this strong desmoplastic staining, the TGF-β expression data generally indicated that altered patterns were not a major feature of most benign or malignant breast lesions.

In contrast, another immunohistochemical analysis of 120 benign and malignant breast tissue samples showed that 38% of invasive carcinomas (82 cases studied) had elevated levels of TGF-β expression in the epithelial component when compared with normal tissue (10 cases) or with carcinoma in situ (4 cases) (Mizukami et al. 1990). The relatively small number of normal samples and the difficulties in quantifying immunohistochemical data make this comparison rather difficult to evaluate. There appeared to be a weak positive correlation between progesterone receptor status and TGF-β expression, and TGF-β-positive tumors were associated with a better prognosis for relapse-free interval (followed over only 2 years) than tumors with no TGF-β staining, although there was no difference in the survival rate. Similarly, a survey of TGF-β_1 mRNA expression in 52 malignant breast biopsies and 15 nonmalignant ones showed that a significantly higher fraction of the malignant samples (76%) showed higher TGF-β_1 mRNA levels than the nonmalignant ones (38%) (Travers et al. 1988). Interpretation of these data is confounded somewhat by the variable cellularity of the samples, although no correlation was observed between TGF-β levels and lymphocytic infiltration. In vitro, little difference in TGF-β expression between normal and malignant cells is seen at the mRNA level with established mammary carcinoma lines (Zajchowski et al. 1988). Normal HMEC and mammary tumor cell lines showed roughly equivalent amounts of TGF-β1 mRNA, although T47D cells appear to be an exception, with very low basal expression of TGF-β_1 mRNA (Zajchowski et al. 1988; Murphy and Dotzlaw 1989; Arrick et al. 1990; Daly and Darbre 1990). No correlation was observed between TGF-β_1 mRNA levels and estrogen receptor status of the tumor line (Daly and Darbre 1990; Daly et al. 1990). Less work has been done with TGF-β_2 and TGF-β_3, but some cell lines have been shown to express the message for these isoforms as well. Thus, T47D expressed mRNA for all three subtypes, ZR-75-1 expressed TGF-β_1 and TGF-β_3, while MDA-MB-231 expressed TGF-β_1 and TGF-β_2 (Arrick et al. 1990). At the protein level, all lines tested secreted TGF-β, although estrogen receptor-negative (ER−) cell lines (MDA-MB-231, MDA-MB-330, Hs578T, and BT20) secreted more TGF-β than did ER-positive (ER+) lines (MCF-7, T47D, and ZR 75-1) (Dickson et al. 1986; Arteaga et al. 1988a). It is unlikely that there is a real correlation between estrogen receptor status and TGF-β expression (Mizukami et al. 1990; McCune et al. 1992). Immortalized, partially transformed, or carcinogen-treated or oncogene-transfected cells have not shown significant differences in TGF-β expression (Valverius et al. 1989).

2.2.3
Data Against TGF-β's as Biologically Important Autocrine Growth Inhibitors of Human Breast Carcinoma Cells

Several reports suggest that the growth inhibitory response of breast cancer cells to antiestrogens may be unrelated to autocrine effects of endogenous TGF-βs. MCF-7 and T47D cell lines can exhibit resistance to TGF-β-mediated growth inhibition despite maintaining tamoxifen sensitivity (Arteaga et al. 1988b; Karey and Sirbasku 1988). In addition, TGF-β-binding sites are low to undetectable in several tamoxifen-sensitive (ER−) breast carcinoma lines (Arteaga et al. 1988b; Murphy and Dotzlaw 1989). The type II TGF-β receptor, which is critical for growth inhibition signaling by TGF-β_1 (Wrana et al. 1992; Inagaki et al. 1993; Sun et al. 1994), is undetectable in antiestrogen-sensitive MCF-7 cells by affinity cross-linking (Inagaki et al. 1993; Arteaga et al. 1996). Finally, antibodies that neutralize all three TGF-β isoforms fail to block tamoxifen-induced retinoblastoma protein dephosphorylation, recruitment in G_1 phase of the cell cycle, and growth inhibition of MCF-7 and T47D breast cancer cells in culture (Arteaga et al. 1996). Paradoxically, the more tumorigenic hormone-independent, ER-negative breast cancer cell lines secrete higher levels of TGF-β activity than other normal and less tumorigenic breast epithelial cells (Knabbe et al. 1987). For example, adriamycin-resistant (ER−) MCF-7 cells secrete a threefold higher level of TGF-β activity into conditioned medium than parental cells. Despite being more sensitive than parental cells to TGF-β_1 in vitro, they rapidly form large tumors in ovariectomized nude mice in the absence of estrogen supplementation (Arteaga et al. 1996). Obviously, measuring pan-TGF-β activity in conditioned medium does not take into account the possible endogenous utilization of secreted TGF-β and/or the identity of the secreted isoform(s). In one study, steady-state levels of TGF-β were roughly similar in a panel of normal and tumor breast epithelial cells (Zajchowski et al. 1988), thus not matching the data discussed before. However, data examining the co-expression of all three isoforms in tumor and nontumor mammary epithelial cell lines are very limited (Arrick et al. 1990). Differences in the level of expression of TGF-β_2 and/or TGF-β_3 may well explain the differences in secreted TGF-β activity. Several published reports argue in favor of a positive association between tumor cells TGF-β overexpression and the progression of breast cancer. MCF-7 cells transfected with v-Ha-*ras* escape estrogen dependence and secreted > fivefold higher levels of TGF-β activity than wild-type cells (Dickson et al. 1987; Kasid et al. 1987). Loss of estrogen dependence in T47D breast cancer cells is associated with an acquired growth stimulatory response to exogenous TGF-β_1 and a dramatic increase in TGF-β_1 mRNA levels (Andres et al. 1989). TGF-β_1 mRNA transcripts are more abundant in highly proliferating tumor than nontumor mammary tissues (Barrett Lee et al. 1990; Coombes et al. 1990). In Western blot analysis, 78% of ER/PgR-negative but only 13% of ER/PgR-positive primary human breast tumors exhibited a high level of TGF-β_1 protein (King et al. 1989). In addition, coexpression of all three TGF-β RNA isoforms in breast tumors has been statistically associated with the presence of lymph node metastases (MacCallum et al. 1994). Exogenous TGF-β_1 can increase the membrane-degrading activity and metastatic ability of mammary tumor cells in the rat (Welch et al. 1990). Immunohistochemical staining of TGF-β_1 in breast tumor cells is greater in invasive than in in situ ductal carcinoma, and is associated with higher tumor clinical stage (Walker and Dearing 1992). Prominent TGF-β_1 staining in breast tumors is strongly associated with lymph node metastases (Walker et al. 1994) as well as with a clinical stage- and ER-independent shorter disease-free interval after mastectomy (Gorsch et al. 1992) compared to tumors with no or low TGF-β_1 reactivity. Finally, Thompson et al. reported that breast tumors unresponsive to the antiestrogen tamoxifen, when rebiopsied, expressed significantly higher levels of TGF-β_1 mRNA than clinically responsive tumors (Thompson et al. 1991), further supporting a positive role for the TGF-βs in the progression of breast carcinoma.

2.2.4 Alterations in Growth Response of Mammary Epithelial Cells by TGF-β

There are conflicting reports in the literature on the responsiveness of mammary tumor-derived cells to growth inhibition by TGF-βs in vitro (Wakefield et al. 1995). Some of these differences may be due to the effects of culture conditions on TGF-β responses (see Sect. 2.2). For example, the inhibitory effect of TGF-β on a number of tumor cell lines in vitro, including MCF-7, has been shown to be dependent on the presence of poly-unsaturated fatty acids (PUFAs) and was reversed by antioxidants or selenite (Newman 1990; Keler et al. 1992; Keler and Sorof 1993). Furthermore, maximal growth inhibition required the presence of retinoids under some conditions. In contrast, growth inhibition of nontransformed epithelial and fibroblastic cells was insensitive to PUFAs. Given the highly variable levels of PUFAs, anti-oxidants, and retinoids in different batches of serum, the author proposed that these observations may explain some of the variability in the responsiveness of malignant cells to TGF-β in culture (Newman 1990). Other differences in the response may be due to genetic drift or epigenetic changes occurring in the various malignant lines during long-term culture in different laboratories. The data for the effects of TGF-β on the growth of mammary tumor cell lines are summarized in Table 1. The controversy mainly centers on whether ER status predicts responsiveness to TGF-β. It seems that (ER−) cell have higher response to TGF-β than (ER+) cell lines, which sometimes correlates with loss of TGF-β receptor expression (Arteaga et al. 1988b).

However, HMECs transformed in vitro retained responsiveness to TGF-β, even in the highly tumorigenic cells generated by transformation with multiple oncogenes (Valverius et al. 1989). Furthermore, anti-TGF-β antibodies stimulated the growth of MDA-MB-231 and Hs578T mammary carcinoma cells in vitro, indicating that these cells have a fully functional negative autocrine loop (Arteaga et al. 1990). These last results are particularly significant, since the cells were shown to secrete TGF-β predominantly (>99%) in the latent form (Arteaga et al. 1990). Thus, the stimulatory effect of the antibodies suggests

Table 1. Effects of TGF-βs on the growth of human breast tumor cell lines

Cell line	ER status	Effect on growth	Growth conditions: Anchorage-independent	Anchorage-dependent	Reference
MCF-7	+	None		+	Aakvaag et al. (1990)
		None		+	Manni et al. (1990)
		None	+		Arteaga et al. (1988b)
		None	+		Zugmaier et al. (1989)
		Inhibition (weak)	+		Zugmaier et al. (1989)
		Inhibition		+	Arrick et al. (1990)
		Inhibition	+		Knabbe et al. (1987)
		Inhibition	+		Roberts et al. (1985)
		Inhibition		+	Ranchalis et al. (1987)
T47D	+	None	+	+	Arteaga et al. (1988b)
		None	+		Arrick et al. (1990)
		None	+		Zugmaier et al. (1989)
		Inhibition (weak)	+		Zugmaier et al. (1989)
		Stimulation		+	Daly et al. (1990)
ZR75-1	+	None	+	+	Arteaga et al. (1988b)
		Inhibition (weak)	+		Zugmaier et al. (1989)
		Inhibition		+	Kerr et al. (1989)
BT20	–	Stimulation		+	Fernandez Pol et al. (1986)
		Inhibition	+		Fernandez Pol et al. (1986)
		Inhibition	+	+	Arteaga et al. (1988b)
HBL100	–	None		+	Fernandez Pol et al. (1986)
		Inhibition		+	Fernandez Pol et al. (1986)
		Inhibition (weak)	+		Fernandez Pol et al. (1986)
Hs578T	–	Inhibition	+		Zugmaier et al. (1989)
		Inhibition	+	+	Arteaga et al. (1988b)
		Inhibition	+		Arteaga et al. (1990)
		Inhibition		+	Arteaga et al. (1990)
MDA-MB-231	–	Stimulation		+	Fernandez Pol et al. (1986)
		Inhibition		+	Fernandez Pol et al. (1986)
		Inhibition	+		Fernandez Pol et al. (1986)
		Inhibition	+	+	Zugmaier et al. (1989)
		Inhibition	+	+	Arteaga et al. (1988b)
		Inhibition	+	+	Kasid et al. (1987)
		Inhibition	+		Arteaga et al. (1990)
		Inhibition			Arteaga et al. (1990)
MDA-330	–	Inhibition (weak)	+	+	Arteaga et al. (1988b)
MDA-MB-435	–	Inhibition		+	Kasid et al. (1987)
MDA-MB-468	–	Inhibition	+		Zugmaier et al. (1989)
SKBr-3	–	None		+	Fernandez Pol et al. (1986)
		Inhibition	+		Fernandez Pol et al. (1986)
		Inhibition	+		Zugmaier et al. (1989)

that these particular tumor cells are probably capable of activating the latent TGF-β they produce. It will be important to determine whether this is also true for other tumor cell lines. In contrast to these results obtained on cell lines that have been maintained long-term in culture, a study using primary cultures of cells that grew out from a fragment (explant) of malignant breast tissue derived from a mastectomy sample demonstrated that TGF-β had no inhibitory effect on the malignant cells. TGF-β inhibited growth of cells derived from nonmalignant regions of the same sample (Takahashi et al. 1990), which underlines the growth-inhibiting role of TGF-β for normal MEC. The results of this type of study may be more meaningful than results obtained on cells after extended periods in culture. However, more samples will have to be analyzed to determine whether loss of inhibition by TGF-β is a general feature of freshly prepared breast tumor cells. Other evidence suggests that long-term culture may indeed restore a response that is not expressed in vivo. Thus, the MDA-MB-231 mammary carcinoma line was potently inhibited by TGF-β_1 and TGF-β_2 in vitro, but chronic administration of TGF-β_1 had no effect on the growth of the tumor in nude mice (Zugmaier et al. 1989).

Furthermore, the data with established breast tumor cell lines in vitro suggest that the malignant phenotype is not strictly coupled to the loss of growth inhibition by TGF-β. However, the dependence of the response to TGF-β on the local cellular microenvironment may make it impossible to determine in vitro whether a tumor cell is refractory to growth inhibition by TGF-β in vivo. The role the extracellular matrix plays in this context is not so clear and, not to be evaluated in in vitro systems. It might be possible that the response of mammary epithelial cells to TGF-β is regulated by the surrounding stroma by modulating its extracellular level.

2.3 TGF-β Is a Paracrine Growth Factor (Effects on Stromal Cells)

As outlined before, TGF-β can act on a paracrine interaction including epithelial as well as stromal cells which play a role in tumor development and regulation of organ development (Daniel and Robinson 1992). Development of a malignant tumor is not dependent on increased cellular proliferation alone, but involves a complex series of changes that allow the malignant cell to escape the confines of the surrounding stroma, to evade the immune surveillance system, and to colonize distant organs successfully. The potential involvement of TGF-β in these processes is considered below.

2.3.1 Extracellular Matrix

Many of the biological activities of the TGF-β appear to be mediated through effects on extracellular matrix (for a review Massague 1990; Massague et al.

1996). In general, TGF-β promotes and stabilizes the deposition of interstitial matrix by multiple mechanisms. TGF-β is shown to increase interstitial matrix expression, which could be evidenced in vivo (Roberts et al. 1986; Silberstein et al, 1990). Interestingly, in infiltrating ductal and lobular breast carcinomas particularly strong extracellular staining for TGF-β_1 has been shown in the stromal matrix of those tumors that showed an intense desmoplastic reaction (McCune et al. 1992). Contrarily, strong immunostaining for TGF-β was observed in the tumor cells of medullary breast carcinomas where no TGF-β expression was found in the stroma, which consisted predominantly of inflammatory cells, and no fibrosis was observed (Dickson et al. 1990). The results are consistent with the elevated levels of TGF-β in the stroma of the desmoplastic tumors driving the increased matrix deposition. It is possible that extracellular TGF-β may be secreted by the tumor cells themselves. Another possibility is that, induced by malignant epithelial cells, surrounding stromal cells secrete more TGF-β, which then stimulates matrix deposition in an autocrine fashion. Indeed, it is possible that TGF-β sequestration by the desmoplastic matrix may actually interrupt a negative autocrine or paracrine control path. Medullary carcinomas have a better clinical prognosis than ductal and lobular carcinomas, and show no such matrix binding. In contrast to the matrix-stabilizing effects of TGF-β on interstitial matrix components, TGF-β may actually promote degradation of the basement membrane matrix components. It has been shown that TGF-β specifically increases the levels of the M_r 72000 type IV collagenase activity in a variety of tumorigenic human cell lines in which the interstitial collagenase was regulated in the opposite direction (Brown et al. 1990; Paulus et al. 1995). An increase in the activity of the type IV collagenase appeared to occur at two levels. TGF-β caused an increased in mRNA levels for type IV collagenase, resulting in an increased expression of the 72-kDa proenzyme, and further enhanced cellular processing of its latent form to active lower molecular weight forms in some cell types. Furthermore, TGF-β specifically decreased the expression of the inhibitor, TIMP-2, which inhibits the type IV collagenase (Stetler Stevenson et al. 1989; Murphy et al. 1993). This suggests that elevated TGF-β expression might facilitate invasion through the basement membrane in vivo. Consistent with this hypothesis, pretreatment of a mammary adenocarcinoma clone MTLn3 with TGF-β caused a two- to threefold increase in the number of lung surface metastases seen when cells were inoculated into syngenic rats (Welch et al. 1990). Increased metastatic potential was associated with increased ability of the cells to extravasate, as determined by their ability to invade through an artificial basement membrane in vitro. This property, in turn, correlated with increases in type IV collagenolytic and heparanolytic activities, and was not due to enhanced association with the basement membrane. Similarly, in a malignant mouse fibrosarcoma model, metastatic potential correlated with an increased expression of active TGF-β compared with the nonmetastatic parent (Schwarz et al. 1990). However, experiments on human mammary adenocarcinoma lines have indicated more complex effects of TGF-β on invasiveness through basement membranes in vitro. In these experiments, TGF-β which had no effect on the migration or invasiveness of MCF-7, T47D, or MDA-MB-468 cells inhibited the invasiveness but not the migration of MDA-MB-436 cells and inhibited both invasiveness and migration of the more malignant, hormone-independent Hs578T and MDA-MB-231 cells. The invasiveness of early passage MCF-7 cells appeared to be stimulated by TGF-β. Similarly, in the oncogene-immortalized A1N4 series of breast epithelial lines (Stampfer and Bartley 1985), TGF-β inhibited the invasiveness of only the most aggressive A1N4-TH cells, while inhibiting the migration of all the cell lines (Thompson et al. 1993). Interestingly, TGF-β increased levels of the type IV pro-collagenase in MDA-MB-436 and A1N4-TH cells, despite inhibiting their invasiveness (Schwarz et al. 1990). This suggests that TGF-β has pleiotropic effects on multiple components of the invasive phenotype and that the positive effects on basement membrane-degrading enzymes may not always dominate (Schor et al. 1989). TGF-β action on the extracellular matrix cannot be regarded as an effect of TGF-β alone, but rather depends on its balance with other growth factors such as the epidermal growth factor (EGF) modulates matrix protein expression regulating cells attachment in vitro and metastatic potential in vivo (Rohde Schulz and Lichtner 1995). TGF-β_1 and TGF-β_2 antagonized the action of MSF on migration and hyaluronic acid synthesis in vitro (Schor et al. 1989). In this case, therefore, TGF-β produced by the epithelium might neutralize an underlying stromal defect. Along the same lines, it will be interesting to see whether TGF-β has any effect on the expression or activity of stromelysin-3, a metalloproteinase that is expressed in all invasive breast cancers. Its expression is restricted to the stromal cells immediately surrounding the neoplastic cells of the invasive, but not the in situ, component of breast carcinomas (Basset et al. 1990).

2.3.2 Stromal-Epithelial Interactions

Experimentally, it is harder to identify defects in paracrine loop than in the autocrine loop. This is due to the fact that monolayer cell cultures are mostly used to study tumorigenesis in vitro. Complicated organ culture (Binas et al. 1992; Brandt et al. 1998) or coculture systems from normal and/or transgenic sources could help to overcome these restrictions. However, this is clearly an important area that merits further study.

Although TGF-β is inhibitory for the growth of endothelial cells in vitro, in vivo it appears to enhance angiogenesis. Indeed, during embryogenesis there is a strong correlation of expression of TGF-β_1 with areas of extensive neovascularization (Heine et al. 1987). TGF-β_1 injected into the nape of the neck of newborn mice evokes the production of granulation tissue with characteristic new blood vessel formation (Roberts et al. 1986). This effect may be indirect due to the recruitment of monocytes and their activation to secrete angiogenic factors, such as tumor necrosis factor-alpha (Wahl et al. 1987). Thus, overproduction of TGF-β by a tumor may promote neovascularization of the tumor by similar mechanisms, and this may be the reason for controversial data about TGF-βs role in growth regulation in MEC and tumor progression.

2.3.3 Implications of Tumor Development, Invasion, and Metastasis

While the induction of interstitial matrix components by TGF-β might decrease the propensity of a tumor to metastasize by physically confining it, other properties of the TGF-β are likely to promote invasion and metastasis. The induction of enzymes that degrade the basement membrane, the induction of components such as tenascin that promote cell detachment and motility, inhibitory effects on the immune surveillance system, and the promotion of neovascularization would all tend to promote metastasis. All these effects are examples of aberrant paracrine communication between epithelium and stroma. Currently, it is not clear which of these effects will dominate at what stage in tumor development. Since the metastatic phenotype arises late in tumorigenesis, elevated TGF-β expression at early stages may tend to have an antitumor effect through increased growth inhibition, while elevated TGF-β at later stages might be deleterious, tending to promote successful invasion and metastasis. Consistent with this, in a malignant mouse fibrosarcoma model, and in benign and malignant rat liver cell lines, metastatic potential was correlated with an increased expression of active TGF-β (Schwarz et al. 1990). Furthermore, TGF-β may actually be growth-stimulatory rather than -inhibitory in some highly malignant carcinoma cell lines (Fernandez Pol et al. 1986; Jetten et al. 1986; Daly and Darbre 1990; Daly et al. 1995).

2.4 Relationship of TGF-β and Steroid Hormones

2.4.1 Storoid Hormones Regulate TGF-β Production by Breast Cancer Cells

A new type of conceptual thinking came with the demonstration that estrogen stimulates the growth of hormone-dependent breast tumor cells in part by autocrine growth stimulation by TGF-alpha (Manni and Wright 1984; Jung Testas and Baulieu 1985; Loser et al. 1985). Conversely, it was proposed that estrogen might reduce the level of growth inhibitors such as TGF-β or that TGF-β expression is increased by antiestrogens (Knabbe et al. 1987, 1991). The first experiments were performed on MCF-7 mammary carcinoma cells where estrogen treatment caused an approximately twofold decrease in TGF-β protein expression under steroid-free conditions (phenol red-free medium) (Knabbe et al. 1987, 1991), whereas the anti-estrogens tamoxifen, LY117018, and 4-hydroxy tamoxifen caused an 8- to 27-fold induction of TGF-β protein (Knabbe et al. 1991). Since no induction of TGF-β was seen in (ER−) mammary epithelial cell lines, it has been concluded that TGF-β induction was mediated through the estrogen receptor. Increasing concentrations of estrogen could antagonize the effect. Interestingly, the biological active form of TGF-β is increased from 5 to 18% (Knabbe et al. 1991). Anti-TGF-β antiserum reversed most of the inhibitory effect of conditioned medium from MCF-7 cells treated with antiestrogens on the (ER−) breast cancer line MDA-MB-231, suggesting that TGF-β was the principal growth-inhibitory activity made by these cells in response to anti-estrogens. Furthermore, the data indicated that TGF-β induced in (ER+) cells by antiestrogen treatment could have a paracrine-inhibitory effect on neighboring (ER−) cells, as well as direct effects on (ER+) cells. Subsequently, numerous other groups examined the effects of estrogen and antiestrogens on TGF-β production by breast cancer cell lines. The results have tended to be complex, and are summarized in Table 2. The regulation of TGF-β by estrogen and antiestrogens appears to be critically dependent on the culture history of the cell and the culture conditions of the experiment. For instance, long-term withdrawal of steroids from (ER+) cell lines in culture can result in changes of baseline expression of the various TGF-β subtypes, as well as changes in the response to estrogens and antiestrogens (Daly and Darbre 1990). These adaptive changes in long-term culture can be variously the result of epigenetic or genetic alterations (Daly et al. 1990). Since regulation of TGF-β_1 by antiestrogens appears to involve posttranscriptional mechanisms in vitro (Colletta et al. 1990, 1994; Knabbe et al. 1991), studies that have only looked at TGF-β_1 mRNA levels could have overlooked a posttranscriptional induction. Where protein levels were determined, there is fairly general agreement that antiestrogens can induce TGF-β protein in (ER+) cell lines. However, neutralizing TGF-β antibodies only partially reversed the growth-inhibitory effects of antiestrogens in some systems, suggesting the involvement of additional growth inhibitors or inhibitory pathways (Wakeling et al. 1984; Wakeling 1993). However, more recent reports discuss that antiestrogens or progestins inhibit mammary epithilial cell proliferation independently of induced TGF-β production (Kalkhoven et al. 1996). A hormone-independent variant of the MCF-7 cell line was shown to have lost its sensitivity to TGF-β during its progression towards an autonomous phenotype, but has preserved its sensitivity to antiestrogens and progestins. Furthermore, T47D cells lacking the TGF-β receptor type II were stable transfected

Table 2. Effects of various steroids on the expression of TGF-βs by human breast tumor cells

Cell line	Steroid/analog	Effects on TGF-β expression			Reference
		Protein	mRNA	TGF-β subtype	
BT-20	Tamoxifen	None	n.d.	β_1	Colletta et al. (1991)
	Gestodene	Increase (3×)	n.d.	n.d.	Colletta et al. (1991)
MCF7	E_2	Decrease (2×)	n.d.	n.d.	Knabbe et al. (1987)
	LY117018	Increase (27×)	None	n.d.	Knabbe et al. (1987)
	Tamoxifen	Increase (5×)	n.d.	n.d.	Knabbe et al. (1987)
	Dexamethasone	Increase (2×)	n.d.	n.d.	Knabbe et al. (1987)
	Gestodene	Increase (10×)	n.d.	β_1	Colletta et al. (1991)
MDA-MB-231	E_2	n.d.	None	β_1, β_2	Arrick et al. (1990)
	Tamoxifen	None	n.d.	β_1	Colletta et al. (1991)
	Gestodene	Increase (6×)	n.d.	β_1	Colletta et al. (1991)
T47D	MPA	n.d.	Decrease	β_1	Murphy and Dotzlaw (1989)
	E_2	Decrease	n.d.	β_2, β_3	Arrick et al. (1990)
	E_2	n.d.	None	β_1	Arrick et al. (1990); Daly and Darbre (1990)
	Tamoxifen	Increase (32×)	None	β_1	Colletta et al. (1991)
	Tamoxifen	n.d.	None	β_1	Arrick et al. (1990)
	Gestodene	Increase (94×)	Increase		Colletta et al. (1991)
ZR-75-1	E_2	n.d.	Decrease	β_2	Arrick et al. (1990)
	Tamoxifen	n.d.	None	β_1, β_3	Arrick et al. (1990)

n.d. = not determined; E_2 = estradiol; MPA = medroxyprogesterone acetate; LY117018 = antiestrogene.

with an expression vector containing the TGF-β type II receptor. In both variants, the estradiol-induced proliferation was equally inhibited by anti-4 hydroxy tamoxifen (Kalkhoven et al. 1996). TGF-β-neutralizing antibodies did not reverse either antiestrogen or progestin-induced growth inhibition (Kalkhoven et al. 1995).

Modulation of TGF-β expression by steroids can also be seen in vivo. It has been demonstrated that estradiol (E_2) withdrawal from MCF-7 breast cancer cells innoculated in nude mice was associated with a threefold increase in mRNA for TGF-β_1 in the tumor (Kyprianou et al. 1991). Conversely, E_2 administration to MCF-7 xenografts in thymectomized irradiated mice caused a decrease in TGF-β_1 mRNA levels (Thompson et al. 1990). Interestingly, some MCF-7 variants respond to E_2 ablation in vivo by increased programmed cell death or apoptosis. The role of E_2 and TGF-β for apoptosis could also be demonstrated for the mouse mammary gland using an organ culture system (Atwood et al. 1995; Chen et al. 1996). Apoptosis is associated with an increase in TGF-β_1 mRNA in the tumor, and TGF-β has been shown to induce apoptosis in prostatic glandular cells and mammary epithelia cells in organ culture (Atwood et al. 1995; Chen et al. 1996; Kyprianou et al. 1991). It will be important to determine how cytotoxic paths may be coupled to TGF-β action in other breast cancer lines that do not show this response.

3 Mammary-Derived Growth Inhibitor (MDGI)

3.1 Purification

The initial description of growth-inhibitory activities derived from the mammary tissue dates back to 1977, when W. Lehmann, P. Langen, and coworkers began to characterize the effects of extracts obtained from ascites fluid of Ehrlich mammary carcinoma cells on the growth of these cells (Lehmann et al. 1977). The assay system had originally been developed as a screening test for antineoplastic agents. Ehrlich cells, harvested from the peritoneal cavity, are cultured in the presence of test samples for 24h, and growth is judged by the estimation of cell number. Since a successful isolation of the inhibitory principle is obviously required, the availability of large amounts of starting material is needed and bovine mammary gland tissue was considered as a source. The "chalone" concept (Bullogh and Laurence 1960), claiming the existence of tissue-specific autocrine growth inhibitors, and the fact that Ehrlich ascites tumor had once been derived from a murine mammary carcinoma guided these studies. Indeed, extracts from lactating bovine mammary tissue displayed inhibitory activities similar to the material previously obtained from ascites and Ehrlich cells (Lehmann et al. 1983). Various purification schemes employing common chromatographic and electrophoretic techniques revealed that the majority of inhibitory activity obtained from lactating bovine mammary tissue is associated with a 13- to 15-kDa protein (Bohmer et al. 1984). Evidence for the identity of this protein with the inhibitory principle provided immunoneutralization experiments (Bohmer et al. 1985). The putative growth inhibitor was purified to homogeneity and the amino sequence of the protein designated mammary-derived growth inhibitor (MDGI) was determined (Bohmer et al. 1987a).

3.1.1 Structural Features and Relationship to Fatty Acid-Binding Proteins

The amino acid sequence of the 14.5-kDa MDGI (Fig. 1) revealed no homology to any of the hitherto known growth inhibitors, such as transforming growth factor-β or the interferons (Bohmer et al. 1987a). A striking homology was,

Homology of bovine MDGI and other fatty acid binding proteins

```
           1        10        20        30          40        50
           |        |         |         |           |         |
MDGI A     VDAFVGTWKLVSSENFDDYMKSLGVGFATRQVGNMA--KPTLIISVNGDTVI
MDGI B     ...........D.K............T--..........T..E.......
(H-FABP)
MP2        SNK.L...........E...A....L...KL..L.--..RV...KK..IIT
CRABP      PN.A....MR......ELL.A...NAML.K.AVA.AS..HVE.RQD..QFY

                   60        70        80        90          100
                   |         |         |         |           |
MDGI A     IKTQSTFKNTEISFKLGVEFDETTADDRKVKSIVNLDEGKLVQVQK----WNGQ
MDGI B     ................................T..G....H...----....
(H-FABP)
MP2        .R.E.P...........Q..E.....N..T..T...AR.S.N....----...N
CRABP      ...ST.VRT...N..V.EG.E.E.V.G..CR.LPTWENENKIHCTQTLLEGD.P

                   110       120       130
                   |         |         |
MDGI A     ETSLVREMVDGKLILTLTHGTAVCTRVYEKQ
MDGI B     .........................T....A
(H-FABP)
MP2        ..TIK.KL....MVVECKMKDV....I...V
CRABP      K.YWT..LANDE.I..FGADDVV...I.V.E
```

Fig. 1. Amino acid sequence of MDGI and closely related bovine forms. For the MDGI-related proteins only nonidentical amino acids are shown. *MDGI* mammary-derived growth inhibitor; *H-FABP* heart fatty acid binding protein; *MP2* myelin P2; CRABP cellular retinoic acid protein

however, evident to members of a large family of hydrophobic ligand-binding proteins comprising fatty acid-binding proteins (FABP) (Matarese et al. 1989; Veerkamp et al. 1993), cellular retinoid-binding proteins, myelin P2, and others (Fig. 1). For the adipocyte, liver, and intestinal FABP, a relationship between their expression and the differentiated stage of the tissue has been documented (Matarese et al. 1989), a property also discussed before for cellular retinoid-binding proteins (Eriksson et al. 1991). The closest structural relationship was found to fatty acid-binding protein(s) of the cardiac type (c-FABP). Indeed, the amino-acid sequence of bovine cardiac FABP predicted from a cDNA sequence (Billich et al. 1988) was found primarily to be 95% homologous to the amino-acid sequence of bovine MDGI. Amino-acid sequencing of MDGI revealed a number of microheterogeneities, i.e., in eight cases (Bohmer et al. 1987a) two residues were found in comparable amounts in the same position of the sequence (Bohmer et al. 1987a). This finding suggests that the material used for sequencing was not entirely homogeneous and contained at least two slightly different isoforms of MDGI (Bohmer et al. 1987a), which could both have contributed to the biological activity of MDGI (MDGI A and MDGI B, Fig. 1). One of them is identical to bovine heart FABP, which has been finally evidenced by electrospray mass spectrometry (Lassen et al. 1995; Specht et al. 1996). Screening of a cDNA library derived from bovine lactating mammary gland revealed only one MDGI form, which turned out to be identical with the cardiac sequence (Kurtz et al. 1990; Fig. 1, MDGI B). In contrast, Specht et al. described that MDGI is not a homogeneous protein but a mix of heart-type and adipocyte-type fatty acid-binding proteins (Specht et al. 1996), both not representing the sequence of MDGI A. Other MDGI species have not been found among nine cDNA clones obtained from the bovine lactating mammary gland library. Furthermore, differential hybridization of PCR fragments amplified from bovine mammary cDNAs from two animals and from bovine genomic DNA with MDGI-specific primers did not give rise to another MDGI form (Kurtz et al. 1990). Taken together, it is more likely that MDGI consists at least of two species (subtypes MDGI A and MDGI B) where MDGI-B is identical with H-FABP and is the abundant form in terminally differentiated mammary gland (MDGI-B, Fig. 1), whereas the other isoform (MDGI-A, Fig. 1) constitutes a minor component in this tissue. Posttranslational mechanisms are not sufficient to explain the eight amino acid differences between the bovine MDGI forms A and B. One could assume that during the development of the mammary gland, two different MDGI forms may become expressed, one of which is abundant during lactation and identical with the cardiac FABP, whereas the other is produced at an earlier stage of development. The hypothesis is supported by two lines of experiments. First, an MDGI form with a primary structure not identical with the cardiac FABP has been detected by screening of a cDNA library that was derived from the mammary gland of the pregnant mouse (Binas et al. 1992). The cloned mouse MDGI cDNA predicts a protein sequence that differs in nine positions from that of the murine cardiac cDNA-derived sequence of FABP (Tweedie and Edwards 1989). The 5′-untranslated regions of the two proteins are completely different. Second, by protein fractionation of the mammary gland of pregnant cow, a slightly "shortened" MDGI form has been detected (Brandt and Grosse 1992). This protein comprises less than 1% of MDGI and is not present in the lactating gland. It is immunologically related to MDGI, however, it does not cross-react with an antibody directed to the C-terminus of MDGI. The amino-acid sequence has not yet been determined, but the available data suggest that this protein is processed from the minor MDGI-A form rather than from an abundant MDGI-B, which is supposed to be present in terminally differentiated tissue. Under nondissociating conditions, the truncated MDGI has been found to be associated with a glycoprotein forming a 70-kDa protein complex which could be separated by elution with N-acetylglucoseamine from WGA-Sepharose (Brandt and Grosse 1992). For the full length MDGI such a protein complex has not yet been identified. The existence of a C-terminal truncated transiently during pregnancy-expressed MDGI is very interesting, due to the fact that a C-terminal peptide of MDGI (see below) can fully mimic the biological effects of the mature protein. Thus, it is thought that the release of a C-terminal MDGI peptide from the tryptic cleavage site can play a regulatory role

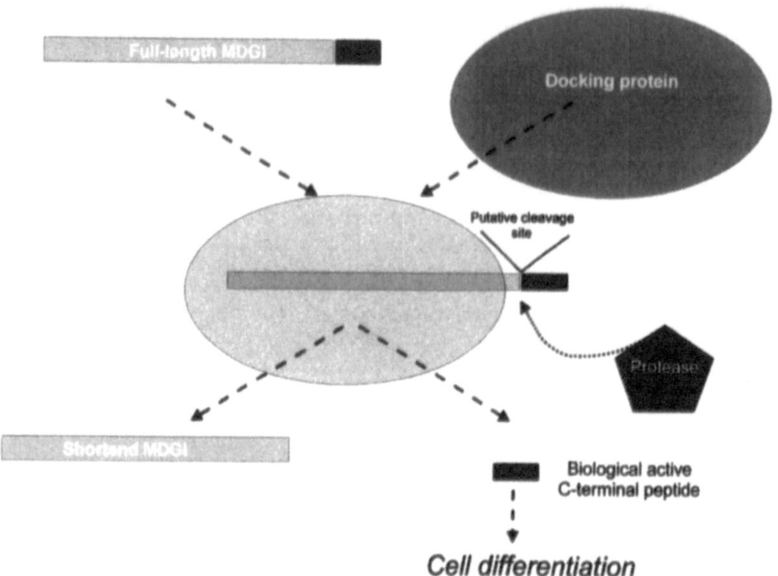

Fig. 2. Scheme showing hypothesized mechanism for MDGI activation. So far, for MDGI no membrane-associated receptor has been identified. We propose that MDGI enters the cytoplasm by a not yet identified mechanism. The full-length MDGI protein binds to approximately 57 kDa *docking protein* that contains by itself protease activity to cleave MDGI at the C-terminus. Additionally, it seems possible that binding of MDGI to the docking protein enables the C-terminal truncation by an unknown protease. Finally, the released C-terminal peptide could activates the MDGI-specific pathway

in induction of mammary gland differentiation. Figure 2 shows a hypothesized mechanistic model for MDGI taking into account the fully active C-terminal peptide.

The homologies of the mouse MDGI with bovine MDGI and bovine cFABP is 80 and 84%, respectively. Three genes located on different chromosomes have been found with a probe for the cardiac FABP in mouse (Heuckeroth et al. 1987). One of them might code for the MDGI transcript. Two mammary gland-derived FABPs have also been isolated from the rat (P.D. Jones et al. 1988). One, which was named mammary gland FABP (Nielsen et al. 1994), has been shown to be identical to MDGI-B by electrospray mass spectrometry of the entire protein and its tryptic peptides. Apart from only nine amino acids not covered by the peptides analyzed, a complete identity was found between the calculated masses for the cardiac FABP and the measured values for the mammary gland FABP (Nielsen et al. 1994). As mentioned before, for some types of FABPs the limited organ distribution and a development-dependent manner of topological tissue expression indicates some role for tissue-specific processes during differentiation (Matarese et al. 1989). Bovine cardiac type-related FABPs have been detected in the heart, mammary gland, skeletal muscle, and brain. Rodent cardiac-type FABP expression is restricted to the heart, mammary gland, adrenal gland, skeletal muscle, kidney, aorta, placenta, and testis. It would be of more general importance to elucidate the identity of the different cFABP/MDGI forms expressed in the different tissues. The pattern of MDGI forms in vivo might be complicated further by post-translational modifications: a potential site for tyrosine phosphorylation exists at Tyr 19 (Fig. 1). The sequence around this tyrosine is rather conserved and is highly homologous to the corresponding sequences of the adipocyte lipid-binding protein p422, aP2, and ALBP (Bernlohr et al. 1984; Fig. 1). For both mammary gland-derived FABP and ALBP, phosphorylation of Tyr 19 by the insulin receptor tyrosine kinase has been shown (Chinander and Bernlohr 1989; Hresko et al. 1988) and has been suggested to be linked with insulin receptor signaling (Chinander and Bernlohr 1989). The phosphorylation in vitro was shown to be activated by the binding of fatty acids to ALBP (Hresko et al. 1990). Fatty acid binding has been suggested to induce a conformational change of ALBP, leading to a facilitated access of the kinase to Tyr 19. For the mammary gland-derived FABP partially phosphorylation on Tyr 19, apparently as a physiological substrate for the insulin receptor tyrosine kinase, has been demonstrated (Nielsen and Spener 1993). Though this has not yet been demonstrated, it seems likely that MDGI can undergo phosphorylation at Tyr 19 and that phosphorylated and nonphosphorylated forms of MDGI may coexist in the cell. Second, the chemical analysis of MDGI preparations revealed the presence of nonextractable, i.e., probably covalently bound, fatty acids (Grosse and Langen 1990). The sites and the exact nature of this possible modification are not yet known. Finally, as outlined later in more detail, MDGI, like its relatives, can bind long-chain fatty acids, creating a further level of posttranslational modification by ligand binding. For all these possible modifications, as for the structural microheterogeneities discussed above, the functional significance remains to be elucidated. The C-terminus of MDGI residues 126–130 are identical to residues 108–112 of bovine growth hormone (Miller et al. 1980; Fig. 1). This stretch of amino acids is part of a sequence of growth hormone that is essential for its biological activity (Yamasaki et al. 1975). The homology raised the question of whether MDGI- or MDGI-derived peptides might compete with growth hormone for binding to the growth hormone or prolactin receptor. This seems, however, not to be the case, since MDGI-derived C-terminal peptides comprising the homologous sequence do not compete with growth hormone or prolactin for binding to their respective receptors. This does not exclude an indirect functional interaction between synthetic MDGI peptides and prolactin-dependent intracellular pathways. The three-dimensional structure of MDGI is most likely very similar to that of the closely homologous proteins for which X-ray crystallography and NMR with and without bound fatty acid has been accomplished (T.A. Jones et al. 1988; Sacchettini et al. 1988;

Muller Fahrnow et al. 1991; Lassen et al. 1993, 1995). The overall structure of the protein is a β-barrel consisting of ten antiparallel β-strands which conform two nearly ontogonal β-sheets of five strands each. Two short helices form a helix-turn-helix motif in the N-terminal region of the polypeptide chain. The fatty acid (palmitic acid) is bound within the protein in a U-shaped confirmation close to the two helices (Lassen et al. 1995). This structure of bovine cardiac FABP presents a fully applicable picture of one of the MDGI forms. Data obtained by analytical ultracentrifugation of MDGI (Behlke et al. 1989) or by theoretical calculations based on atomic coordinates (Suessmilch et al. 1990) are in accordance with the crystallographic data. As in the case of the structural relatives of MDGI, the sequences of the cloned cDNA for bovine MDGI-B and murine MDGI lack a signal sequence for membrane translocation (Kurtz et al. 1990; Binas et al. 1992). This finding raises several questions with regard to the physiological significance of the observed biological effects of MDGI added extracellularly to target cells. If MDGI is supposed to act in an autocrine manner as a growth inhibitor, how can it reach the putative receptor on the target cells if it is not secreted? If most of MDGI has an intracellular localization, what is the function of this protein inside the cell? We will come to this later. However, with regard to secretion, an analogy might exist to other growth factors that also lack a signal sequence, like FGF and PG-ECGF (Abraham et al. 1986; Ishikawa et al. 1989). It should be noted that the currently available data do not allow the exclusion of some "leakage" of MDGI into the extracellular compartment. In cryosections of mouse mammary glands, MDGI-related immunostaining has been observed over the extracellular matrix (F. Vogel, unpubl. observation). For rat cardiac FABP small amounts were found to be continuously present in serum with an increase in experimental ischemia (Knowlton et al. 1989). Also, a more distant structural relative of MDGI, gastrotropin (Walz et al. 1988), is supposed to be present and to act physiologically in the extracellular compartment. In addition, an immunologically cross-reacting growth-inhibitory protein, FGRs (Hsu and Wang 1986; Bohmer et al. 1987b), has been purified from conditioned media of 3T3 fibroblasts (Hsu and Wang 1986). Alternatively, the physiological site of action of MDGI might be the cell interior. So far, a surface receptor for MDGI has not been demonstrated. Experiments with ^{125}I-labeled MDGI and Ehrlich cells revealed that at least a part of the protein is taken up by the cells and remains stable as a 14.5-kDa molecule over several hours of incubation (T. Müller, F-D Böhmer, F. Vogel et al. pers. comm.). With regard to the high amounts of MDGI in lactating tissue (in contrast to very small levels in all other situations, for example, target cells in vitro!), the key to understanding might be a dissection of the various possible forms of MDGI (see above). Only a minor form of MDGI (MDGI-A) might participate in regulation, whereas the bulk form(s) could serve a distinct function, possibly in fatty acid mobilization and transport, as discussed for FABPs in reviews (Matarese et al. 1989; Spener et al. 1990). As expected from the structural homologies to lipid-binding proteins, MDGI was revealed to be a lipid-binding protein as well (Bohmer et al. 1988). Among the various potential ligands tested, only long-chain fatty acids were found to become bound to MDGI with characteristics very similar to other fatty acid-binding proteins (Wallukat et al. 1991a; Table 1). Therefore MDGI, or rather its predominant form (MDGI-B) in the lactating mammary gland, could well be referred to also as bovine mammary fatty acid-binding protein. An obviously interesting question is whether ligand binding could be related to the biological activity of MDGI. In principle, transport of a physiologically active (growth-inhibitory) ligand or depletion of a ligand essential for cell growth could be putative ways of MDGI-ligand interactions in cell growth inhibition. So far the search for a biologically potent ligand has failed. In particular, no binding of any of the tested eicosanoids could be detected. In contrast, for rat liver FABP, binding of eicosanoids (15-HPETE, S-HETE, S-HPETE, PGA_1, PGA_2) with a relatively high affinity has been demonstrated (Raza et al. 1989; Khan and Sorof 1990) and has been linked to a role of this protein in mitosis and as a target for carcinogens (Bassuk et al. 1987). Long-chain fatty acids, however, added to cell cultures at concentrations as low as those of MDGI in the growth assay (Bohmer et al. 1984, 1985, 1987a), failed to inhibit cellular proliferation. Also depletion of essential fatty acids by MDGI is unlikely to form the basis for the inhibitory effect, considering the relatively low affinity for fatty acid binding. In addition, direct uptake measurements revealed that long-chain fatty acids bound to MDGI are even more easily taken up by the target cells than those added in free form (Bohmer et al. 1988). As outlined in detail below, we found that synthetic peptides derived from the MDGI sequence have biological activity similar to MDGI. These peptides are, however, unable to bind fatty acids (Wallukat et al. 1991a). We therefore conclude that the MDGI-ligand interaction is not directly involved in its biological activity. It is likely that the affinity of MDGI to fatty acids is associated with the synthesis and secretion of lipids into the milk. Indeed MDGI is abundantly present in "milk fat globules" (Brandt et al. 1988) and might have some protective function for the descendants. It remains a possibility that an indirect effect of ligand binding, in terms of a structural change of MDGI upon ligand binding, might be necessary for full functional activity. Evidence for such a structural change has been obtained by analytical ultracentrifugation studies (Behlke et al. 1989) and structural analysis of cardiac FABP (Lassen et al. 1993, 1995; Muller Fahrnow et al. 1991). An analogy can be drawn to that suggested to form the basis of the facilitated access of tyrosin kinase to Tyr 19 in ALBP upon ligand binding.

Some remaining questions have to be investigated. (1) How is MDGI secreted to the extracellular matrix and how is it transmitting the growth-inhibiting signal into the nucleus? Does any MDGI receptor exist? (2) What biological role, if any, does the MDGI C-terminal peptide play? Is it naturally released from the MDGI-A form to regulate or stabilize MDGI function for growth inhibition and differentiation of the mammary epithelium? It has been demonstrated that MDGI induces it own expression during pregnancy and lactation, but it is unclear which form of MDGI is synthesized because of lacking different mRNA species. Might it be possible that MDGI-A induces MDGI-B in late pregnancy stabilizing mammary gland lactation and modulates fatty acid metabolism, as has been postulated for the cardiac type of bovine FABP in cardiac myocytes (Spener et al. 1989)?

3.2 Biological Activities

3.2.1 Growth Inhibition of Mammary Epithelial Cells

Ehrlich cells taken from the stationary in vivo growth phase were first used to assay growth-inhibitory activities in vitro (Lehmann et al. 1977). As outlined in detail previously, in these cells half-maximal inhibition was obtained with about 10^{-10}M MDGI (1 ng/ml). Inhibition was abolished by simultaneously adding MDGI with EGF, insulin, or 2′-deoxycytidine (Bohmer et al. 1984; Langen et al. 1984). Ehrlich cells taken from the logarithmic phase of in vivo growth do not respond to MDGI. This is in accordance with pulse cytophotometry measurements indicating that cells late in the G_1-phase were more sensitive towards MDGI action than S- or G_2-phase cells (Lehmann et al. 1977). The

Table 3. Inhibition of DNA sysnthesis by MDGI in normal and transformed cells: DNA was quantified by ^{3}H-thymidin pulse labeling. (Bohmer et al. 1984, 1987a; Bielka et al. 1986)

Cell line	Tissue	% Inhibition
MATU	Human ductal carcinoma	40–50
T47D	Human ductal carcinoma	40
MCF-7	Human ductal carcinoma	0–10
mMa Ca	Human ductal carcinoma	30
Normal human MEC	Ductular mammary epithelium	
2nd passage		0
3rd passage		23
5th passage		30
6th passage		60
12th passage		100
14th passage		100
NRK	Rat kidney fibroblast	0
A431	Squamous epidermoid carcinoma	0
P19	Mouse embryonic stem cells	0

response to MDGI of permanent mammary carcinoma cell lines, normal human mammary epithelial cells, and nonepithelial cells is compared in Table 3. The proliferation assay is based on serum starvation, followed by a restimulation period with fresh medium. MDGI was present during the restimulation period for 16–20 h. Flow cytophotometric measurements proved them to be arrested in G_1/G_0, which was a prerequisite for growth inhibition. The data indicate that various mammary epithelial cells are sensitive to the growth-inhibitory action of MDGI. However, there is no obvious relationship between sensitivity towards MDGI and growth parameters of human mammary carcinoma cells (such as their ability to clonogenic growth) or origin (murine or human). In a soft agar growth assay with MATU mammary carcinoma cells, treatment with MDGI for 6 days did not exceed growth inhibition obtained in monolayer culture (unpubl. data). The enhanced sensitivity of normal mammary epithelial cells towards MDGI (Lehmann et al. 1989) is interesting. While the inhibition of growth of Ehrlich cells could be prevented completely by low concentrations of fetal calf serum, insulin, or EGF (Lehmann et al. 1984), restimulation of normal mammary epithelial cells was performed in the presence of 10% fetal calf serum, EGF, insulin, and pituitary extract. The degree of inhibition increased with the passage number. This finding gives rise to speculations that MDGI might act on normal, rather than on transformed, cells, and that in later passages cells may have lost their proliferative capacity, and/or some differentiated functions may have been selected in certain responsive cell clones.

3.2.2
Stimulation of Differentiation

As discussed before, in the mammary gland from pregnant mouse, an MDGI form was detected that differs from murine-cardiac FABP. Expression studies clearly indicate a relationship between the onset of endogenous MDGI expression and differentiation of the mouse and bovine mammary gland (Kurtz et al. 1990; Binas et al. 1992). Therefore, it seems reasonable to assume that the protein may fulfill a function related to differentiation. In order to prove directly MDGI's function during differentiation, an organ explant culture system with mouse mammary glands has been established (Binas et al. 1992). Furthermore, a synthetic peptide comprising the last 11 C-terminal amino acids (121–131) (to prevent dimerization and subsequent loss of activity Cys was replaced by Ser) of MDGI has been taken in advantage to study its biological function. This peptide (P108) behaves in several aspects as MDGI. It partially blocks insulin and 2′-deoxycytidine (Grosse and Langen 1990), and it reverses the supersensitivity of beta-adrenergic receptors towards isoprenaline (Wallukat et al. 1991a) (see below). If mammary explants from mid- to late-pregnant Balb c-mice were cultured for several days in a serum-free medium containing prolactin, insulin, and hydrocortisone (lactogenic medium), lobuloalveoli expressed the typical phenotype of a functionally differentiated mammary gland, i.e., the alveolar cells are vacuolized, secretory active, and they synthesize beta-casein mRNA (Binas et al. 1992). Under these conditions, the proliferative rate, as indicated by the labeling index, is very low. The continuous presence of EGF during culture caused a clear suppression of the differentiated phenotype of alveolar cells. The vacuolization index was chosen as a parameter to quantitatively characterize the degree of morphological differentiation. The EGF-exerted inhibition of functional differentiation could be reversed by the additional presence of the synthetic MDGI peptide. P108 reversed the vacuolization index to the normal control level. EGF-dependent inhibition of ductal growth and functional differentiation have been described earlier (Coleman and Daniel 1990). The underlying mechanism is unclear, though it is most likely not associated with the mitogenic activity of EGF (Coleman and Daniel 1990). EGF probably interferes with transcriptional mechanisms involved in the maintenance of functional differentiation in the mammary gland of the pregnant mouse (Taketani and Oka 1983).

3.2.3 Desensitization of Beta-Adrenergic Receptors

The close structural relationship between MDGI and heart FABP prompted to test MDGI on cardiac myocytes (Wallukat et al. 1991a,c). It turned out that in neonatal cardiac myocytes MDGI exhibited an effect that at first glance seemed unrelated to growth inhibition. These cells, when taken into primary culture, beat with a basal frequency of 140–160 beats/min. In the standard system developed by G. Wallukat et al. (Haase et al. 1987) and used for the tests, β-adrenergic agents increase the beating rate in a characteristic dose-dependent manner, with the first response detectable at about 10^{-9} to 10^{-11} M isoprenaline. The authors found that this sensitivity can be greatly increased by pretreating the cells with L(+)-lactate (Wallukat et al. 1991c). In this way, the cells respond to isoprenaline at 10^{-10} M, probably on account of an increased responsiveness or accessibility of a subpopulation of β-adrenergic receptors to the agonist. Circumstantial evidence points to a metabolic pathway involving activation of phospholipase A_2, the generation of arachidonic acid, and conversion of arachidonic acid to 15S-hydroxyeicosatetraenoic acid (15S-HETE) or 11S-HETE by a lipoxygenase (Wallukat et al. 1991b,c); this may be the mechanism of the sensitization. This concept is also supported by findings that the supersensitivity can be induced directly with 15S-HETE or 11-HETE, but not with 15R-HETE (Wallukat et al. 1991a,c). MDGI strikingly inhibited the induction of supersensitivity by all agents. Moreover, the C-terminal peptide (MDGI 121–131 or the Ser-derivative P108) blocked arachidonic acid and 15S-HETE-induced supersensitivity. This means that a mechanism of interference by MDGI subsequent to the generation of 15S-HETE is involved (Wallukat et al. 1991c). The data have several interesting implications. In particular, they draw the attention to the possibility that the β-adrenergic system might be important for mammary epithelial cell growth and differentiation, and that the effects of MDGI on growth and β-adrenergic sensitivity might have a common and physiologically important mechanism. Indeed, several reports point to an involvement of the beta-adrenergic receptors, PGE_2 and HETEs, for growth and developmental changes in the mammary gland (Marchetti et al. 1990). The possible effects of MDGI or MDGI-derived peptides on related metabolic mechanisms may help to elucidate the mechanisms of normal proliferation and differentiation in the mammary gland. Second, one function of heart FABP could be to protect the heart under pathophysiological conditions from lipoxygenase metabolites causing supersensitivity of β-adrenergic receptors. Finally, the data indicate common targets to MDGI and lipoxygenase products in the modulation of cellular functions. In this respect, it is interesting to note that some data show an involvement of the lipoxygenase pathway in the signaling of growth stimulation of mammary epithelial cells (Rose and Connolly 1990), as well as of growth inhibition by TGF-β (Datta et al. 1990; Newman 1990), interferon (Hannigan and Williams 1991), and TNF-alpha (Haliday et al. 1991).

3.3 Expression of MDGI

3.3.1 Developmental Regulation in the Bovine Mammary Gland

MDGI expression monitored at the protein and mRNA levels is dependent on the developmental stage of bovine mammary tissue. This was shown by immunochemical methods (Muller et al. 1989), by cell-free, in-vitro translation experiments, and by a Northern blot analysis (Muller et al. 1989). MDGI is not detectable in virgin mammary tissue. The onset of MDGI expression was detected during pregnancy with increasing amounts towards terminal differentiation of the gland (Bohmer et al. 1985; Muller et al. 1989). In the midpregnant mammary tissue, which is characterized by a high proliferation rate and ongoing differentiation, the amount of MDGI reaches 0.1% of the total protein and mRNA. Maximal MDGI expression was obtained in the lactating, terminally differentiated mammary gland with 0.5–1% of total protein and RNA (Muller et al. 1989). MDGI is also secreted into bovine and human milk (Brandt et al. 1988). Its level increases with lactation (Brandt et al. 1988). MDGI transcripts and protein were analyzed in tissue sections from embryonic, virgin, pregnant, and lactating mammary gland. In the mammary anlagen of a 5-month-old female embryo, MDGI transcripts and protein are clearly detectable. MDGI is evenly distributed over the proximal and distal epithelial cells of the mammary rudiment and the epidermis. Mesenchymal tissue was not labeled. β-Casein expression was also found, but was restricted to the mammary epithelial rudiment. The coexpression of β-casein and MDGI is probably due to the presence of maternal lactogenic hormones in the embryo. In the immature, resting mammary gland of the virgin animal, MDGI is not expressed. During midpregnancy, ductal and alveolar epithelial cells, in combination with myoepithelial cells, form the lobuloalveolar gland. During this stage of development ductal cells express MDGI at a rather low level, whereas MDGI is clearly detectable over the alveoli. Other cell types, such as myoepithelial cells and connective tissue, do not transcribe the MDGI gene. It has been found that early alveolar differentiation is coupled with an increase in MDGI transcription. Interestingly, in those alveolar epithelial cells bordering the connective tissue, MDGI transcripts are augmented (Kurtz et al. 1990). These are the epithelial cells with an increased proliferation rate that first start to transit into functional differentiation. The presence of mesenchymal components has been shown to be important for the induction of limited functional differentiation (Oka and Yoshimura 1986). The induction of MDGI expression by paracrine-acting connective tissue signals could be accompanied by a feedback reaction causing inhibition of growth and the onset of functional differentiation of the mammary epithelium. On the other hand, it might also be possible that MDGI expression is a necessary prerequisite for the alveolar outgrowth into the regressing stroma. In contrast to the pregnant stage, in the differentiated gland both the alveolar and the ductal epithelial cells express MDGI. Along the duct a striking increase in the level of MDGI expression was obtained with the highest levels in the terminal part. In lactation, MDGI is expressed at higher levels in the ductal epithelial cells than in the alveolar epithelial cells. Myoepithelial cells and fibroblasts again do not express MDGI. In the lactating mammary gland, MDGI expression follows an anatomical-histological gradient. In the less differentiated distal parts of the gland, the MDGI level is clearly reduced compared to the proximal regions. In all stages of mammary-gland development, the MDGI-mRNA level parallels the protein level. This was also confirmed by an electron microscopical analysis of the intracellular distribution of MDGI in the mammary gland of different functional states (Muller et al. 1989). The apparent close parallel between the expression of MDGI and its mRNA during normal mammary gland development suggest that transcriptional control mechanisms are a major regulator. Those control mechanisms could be induced by lactogenic hormones, such as prolactin, leading to the enhanced MDGI expression necessary for local control of proliferation and differentiation.

3.3.2
Hormonal Dependence of Expression in Mouse Organ Culture

In order to obtain access to the more complex mechanisms of MDGI expression, advantage has been taken of the organ-culture system employing explants from abdominal mammary glands of primed virgin and pregnant mice. The mammary gland of 4-week-old virgin Balb/c-mice primed with estradiol and progesterone consists of a system of sparsely branching ducts embedded in adipose tissue. Upon cultivation with aldosterone, prolactin, insulin, and hydrocortisone (APIH medium), lobuloalveolar structures develop that exhibit cubic epithelial cells with fat vacuoles and eosinophilic secretory material in their lumina (Binas et al. 1992). It has been demonstrated that both β-casein and MDGI transcripts, which are absent in the mammary tissue of primed virgin mice, become expressed in the terminally differentiated mammary tissue from lactating mice. An identical expression pattern was demonstrated for mammary glands cultured for 9 days in APIH medium (Binas et al. 1992). Whole-mount stains of the cultured glands represented fully branching of the ducts and developed lobuloalveoli, as is known from lactating gland in vivo (Binas et al. 1992). Thus, local expression of β-casein and MDGI mRNA can be induced by a combination of systemic hormones. The tissue distribution of β-casein and MDGI has been shown by immunostaining to be different. β-Casein becomes clearly expressed upon culture in the APIH medium in alveoli, ductules, and the lumina if compared to the control, where only MDGI was detected. The use of organ explants from primed virgin mice for studying in vivo morphogenesis implies that lobuloalveolar development and functional differentiation cannot be separated from each other in culture. In order to study a developmental stage associated with functional differentiation rather than with lobuloalveolar morphogenesis, it has been taken advantage of organ explants derived from late-pregnant mice. They were cultured in a medium containing prolactin, cortisol, and insulin (PIH medium). In addition, to monitor the influence of prolactin on MDGI expression, either the hormone concentration was reduced or prolactin and cortisol were replaced by EGF. At optimal prolactin concentrations, the β-casein mRNA level was only slightly enhanced, while at the same time the MDGI mRNA level clearly declined. At a reduced prolactin concentration, transcription of the MDGI gene ceased completely, while some casein expression still took place. It should be emphasized that under these conditions the typical phenotype of differentiated alveoli was maintained, although lowering the prolactin concentrations led to suppression of secretory alveolar activity. If prolactin and cortisol were substituted by EGF (EI-medium) MDGI expression was completely inhibited. However, by subsequently culturing the EI-preteated explants in the medium containing prolactin, insulin, and hydrocortison (PIH medium), MDGI transcription could be reinduced and reached the PIH control (Binas et al. 1992). Finally, MDGI behaves as a differentiation marker for the mammary gland, differing in some aspects from casein. The cellular patterns of expression are variable in their dependence on the developmental stage and seem to be regulated by different mechanisms. The data support a role for MDGI during functional differentiation in the mammary gland of pregnant mice. This role is at least dependent on prolactin.

3.4 Nuclear Localization

Immunocytochemistry of Lowicryl K4M embedded sections, as well as of cryosections from lactating bovine mammary gland tissue, showed the euchromatic regions of nuclei from mammary epithelial cells to be densely labeled with the immuno-gold complex (Binas et al. 1992). Since then, other members of the FABP family have also been shown to be present in the nuclei of hepatocytes and myocytes (Borchers et al. 1989). In order to prove the molecular identity of the nuclear antigen, lactating mammary tissue was homogenized and a nuclear fraction was enriched. By Western blotting and immunostaining with affinity-purified anti-MDGI antibodies, only tiny amounts of the nuclear localized MDGI could be detected, much less than expected from the immunocytochemistry (Vogel et al. 1990).

Using an MDGI-specific antibody, another cross-reactive 70-kDa protein has been detected. This protein was present neither in the soluble cytoplasmic fraction nor in the mitochondrial or microsomal fraction. Thus it seemed to be truly nuclear. The 70-kDa antigen was also recognized by antibodies raised against synthetic peptides of the MDGI sequence. Comparing the nuclear fractions from several tissues with respect to the presence of p70 indicates the mammary gland specificity, in keeping with the MDGI distribution. The MDGI-related 70-kDa protein was purified by a combination of steps, including salt extraction of nuclei, ammonium sulfate precipitation, selective solubilization, passage through Q-Sepharose, mono S-chromatography, and RP-chromatography in formic acid. Several peptides comprising 120 amino acids have been sequenced. They showed no significant homology, either with MDGI or with any protein in a sequence data bank. From the sequencing data it is apparent that p70 constitutes a protein that is different from MDGI. As outlined before, a truncated form of MDGI has been found complexed with a glycoprotein eluting as an about 70-kDa complex from Sepharose. Thus the nuclear-detected immunoreactivity may be related to the former described protein complex. Furthermore, it is possible that this p70 can function as protease, regulating the release of the C-terminal domain of MDGI, which is shown to develop fully biological activity. Alternatively, it could be an entirely different protein, sharing a structural and/or functional domain with MDGI; for example, a ligand-binding domain common to MDGI. Taking into account its association with chromatin, purification behavior, yield, and DNA-binding capacity, it is tempting to speculate about a role of p70 as a transcription factor or transport function shuttling MDGI into the nucleus and exerting tumor suppressor activity (Huynh et al. 1995).

4 Tumor Necrosis Factor Alpha (TNF-α)

4.1 TNF-α Structure and Function in Breast Tissue

Tumor necrosis factor (TNF), a homotrimer of 17-kDa subunits, is a potent macrophage-derived multifunctional cytotoxic cytokine which plays a possible crucial role in apoptosis or killing of tumor cells by activated effector cells of the human immune system. Several other cell types also produce TNF.

TNF seems to be pleiotrophic, causes the necrosis of different solid tumors in vivo, and exhibits direct cytotoxic and/or cytostatic effects on several cell lines. Furthermore, it induces cachexia, septic shock, shows antiviral, antiinflammatory, and immunoregulatory activities, but also may stimulate the growth of different cells. As in several other cell types in vivo, in mammary epithelial cells two different receptors (TNFR-I or p55 or p60, and TNFR-II or p80 or p75) bind TNF with a similar high affinity. The extracellular domain of both TNFR-I and TNFR-II share 28% sequence identity with each other and other members of the TNF/NGFR-receptor family. Little homology exists between the cytoplasmatic domains of TNFR-I and TNFR-II. The cytoplasmic domain of TNF-R I contains a region referred to as a death domain (DD), which is found in proteins involved in apoptosis.

TNF-α has a functional role in the mammary epithelium such as phosphorylation of epidermal growth factor (EGF) receptor (Donato et al. 1989, 1992) hsp27 and hsp70 and other proteins (Minshull et al. 1990; Alberts et al. 1993; Jaattela et al. 1993). Overexpression of these proteins protects cells against the cytotoxic effects of TNF-α, which can be understood as a feedback mechanism (Landry et al. 1989). Additionally, TNF-α induces phosphorylation of the nuclear factor-κB (NF-KB) and the eukaryotic initiation factor EIF 4E (Duh et al. 1989; Marino et al. 1991). The cellular responses to TNF-α are necrotic or apoptotic killing, resulting in DNA fragmentation, lipid peroxidation, changes in arachidonic acid metabolism, increased production of prostaglandin E and A, and inhibition of mitochondrial electron transfer resulting in a collapse of mitochondrial membrane potential, an early marker of apoptosis (Jeoung et al.

1995). In parallel, TNF induces the synthesis of protective proteins in target cells, e.g., the Mn-superoxide dismutase in MCF-7 breast cancer cells (Zyad et al. 1994). Transfection of human tumor cells with the c-*sis* gene leads to downmodulation of TNF receptors and also to a decrease of intracellular gluthatione levels. Through transfection of NIH 3T3 cells with platelet-derived growth factor (PDGF-Blc-sis) fusion gene, Aggarwal et al. (1995) demonstrated that overexpression of PDGF-Blc-sis in certain tumor cells leads to their protection from the anti-cellular effects of TNF. Interestingly, the cytotoxic effects of TNF-α on human MCF-7 cells may be potentiated by 1,25-dihydroxyvitamin D_3, which stimulates the antitumor activity of the immune system (Rocker et al. 1994).

TNF-α shows a dose- and time-dependent antimitogenic activity in (ER+) MCF-7 cells accompanied by a decreased number of cells in the S-phase and arresting cells in the G_1 phase of the cell cycle. Furthermore, the response of cell cycle-regulated proteins such as the Retinoblastoma protein (Rb) was decreased after TNF-α treatment of mammary epithelial cells, whereas p53 and p21 expression levels increased in response to TNF-α. It has been concluded that TNF-α-mediated anti-mitogenicity of MCF-7 cells specifically involves regulation of Rb and p53 protein expression. However, G_1 cyclins, like cyclin D, did not show decreased expression levels in response to TNF-α treatment under normal culture conditions using asynchronous cells, whereas expression levels of cyclin A and B did not change significantly in response to TNF-α (Jeoung et al. 1995). TNF-α treatment of MEC results in increased p53 expression which, in turn, induces p21 expression whereas Rb finally is decreased. TNF-α has been shown to mediate its action through the activation of several other transcriptional factors, including c-*Jun*/AP-1, c-*Fos*, c-*Myc*, IRF-1, and the early growth response gene Egr-1 (Aggarwal and Natarajan 1996). Most of these factors have been implicated in cell proliferation. Cai et al. (1997) emphasized that nuclear factors distinct from NF-κB are capable of mediating TNF-α-induced cell death. The zinc finger protein A20, a TNF-α and IL-1-inducible gene product in endothelial cells, specifically inhibits signal transduction pathways induced by TNF and IL-1 in breast cancer cell lines, suggesting that it functions as a negative regulator of cytokine response (Jaattela et al. 1996). Overexpression of A20 in MCF-7 cells inhibits TNF-α-induced apoptosis, whereas cytotoxicity induced by anti-Fas antibodies, lymphokine-activated killer (LAK) cells, serum starvation, oxidative stress, or okadaic acid is not inhibited. Overexpression of A20, similar to TNF-induced apoptosis, inhibits the TNF-α-induced activation of phospholipase A_2 in a dose-dependent manner, whereas the anti-Fas activation of phospholipase A_2 was not affected. However, A20 blocks TNF-induced signal transduction to the nucleus, such as the activation of the transcription factors NF-κB and AP-1, but binding of TNF to its surface receptors is not affected (Jaattela et al. 1996).

4.2
Hormone Dependence of TNF-Related Effects in Mammary-Derived Cells

Estradiol (E_2) plays a well-established role in the etiology of breast cancer. It is reversibly formed via estradiol 17-β-hydroxysteroid dehydrogenase (E_2DH) from the biological inactive estrogen (E_1). E_2DH has a pivotal role for the regulation of E_2 concentrations in normal and malignant breast tissues. Additionally, the effect of TNF-α, IL-1β and IL-6 on E_2DH activity in MCF-7 ER-positive breast cancer cells has been investigated (Duncan et al. 1994). Treatment of MCF-7 cells with TNF-α and IL-1β significantly increased reductive E_2DH activity, e.g., the conversion of E_1 to E_2. A combination of TNF-α, IL-1β, and IL-6 stimulated reductive E_2DH activity synergistically. No cytokine-affected oxidative (E_2 to E_1) activity has been observed. It has been concluded that TNF-α, IL-1β and IL-6 might play an important role in regulating E_2DH activity in breast cancer cells and might act synergistically in vivo to enhance the formation of E_2 in breast tumors. In situ estrogen synthesis makes an important contribution to the high estrogen concentrations which was found in breast tumors (Macdiarmid et al. 1994). TNF-α stimulates aromatase activity in the presence of dexamethasone, not only in breast tumor-derived fibroblasts, but also in fibroblasts derived from normal breast tissue which, in postmenopausal women, contains mainly of adipose tissue. Formation of estrone via the sulphatase pathway may be the major route of estrogen synthesis in tumors. TNF-α stimulates the sulphatase activity in a dose-dependent manner (Macdiarmid et al. 1994). TNF-α significantly stimulated fibroblast aromatase activity in a dose-dependent manner in the presence of stripped fetal calf serum and in the presence of dexamethasone. Interestingly, IL-1 or IL-6 had no marked synergism with TNF-α. Using a specific radioimmunoassay (RIA), significant concentrations of TNF-α were detected in samples of breast cyst fluid (BCF) and breast tumor cytosol, which had previously been shown to stimulate aromatase activity, but was not detected in conditioned medium from breast tumor-derived fibroblasts. Taken together, TNF-α may be potentially expressed and produced in the adipose breast tissue components, and may play an important role in regulating estrogen synthesis in normal and malignant breast tissues (Macdiarmid et al. 1994).

As outlined before, the growth of hormone-dependent human MCF-7, ZR-75-1 and T47-D breast cancer cells was inhibited by TNF-α with an IC_{50} of 0.25 nM (Mueller et al. 1996). In contrast, the growth of hormone-independent cell lines MDA-MB 231 and HS578T was not affected by TNF-α alone, but a synergistic inhibition was observed when using IFN-γ and TNF-α together. The mRNA for the proto-oncogene c-*myc*, as an intracellular indicator of cell activation, was significantly increased in MCF-7 cells treated with TNF-α. No change in the number of TNF-R could be observed. The combinatorial treatment of breast tumors with TNF-α and other cytokines, like IFN-γ, may provide a successful approach to overcome the cellular heterogeneity of advanced breast carcinomas (Mueller et al. 1996).

4.3 Expression of TNF-α in Breast Cancer

TNF-α expression was correlated with proliferative activity, neovascularization, lymph node status, grading and the degree of lymphoid infiltration. Expression of TNF-α or TNF-R p75 was not detectable in normal or nonmalignant breast tissue adjacent to carcinomas (Pusztai et al. 1994). The TNF receptor p55 was expressed by occasional stromal cells in normal tissue. TNF-α was expressed focally in 50% of all carcinomas, being largely localized to macrophage-like cells in the stroma. A population of stromal cells in all tumors and a varying proportion of neoplastic cells expressed TNF-R p55 in 75% of the examined cases. TNF-R p75 was detected in stromal cells in about 70% of examined tumors. No association between the TNF-R expression and clinically relevant parameters, such as lymph node status, tumor grade, labeling index, or degree of angiogenesis (CD31) was found. However, there was a correlation between the expression of TNF-R p55 in blood vessels and the number of leukocytes. It is possible that TNF-α and its receptors play an important role in the progression of preclinical breast cancer rather than in clinical manifestation or in advanced tumors (Pusztai et al. 1994). Other studies claim that the expression of TNF-α mRNA was scant and focal, whereas TNF-α protein expression was more widespread and frequent (Miles et al. 1994). These findings may reflect the relative instability of TNF-α mRNA as compared to the proposed longer half-life of the protein. TNF-α mRNA and protein colocalized to the tumor stroma, particularly in stromal areas adjacent to areas of invasive carcinoma (Miles et al. 1994). Cells of the mononuclear infiltrate, in particular T cells, expressed the TNF receptor p75. The p55 receptor was expressed to a lesser extent. Interestingly, breast epithelial cells and tumor cells were negative for both TNF receptors in all cases. The lack of TNF receptors in MEC would suggest that TNF-α may not be responsible for direct cytolysis of primary breast cancer tumor cells (Miles et al. 1994). However, TNF-α may be important for indirect antitumor mechanisms including cytotoxic T-cell response and the augmentation of non-MEC-restricted cytotoxic mechanisms such as NK-cell activity and lymphokine-activated killer-cell induction. TNF-α may also be involved in the autocrine induction of macrophage cytotoxicity (Ranges et al. 1987; Owen Schaub and Grimm 1989; Lake and Riches 1992). Additionally, the antitumor activity of TNF-α may be mediated by effects on the tumor vascular endothelium (Miles et al. 1994). The signal transduction cascade of TNF-α into the nucleus as well as the specificity of the effects is not yet clear. Recently, two intracellular TNF receptor-associated factors TRAF1 and TRAF2, which interact with type 2 TNF receptor and the TNF receptor 1-associated death domain protein, TRADD, has been cloned.

Most likely these proteins regulated specificity of the TNF-a action by switching on different transduction pathways (Rothe et al. 1994; Hsu et al. 1995). However, protein kinase C phosphorylation is specifically associated with the cytoplasmatic domain of the p60 form of the TNF receptor (Darnay et al. 1994). TNF-α stimulation of MCF-7 for 5–15 min induced approximately 50–240% increases in phosphorylation of p60CDΔ1-GST fusion protein. This p60 TNF receptor-associated protein and the associated kinase are referred to as p60-TRAP and p60-TRAK, respectively (Darnay et al. 1994).

The nuclear factor-kB (NF-κB) activity is induced by TNF-α and its role in signaling TNF-α-induced cell death remains controversial. TNF-α is one of the cytokines known to activate NF-κB within minutes, leading to the transcriptional activation of various important cellular genes (Aggarwal and Natarajan 1996). The activation of NF-κB is considered integral to the transfer of the TNF signal to the nucleus (Thanos and Maniatis 1995). Both types of TNF receptors independently mediate NF-κB activation after TNF binding (Hsu et al. 1995; Rothe et al. 1995). TNF-α activates phosphatidylcholine-specific phosphlipase C. This leads to the sequential activation of an acidic sphingo-myelinase and the production of ceramide, which in breast tumors causes the NF-κB activation and induces apoptosis (Bielawska et al. 1992; Wiegmann et al. 1994; Jaattela et al. 1995). However, Cai et al. demonstrated that NF-κB activation is not required for induction of apoptosis by TNF (Cai et al. 1997). Several data indicate that TNF-R1-associated DD protein (TRADD) directly interacts with one of the TNF-R2-associated factors (TRAF2) and the Fas-associated factor (FADD) to induce NF-κB activation and apoptosis, respectively (Hsu et al. 1995, 1996). Conversely, a dominant-negative FADD mutant inhibited TNF-induced apoptosis, but did not inhibit NF-κB activation. Thus, it is suggested that TNF-R1 may utilize distinct TRADD-dependent mechanisms to activate signaling pathways for NF-κB activation and apoptosis. TNF has been shown to mediate FADD-DN, a dominant-negative derivative of FADD, which blocked TNF-induced apoptosis while not affecting NF-κB activation (Chinnaiyan et al. 1995).

Human breast cancer progressively grows despite the presence of extensive lymphocytic infiltration and specific antitumor immune recognition, thereby calling into question the competency of breast cancer tumor-infiltrating lymphocytes (TIL). Compared to lymphocytes obtained from benign lesions, a significant increase of TNF-α and IL-2 in breast cancer TILs has been observed (Camp et al. 1996) which could not be correlated to disease-free survival.

5 Conclusion

In this chapter we described polypeptidic growth inhibitors for the mammary gland. Excluding tumor necrosis factor alpha, the others are multifunctional in the sense of mammary gland development, which means that they are not purely inhibitors of growth. So we can learn that the transforming growth factor beta (TGF-β) has a regulating role for the nontransformed mammary epithelium rather than a classical growth inhibition function. Data describing TGF-β as a growth inhibitor were mostly obtained using monolayer cell cultures on coated or plastic culture dishes. These culture conditions are nonphysiological, thus it is not surprising that often these data are opposite to data obtained under more physiological conditions (Vonderhaar 1987; Cardiff et al. 1991; Jhappan et al. 1993; Atwood et al. 1995; Kordon et al. 1995). Furthermore, TGF-β's function strongly depends on the physiological stage of the mammary epithelial cell (autocrine interaction) and on the quality of the microenvironment (paracrine interaction). The report of Pierce et al. (1993) strongly supports this argument. Even after extreme retardation of the ductal development by overexpression of TGF-β during pregnancy, transgenic animals restored the ability to feed litters routinely. Data obtained from single-type monolayer cell culture should be extrapolated into the in vivo situation very carefully. In particular, it will be interesting to clarify and exploit possible mechanistic links between TGF-β and the process of programmed cell death. TNF-α, a growth factor exploring an indirect effect on mammary epithelium, could be of main interest in this sense. Depending on the physiological state of the mammary gland, TNF-α induces apoptosis while acting in the stroma of the mammary gland. So far there are no studies examining possible synergistic effects of TGF-β and TNF-α, both stored in the stroma. Studying this synergism on double trangenic mice overexpressing both TGF-β and TNF-α in the mammary would be a good model to find an answer to this question.

In the second part of the chapter we discussed the role of the mammary-derived growth inhibitor (MDGI) for the development of the mammary epithelium. Summarized from the organ culture data, MDGI most likely acts contrarily in vitro and in vivo. The tissue- and cell-specific, developmentally and hormone-dependent expression; the preferential inhibitory action on the in vitro proliferation of normal mammary epithelial cells; the differentiation-modulating activity in organ cultures; and the antagonistic interaction with the "mammogenic" EGF indicate a role for MDGI in the local regulation of the growth and development of the mammary gland. It remains to be elucidated whether different forms of MDGI, encoded by different genes, regulate different functions, such as lipid transport or growth inhibition, or whether the same molecule acts on different events at different stages of tissue development. What are the functions of the closely related MDGI forms in the heart or brain? The comparison of the intron/exon structure of genes of the protein family supports a common ancestral origin (Matarese et al. 1989). Future work will show whether certain differences in the promoter regions and other regulatory elements reflect divergent evolution to accomplish specialized functions in tissue-specific growth and development. The main questions to be answered in future are which elements direct the expression of the MDGI in the mammary gland and cFABP in cardiac tissue? What is the molecular basis for hormone-dependent MDGI expression during differentiation? what is the mechanism switching off MDGI expression in normal proliferating and mammary carcinoma cells? In this context it is of main interest which role other growth factors like EGF and TGF-β might play. It has not yet been identified a pathway leading to the mechanisms underlying the growth-inhibitory or differentiation-associated activities of MDGI. A primarily intracellular mechanism of action must be considered, as so far no membrane receptor for MDGI has been identified. It might be possible that the functional receptor binds only to the C-terminal peptide released for activating the function of MDGI. We hypothesize, that the C-terminal of MDGI, mimicking exclusively all biological effects of the major protein could be a key element in regulating growth and differentiation of the mammary epithelium during embryogenesis and pregnancy (Fig. 2). MDGI is a member of the FABP proetin family. The involvement of eicosanoids in the regulation of proliferation and differentiation has been discussed in terms of the stimulation of growth of the mammary epithelium (Imagawa et al. 1990). It is exciting to presume that this pathway is affected by MDGI because it could lead us to new and physiologically important regulators of growth and differentiation. Finally, MDGI expression is dependent on systemic hormones. Therefore, the study of its action might serve as a model indicating how the interplay between endocrine and locally acting factors controls growth and development in the normal mammary gland, and identifying steps abrogated in mammary carcinoma cells.

In future, it will be necessary to study growth and differentiation of epithelial tissue in a more complex manner. Using culture models close to the in vivo situation, e.g., serum-free whole-mount organ culture systems, might be a step in the right direction. It would be of interest to prepare mammary glands from genetically modified mice to study effects of growth factors under serum-free culture conditions. The combination of novel knockout techniques will allow genes to be switched on/off conditionally in adult mice; – a combination of classical primary MEC cultures or organ cultures will open a new, modern era to study growth regulation and differentiation of the mammary epithelium.

Acknowledgements. We would like to thank Drs. David S. Salomon and Richard Grosse, who generously communicated ideas and experimental results for the MDGI data. Also we would like to thank Stephen Barnett for the help in preparing the final manuscript. The article was supported by DFG-grant EB152/4-1.

References

Aakvaag A, Utaaker E, Thorsen T, Lea OA, Lahooti H (1990) Growth control of human mammary cancer cells (MCF-7 cells) in culture: effect of estradiol and growth factors in serum-containing medium. Cancer Res 50:7806–7810

Abraham JA, Mergia A, Whang JL, Tumolo A, Friedman J, Hjerrild KA, Gospodarowicz D, Fiddes JC (1986) Nucleotide sequence of a bovine clone encoding the angiogenic protein, basic fibroblast growth factor. Science 233:545–548

Aggarwal BB, Natarajan K (1996) Tumor necrosis factors: developments during the last decade. Eur Cytokine Netw 7:93–124

Aggarwal BB, Pocsik E, Totpal K, Ali Osman F (1995) Suppression of antiproliferative effects of tumor necrosis factor by transfection of cells with human platelet-derived growth factor B/c-sis gene. FEBS Lett 357:1–6

Alberts AS, Thorburn AM, Shenolikar S, Mumby MC, Feramisco JR (1993) Regulation of cell cycle progression and nuclear affinity of the retinoblastoma protein by protein phosphatases (published erratum appears in Proc Natl Acad Sci USA 1993 Mar 15;90(6):2556). Proc Natl Acad Sci USA 90:388–392

Andres J, Stanley K, Cheifetz S, Massague J (1989) Membrane-anchored and soluble forms of beta glycan, a polymorphic proteoglycan that binds transforming growth factor-beta. J Cell Biol 109:3137–3145

Arrick BA, Korc M, Derynck R (1990) Differential regulation of expression of three transforming growth factor beta species in human breast cancer cell lines by estradiol. Cancer Res 50:299–303

Arteaga CL, Hanauske AR, Clark GM, Osborne CK, Hazarika P, Pardue RL, Tio F, Von Hoff DD (1988a) Immunoreactive α transforming growth factor activity in effusions from cancer patients as a marker of tumor burden and patients prognosis. Cancer Res 48:5023–5028

Arteaga CL, Tandon AK, Von Hoff DD, Osborne CK (1988b) Transforming growth factor beta: potential autocrine growth inhibitor of estrogen receptor-negative human breast cancer cells. Cancer Res 48:3898–3904

Arteaga CL, Coffey RJ Jr, Dugger TC, McCutchen CM, Moses HL, Lyons RM (1990) Growth stimulation of human breast cancer cells with anti-transforming growth factor beta antibodies: evidence for negative autocrine regulation by transforming growth factor beta. Cell Growth Differ 1:367–374

Arteaga CL, Dugger TC, Hurd SD (1996) The multifunctional role of transforming growth factor (TGF)-beta s on mammary epithelial cell biology. Breast Cancer Res Treat 38:49–56

Atwood CS, Ikeda M, Vonderhaar BK (1995) Involution of mouse mammary glands in whole organ culture: a model for studying programmed cell death. Biochem Biophys Res Commun 207:860–867

Barrett Lee P, Travers M, Luqmani Y, Coombes RC (1990) Transcripts for transforming growth factors in human breast cancer: clinical correlates. Br J Cancer 61:612–617

Basset P, Bellocq JP, Wolf C, Stoll I, Hutin P, Limacher JM, Podhajcer OL, Chenard MP, Rio MC, Chambon P (1990) A novel metalloproteinase gene specifically expressed in stromal cells of breast carcinomas. Nature 348:699–704

Bassuk JA, Tsichlis PN, Sorof S (1987) Liver fatty acid binding protein is the mitosis-associated polypeptide target of a carcinogen in rat hepatocytes. Proc Natl Acad Sci USA 84:7547–7551

Behlke J, Mieth M, Bohmer FD, Grosse R (1989) Hydrodynamic and circular dichroic analysis of mammary-derived growth inhibitor (MDGI). Biochem Biophys Res Commun 161:363–370

Bernlohr DA, Angus CW, Lane MD, Bolanowski MA, Kelly TJ Jr (1984) Expression of specific mRNAs during adipose differentiation: identification of an mRNA encoding a homologue of myelin P2 protein. Proc Natl Acad Sci USA 81:5468–5472

Bielawska A, Linardic CM, Hannun YA (1992) Modulation of cell growth and differentiation by ceramide. FEBS Lett 307:211–214

Billich S, Wissel T, Kratzin H, Hahn U, Hagenhoff B, Lezius AG, Spener F (1988) Cloning of a full-length complementary DNA for fatty-acid-binding protein from bovine heart. Eur J Biochem 175:549–556

Binas B, Spitzer E, Zschiesche W, Erdmann B, Kurtz A, Muller T, Niemann C, Blenau W, Grosse R (1992) Hormonal induction of functional differentiation and mammary-derived growth inhibitor expression in cultured mouse mammary gland explants. In Vitro Cell Dev Biol 28A:625–634

Bohmer FD, Lehmann W, Schmidt HE, Langen P, Grosse R (1984) Purification of a growth inhibitor for Ehrlich ascites mammary carcinoma cells from bovine mammary gland. Exp Cell Res 150:466–476

Bohmer FD, Lehmann W, Noll F, Samtleben R, Langen P, Grosse R (1985) Specific neutralizing antiserum against a polypeptide growth inhibitor for mammary cells purified from bovine mammary gland. Biochim Biophys Acta 846:145–154

Bohmer FD, Kraft R, Otto A, Wernstedt C, Hellman U, Kurtz A, Muller T, Rohde K, Etzold G, Lehmann W et al. (1987a) Identification of a polypeptide growth inhibitor from bovine mammary gland. Sequence homology to fatty acid- and retinoid-binding proteins. J Biol Chem 262:15137–15143

Bohmer FD, Sun Q, Pepperle M, Muller T, Eriksson U, Wang JL, Grosse R (1987b) Antibodies against mammary derived growth inhibitor (MDGI) react with a fibroblast growth inhibitor and with heart fatty acid binding protein. Biochem Biophys Res Commun 148:1425–1431

Bohmer FD, Mieth M, Reichmann G, Taube C, Grosse R, Hollenberg MD (1988) A polypeptide growth inhibitor isolated from lactating bovine mammary gland (MDGI) is a lipid-carrying protein. J Cell Biochem 38:199–204

Borchers T, Unterberg C, Rudel H, Robenek H, Spener F (1989) Subcellular distribution of cardiac fatty acid-binding protein in bovine heart muscle and quantitation with an enzyme-linked immunosorbent assay. Biochim Biophys Acta 1002:54–61

Bouchard L, Lamarre L, Tremblay PJ, Jolicouer P (1989) Stochastic appearance of mammary tumours in transgenic mice carrying the MMTV/c-neu oncogene. Cell 57:931–936 (Abstract)

Brandt R, Grosse R (1992) Purification of a mammary-derived growth inhibitor (MDGI) related polypeptide expressed during pregnancy. Biochem Biophys Res Commun 189:406–413

Brandt R, Pepperle M, Otto A, Kraft R, Bohmer FD, Grosse R (1988) A 13-kilodalton protein purified from milk fat globule membranes is closely related to a mammary-derived growth inhibitor. Biochemistry 27:1420–1425

Brandt R, Eisenbrandt R, Zschiesche W, Binas B, Juergensen C, Theuring F (1997) Mammary gland-specific hEGF receptor transgene expression induces neoplasia. (unpublished)

Brown PD, Levy AT, Margulies IM, Liotta LA, Stetler Stevenson WG (1990) Independent expression and cellular processing of Mr 72,000 type IV collagenase and interstitial collagenase in human tumorigenic cell lines. Cancer Res 50:6184–6191

Bullogh WS, Laurence EB (1960) The control of epidermal mitotic activity in the mouse. Proc R Soc Lond Ser B 151:990–993

Cai Z, Korner M, Tarnatino N, Chouaib S (1997) IkB-alpha overexpression in human breast carcinoma MCF-7 cells inhibits nuclear factor kB activation but not tumor necrosis factro-alpha-induced apoptosis. Br J Cancer 73:1356–1361

Camp BJ, Dyhrman ST, Memoli VA, Mott LA, Barth RJ Jr (1996) In situ cytokine production by breast cancer tumor-infiltrating lymphocytes. Ann Surg Oncol 3:176–184

Cardiff RD, Sinn E, Muller W, Leder P (1991) Transgenic oncogene mice. Tumor phenotype predicts genotype. Am J Pathol 139:495–501

Chen HM, Tritton TR, Kenny N, Absher M, Chiu JF (1996) Tamoxifen induces tgf-beta-1 activity and apoptosis of human mcf-7 breast cancer cells in vitro. J Cell Biochem 61:9–17

Chinander LL, Bernlohr DA (1989) Cloning of murine adipocyte lipid binding protein in *Escherichia coli*. Its purification, ligand binding properties, and phosphorylation by the adipocyte insulin receptor. J Biol Chem 264:19564–19572

Chinnaiyan AM, O'Rourke K, Tewari M, Dixit VM (1995) FADD, a novel death domain-containing protein, interacts with the death domain of Fas and initiates apoptosis. Cell 81:505–512

Coleman S, Daniel CW (1990) Inhibition of mouse mammary ductal morphogenesis and down-regulation of the EGF receptor by epidermal growth factor. Dev Biol 137:425–433

Colletta AA, Wakefield LM, Howell FV, van Roozendaal KE, Danielpour D, Ebbs SR, Sporn MB, Baum M (1990) Anti-oestrogens induce the secretion of active transforming growth factor beta from human fetal fibroblasts. Br J Cancer 62:405–409

Colletta AA, Wakefield LM, Howell FV, Danielpour D, Baum M, Sporn MB (1991) The growth inhibition of human breast cancer cells by a novel synthetic progestin involves the induction of transforming growth factor beta. J Clin Invest 87:277–283

Colletta AA, Benson JR, Baum M (1994) Alternative mechanisms of action of anti-oestrogens. Breast Cancer Res Treat 31:5–9

Coombes RC, Barrett Lee P, Luqmani Y (1990) Growth factor expression in breast tissue. J Steroid Biochem Mol Biol 37:833–836

Daly RJ, Darbre PD (1990) Cellular and molecular events in loss of estrogen sensitivity in ZR-75-1 and T-47-D human breast cancer cells. Cancer Res 50:5868–5875

Daly RJ, King RJ, Darbre PD (1990) Interaction of growth factors during progression towards steroid independence in T-47-D human breast cancer cells. J Cell Biochem 43:199–211

Daly RJ, Carrick N, Darbre PD (1995) Progression to steroid autonomy is accompanied by altered sensitivity to growth factors in S115 mouse mammary tumour cells. J Steroid Biochem Mol Biol 54:21–29

Daniel CW, Robinson SD (1992) Regulation of mammary growth and function by TGF-beta. Mol Reprod Dev 32:145–151

Daniel CW, Silberstein GB, Van Horn K, Strickland P, Robinson S (1989) TGF-beta 1-induced inhibition of mouse mammary ductal growth: developmental specificity and characterization. Dev Biol 135:20–30

Dardick I, Burford Mason AP, Garlick DS, Carney WP (1992) The pathobiology of salivary gland. II. Morphological evaluation of acinic cell carcinomas in the parotid gland of male transgenic (MMTV/v-Ha-ras) mice as a model for human tumours. Virchows Arch A Pathol Anat Histopathol 421:105–113

Darnay BG, Reddy SA, Aggarwal BB (1994) Identification of a protein kinase associated with the cytoplasmic domain of the p60 tumor necrosis factor receptor. J Biol Chem 269:20299–20304

Datta H, Sullivan M, Ohri S, Macintyre I (1990) Transforming growth factor-beta 1 inhibits lipoxygenase and epoxygenase eicosanoid production by osteosarcoma cells. Biochem Soc Trans 18:1259–1260

Dickson MC, Martin JS, Cousins FM, Kulkarni AB, Karlsson S, Akhurst RJ (1995) Defective haematopoiesis and vasculogenesis in transforming growth factor-beta 1 knock out mice. Development 121:1845–1854

Dickson RB, Bates SE, McManaway ME, Lippman ME (1986) Characterization of estrogen responsive transforming activity in human breast cancer cell lines. Cancer Res 46:1707–1713

Dickson RB, Kasid A, Huff KK, Bates SE, Knabbe C, Bronzert D, Gelmann EP, Lippman ME (1987) Activation of growth factor secretion in tumorigenic states of breast cancer induced by 17 beta-estradiol or v-Ha-ras oncogene. Proc Natl Acad Sci USA 84:837–841

Dickson RB, Thompson EW, Lippman ME (1990) Regulation of proliferation, invasion and growth factor synthesis in breast cancer by steroids. J Steroid Biochem Mol Biol 37:305–316

Donato NJ, Gallick GE, Steck PA, Rosenblum MG (1989) Tumor necrosis factor modulates epidermal growth factor receptor phosphorylation and kinase activity in human tumor cells. Correlation with cytotoxicity. J Biol Chem 264:20474–20481

Donato NJ, Rosenblum MG, Steck PA (1992) Tumor necrosis factor regulates tyrosine phosphorylation on epidermal growth factor receptors in A431 carcinoma cells: evidence for a distinct mechanism. Cell Growth Differ 3:259–268

Dublin EA, Barnes DM, Wang DY, King RJ, Levison DA (1993) TGF alpha and TGF beta expression in mammary carcinoma. J Pathol 170:15–22

Duh EJ, Maury WJ, Folks TM, Fauci AS, Rabson AB (1989) Tumor necrosis factor alpha activates human immunodeficiency virus type 1 through induction of nuclear factor binding to the NF-kappa B sites in the long terminal repeat. Proc Natl Acad Sci USA 86:5974–5978

Duncan LJ, Coldham NG, Reed MJ (1994) The interaction of cytokines in regulating oestradiol 17 beta-hydroxysteroid dehydrogenase activity in MCF-7 cells. J Steroid Biochem Mol Biol 49:63–68

Eriksson V, Hansson E, Nilsson K, Jansson H, Sundelin J, Peterson P (1991) Structural and functional features of different types of cytoplasmic fatty acid binding proteins. Biochim Biophys Acta 1081:1–24

Ervin PR Jr, Kaminski MS, Cody RL, Wicha MS (1989) Production of mammastatin, a tissue-specific growth inhibitor, by normal human mammary cells. Science 244:1585–1587

Ethier SP, Van de Velde RM (1990) Secretion of a TGF-beta-like growth inhibitor by normal rat mammary epithelial cells in vitro. J Cell Physiol 142:15–20

Fernandez Pol JA, Klos DJ, Grant GA (1986) Purification and biological properties of type beta transforming growth factor from mouse transformed cells. Cancer Res 46:5153–5161

Gatherer D, ten Dijke P, Baird DT, Akhurst RJ (1990) Expression of TGF-beta isoforms during first trimester human embryogenesis. Development 110:445–460

Gorsch SM, Memoli VA, Stukel TA, Gold LI, Arrick BA (1992) Immunohistochemical staining for transforming growth factor beta 1 associates with disease progression in human breast cancer. Cancer Res 52:6949–6952

Grosse R, Langen P (1990) Mammary-derived growth inhibitor. In: Sporn MB, Roberts A (eds) Peptide growth factors and their receptors. Springer, Berlin Heidelberg New York, pp 249–265

Haase H, Wallukat G, Vetter R, Will H (1987) Characterization of calcium antagonist receptors in highly purified procine cardiac sarcolemma. Biomed Biochim Acta 46:S363–S369

Haliday EM, Ramesha SS, Ringold G (1991) TNF induces *c-fos* via a novel pathway requirring conversion of arachidonic acid to a lipoxigenase metabolite. EMBO J 10:109–115

Hannigan G, Williams B (1991) Signal transduction by interferon alpha through arachidonic acid metabolism. Science 205:204–206

Heine UI, Munoz EF, Flanders KC, Ellingsworth LR, Lam HY, Thompson NL, Roberts AB, Sporn MB (1987) Role of transforming growth factor-beta in the development of the mouse embryo. J Cell Biol 105:2861–2876

Heuckeroth RO, Birkenmeier EH, Levin MS, Gordon JI (1987) Analysis of the tissue-specific expression, developmental regulation, and linkage relationships of a rodent gene encoding heart fatty acid binding protein. J Biol Chem 262:9709–9717

Hosobuchi M, Stampfer MR (1989) Effects of transforming growth factor beta on growth of human mammary epithelial cells in culture. In Vitro Cell Dev Biol 25:705–713

Hresko RC, Bernier M, Hoffman RD, Flores Riveros JR, Liao K, Laird DM, Lane MD (1988) Identification of phosphorylated 422(aP2) protein as pp15, the 15-kilodalton target of the insulin receptor tyrosine kinase in 3T3-L1 adipocytes. Proc Natl Acad Sci USA 85:8835–8839

Hresko RC, Hoffman RD, Flores Riveros JR, Lane MD (1990) Insulin receptor tyrosine kinase-catalyzed phosphorylation of 422(aP2) protein. Substrate activation by long-chain fatty acid. J Biol Chem 265:21075–21085

Hsu H, Xiong J, Goeddel DV (1995) The TNF receptor 1-associated protein TRADD signals cell death and NF-kappa B activation. Cell 81:495–504

Hsu H, Shu HB, Pan MG, Goeddel DV (1996) TRADD-TRAF2 and TRADD-FADD interactions define two distinct TNF receptor 1 signal transduction pathways. Cell 84:299–308

Hsu YM, Wang JL (1986) Growth control in cultured 3T3 fibroblasts. V. Purification of an Mr 13,000 polypeptide responsible for growth inhibitory activity. J Cell Biol 102:362–369

Hubbs AF, Hahn FF, Thomassen DG (1989) Increased resistance to transforming growth factor beta accompanies neoplastic progression of rat tracheal epithelial cells. Carcinogenesis 10:1599–1605

Huynh HT, Larsson C, Narod S, Pollak M (1995) Tumor suppressor activity of the gene encoding mammary-derived growth inhibitor. Cancer Res 55:2225–2231

Imagawa W, Bandyopadhyay GK, Nandi S (1990) Regulation of mammary epithelial cell growth in mice and rats. Endocr Rev 11:494–523

Inagaki M, Moustakas A, Lin HY, Lodish HF, Carr BI (1993) Growth inhibition by transforming growth factor beta (TGF-beta) type I is restored in TGF-beta-resistant hepatoma cells after expression of TGF-beta receptor type II cDNA. Proc Natl Acad Sci USA 90:5359–5363

Ishikawa F, Miyazono K, Hellman U, Drexler H, Wernstedt C, Hagiwara K, Usuki K, Takaku F, Risau W, Heldin CH (1989) Identification of angiogenic activity and the cloning and expression of platelet-derived endothelial cell growth factor. Nature 338:557–562

Jaattela M, Wissing D, Bauer PA, Li G (1993) Major heat shock protein hsp70 protects tumor cells from tumor necrosis factor cytotoxicity. EMBO 11:3507–3512

Jaattela M, Benedict M, Tewari M, Shayman JA, Dixit VM (1995) Bcl-x and Bcl-2 inhibit TNF and Fas-induced apoptosis and activation of phospholipase A2 in breast carcinoma cells. Oncogene 10:2297–2305

Jaattela M, Mouritzen H, Elling F, Bastholm L (1996) A 20 zinc finger protein inhibits TNF and IL-1 signaling. J Immunol 156:1166–1173

Jeoung DI, Tang B, Sonenberg M (1995) Effects of tumor necrosis factor-alpha on antimitogenicity and cell cycle-related proteins in MCF-7 cells. J Biol Chem 270:18367–18373

Jetten AM, Shirley JE, Stoner G (1986) Regulation of proliferation and differentiation of respiratory tract epithelial cells by TGF beta. Exp Cell Res 167:539–549

Jhappan C, Geiser AG, Kordon EC, Bagheri D, Hennighausen L, Roberts AB, Smith GH, Merlino G (1993) Targeting expression of a transforming growth factor beta 1 transgene to the pregnant mammary gland inhibits alveolar development and lactation. EMBO J 12:1835–1845

Jones PD, Carne A, Bass NM, Grigor MR (1988) Isolation and characterization of fatty acid binding proteins from mammary tissue of lactating rats. Biochem J 251:919–925

Jones TA, Bergfors T, Sedzik J, Unge T (1988) The three-dimensional structure of P2 myelin protein. EMBO J 7:1597–1604

Jung Testas I, Baulieu EE (1985) Effects of steroid hormones and antihormones in cultured cells. Exp Clin Endocrinol 86:151–164

Kalkhoven E, Kwakkenbos Isbrucker L, Mummery CL, de Laat SW, van den Eijnden van Raaij AJ, van der Saag PT, Van Der Burg B (1995) The role of TGF-beta production in growth inhibition of breast-tumor cells by progestins. Int J Cancer 61:80–86

Kalkhoven E, Beraldi E, Panno ML, De Winter JP, Thijssen JH, Van Der Burg B (1996) Growth inhibition by anti-estrogens and progestins in TGF-beta-resistant and -sensitive breast-tumor cells. Int J Cancer 65:682–687

Karey KP, Sirbasku DA (1988) Differential responsiveness of human breast cancer cell lines MCF-7 and T47D to growth factors and 17 beta-estradiol. Cancer Res 48:4083–4092

Kasid A, Knabbe C, Lippman ME (1987) Effect of v-rasH oncogene transfection on estrogen-independent tumorigenicity of estrogen-dependent human breast cancer cells. Cancer Res 47:5733–5738

Keler T, Sorof S (1993) Growth promotion of transfected hepatoma cells by liver fatty acid binding protein. J Cell Physiol 157:33–40

Keler T, Barker CS, Sorof S (1992) Specific growth stimulation by linoleic acid in hepatoma cell lines transfected with the target protein of a liver carcinogen. Proc Natl Acad Sci USA 89:4830–4834

Kerr DJ, Pragnell IB, Sproul A, Cowan S, Murray T, George D, Leake R (1989) The cytostatic effects of alpha-interferon may be mediated by transforming growth factor β. J Mol Endocrinol 131–136

Khan SH, Sorof S (1990) Preferential binding of growth inhibitory prostaglandins by the target protein of a carcinogen. Proc Natl Acad Sci USA 87:9401–9405

King RJ, Wang DY, Daly RJ, Darbre PD (1989) Approaches to studying the role of growth factors in the progression of breast tumours from the steroid sensitive to insensitive state. J Steroid Biochem 34:133–138

Knabbe C, Lippman ME, Wakefield LM, Flanders KC, Kasid A, Derynck R, Dickson RB (1987) Evidence that transforming growth factor-beta is a hormonally regulated negative growth factor in human breast cancer cells. Cell 48:417–428

Knabbe C, Zugmaier G, Schmahl M, Dietel M, Lippman ME, Dickson RB (1991) Induction of transforming growth factor beta by the antiestrogens droloxifene, tamoxifen, and toremifene in MCF-7 cells. Am J Clin Oncol 14 Suppl 2:S15–20

Knowlton AA, Apstein CS, Saouf R, Brecher P (1989) Leakage of heart fatty acid binding protein with ischemia and reperfusion in the rat. J Mol Cell Cardiol 21:577–583

Kordon EC, Mcknight RA, Jhappan C, Hennighausen L, Merlino G, Smith GH (1995) Ectopic TGF beta 1 expression in the secretory mammary epithelium induces early senescence of the epithelial stem cell population. Dev Biol 168:47–61

Kurtz A, Vogel F, Funa K, Heldin CH, Grosse R (1990) Developmental regulation of mammary-derived growth inhibitor expression in bovine mammary tissue. J Cell Biol 110:1779–1789

Kyprianou N, English HF, Davidson NE, Isaacs JT (1991) Programmed cell death during regression of the MCF-7 human breast cancer following estrogen ablation. Cancer Res 51:162–166

Lake FR, Riches DWH (1992) Hyaluronate activation of CD44 induces insulin-like growth factor-1 expression by a tumor necrosis-alpha-dependent mechanism in murine macrophages. FASEB J 6:1614

Landry J, Chretien P, Lambert H, Hickey E, Weber LA (1989) Heat shock resistance conferred by expression of the human HSP27 gene in rodent cells. J Cell Biol 109:7–15

Langen P, Lehmann W, Graetz H, Bohmer FD, Grosse R (1984) Is ribonucleotide reductase in Ehrlich ascites mammary tumour cells the target of a growth inhibitor purified from bovine mammary gland? Biomed Biochim Acta 43:1377–1383

Lassen D, Lucke C, Kromminga A, Lezius A, Spener F, Ruterjans H (1993) Solution structure of bovine heart fatty acid-binding protein (H-FABPc). Mol Cell Biochem 123:15–22

Lassen D, Lucke C, Kveder M, Mesgarzadeh A, Schmidt JM, Specht B, Lezius A, Spener F, Ruterjans H (1995) Three-dimensional structure of bovine heart fatty-acid-binding protein with bound palmitic acid, determined by multidimensional NMR spectroscopy. Eur J Biochem 230:266–280

Lehmann W, Graetz H, Schütt M, Langen P (1977) Chalone-like inhibition of Ehrlich ascites cell proliferation in vitro by an ultrafiltrate obtained from ascites fluid. Acta Biol Med Germ 36:43–52

Lehmann W, Samtleben R, Graetz H, Langen P (1983) Purification of a chalone-like inhibitor for Ehrlich ascites mammary carcinoma cells from bovine mammary gland. Eur J Cancer Clin Oncol 19:101–107

Lehmann W, Graetz H, Widmayer R, Langen P (1984) Fetal calf serum and epidermal growth factor prevent the inhibitory action of a chalone-like growth inhibitor for Ehrlich ascites mammary carcinoma cells in vitro. Biomed Biochim Acta 43:971–974

Lehmann W, Widmaier R, Langen P (1989) Response of different mammary epithelial cell lines to a mammary derived growth inhibitor (MDGI). Biomed Biochim Acta 48:143–151

Loser R, Seibel K, Roos W, Eppenberger U (1985) In vivo and in vitro antiestrogenic action of 3-hydroxytamoxifen, tamoxifen and 4-hydroxytamoxifen. Eur J Cancer Clin Oncol 21:985–990

MacCallum J, Bartlett JM, Thompson AM, Keen JC, Dixon JM, Miller WR (1994) Expression of transforming growth factor beta mRNA isoforms in human breast cancer [see comments]. Br J Cancer 69:1006–1009

Macdiarmid F, Wang D, Duncan LJ, Purohit A, Ghilchick MW, Reed MJ (1994) Stimulation of aromatase activity in breast fibroblasts by tumor necrosis factor alpha. Mol Cell Endocrinol 106:17–21

Manni A, Wright C (1984) Reversal of the antiproliferative effect of the antiestrogen tamoxifen by polyamines in breast cancer cells. Endocrinol 114:836–839

Manni A, Wright C, Badger B, Bartholomew M, Herlyn M, Mendelsohn J, Masui H, Demers L (1990) Role of transforming growth factor α related peptides in the autocrine/paracrine control of experimental breast cancer growth in vitro by estradiol, prolactin, and progesterone. Breast Cancer Res Treat 15:73–83

Marchetti B, Fortier MA, Poyet P, Follea N, Pelletier G, Labrie F (1990) Beta-adrenergic receptors in the rat mammary gland during pregnancy and lactation: characterization, distribution, and coupling to adenylate cyclase. Endocrinology 126:565–574

Marino MW, Feld LJ, Jaffe EA, Pfeffer LM, Han HM, Donner DB (1991) Phosphorylation of the proto-oncogene product eukaryotic initiation factor 4E is a common cellular response to tumor necrosis factor. J Biol Chem 266:2685–2688

Martikainen P, Kyprianou N, Isaacs JT (1990) Effect of transforming growth factor-beta 1 on proliferation and death of rat prostatic cells. Endocrinology 127:2963–2968

Massague J (1990) The transforming growth factor-beta family. Annu Rev Cell Biol 6:597–641

Massague J, Cheifetz S, Laiho M, Ralph DA, Weis FM, Zentella A (1996) Transforming growth factor-β. Cancer Surv 12:81–103

Masui T, Wakefield LM, Lechner JF, LaVeck MA, Sporn MB, Harris CC (1986) Type beta transforming growth factor is the primary differentiation-inducing serum factor for normal human bronchial epithelial cells. Proc Natl Acad Sci USA 83:2438–2442

Matarese V, Stone RL, Waggoner DW, Bernlohr DA (1989) Intracellular fatty acid trafficking and the role of cytosolic lipid binding proteins. Prog Lipid Res 28:245–272

Matsui Y, Halter S, Holt JT, Hogan BLM, Coffey RJ (1990) Development of mammary hyperplasia and neoplasia in MMTV-TGFα transgenic mice. Cell 61:1137–1146 (Abstr)

McCune BK, Mullin BR, Flanders KC, Jaffurs WJ, Mullen LT, Sporn MB (1992) Localization of transforming growth factor-beta isotypes in lesions of the human breast [see comments]. Hum Pathol 23:13–20

Mieth M, Boehmer FD, Ball R, Groner B, Grosse R (1990) Transforming growth factor-beta inhibits lactogenic hormone induction of beta-casein expression in HC11 mouse mammary epithelial cells. Growth Factors 4:9–15

Miettinen PJ, Ebner R, Lopez AR, Derynck R (1994) TGF-beta induced transdifferentiation of mammary epithelial cells to mesenchymal cells: involvement of type I receptors. J Cell Biol 127:2021–2036

Miles DW, Happerfield LC, Naylor MS, Bobrow LG, Rubens RD, Balkwill FR (1994) Expression of tumour necrosis factor (TNF alpha) and its receptors in benign and malignant breast tissue. Int J Cancer 56:777–782

Millan FA, Denhez F, Kondaiah P, Akhurst RJ (1991) Embryonic gene expression patterns of TGF beta 1, beta 2 and beta 3 suggest different developmental functions in vivo. Development 111:131–143

Miller W, Mortial JA, Baxter JD (1980) Molecular cloning of DNA complementary to bovine growth hormone mRNA. J Biol Chem 255:7512–7524

Minshull J, Golsteyn R, Hill CS, Hunt T (1990) The A- and B-type cyclin associated cdc2 kinases in *Xenopus* turn on and off at different times in the cell cycle. EMBO J 9:2865–2875

Nielsen SU, Spener F (1993) Fatty acid-binding protein from rat heart is phosphorylated on Tyr19 in response to insulin stimulation. J Lipid Res 34:1355–1366

Nielsen SU, Rump R, Hojrup P, Roepstorff P, Spener F (1994) Differentiational regulation and phosphorylation of the fatty acid-binding protein from rat mammary epithelial cells. Biochim Biophys Acta 1211:189–197

Oka T, Yoshimura M (1986) Paracrine regulation of mammary gland growth. Clin Endocrinol Metab 15:79–97

Owen Schaub LB, Grimm EA (1989) Lymphokine-activated killer lymphocytes: evidence for regulation of induction and function by multiple cytokines. Immunol Ser 48:19–35

Paulus W, Baur I, Huettner C, Schmausser B, Roggendorf W, Schlingensiepen KH, Brysch W (1995) Effects of transforming growth factor-beta 1 on collagen synthesis, integrin expression, adhesion and invasion of glioma cells. J Neuropathol Exp Neurol 54:236–244

Pierce DF Jr, Johnson MD, Matsui Y, Robinson SD, Gold LI, Purchio AF, Daniel CW, Hogan BL, Moses HL (1993) Inhibition of mammary duct development but not alveolar outgrowth during pregnancy in transgenic mice expressing active TGF-beta 1. Genes Dev 7:2308–2317

Pierce DF Jr, Gorska AE, Chytil A, Meise KS, Page DL, Coffey RJ Jr, Moses HL (1995) Mammary tumor suppression by transforming growth factor beta 1 transgene expression. Proc Natl Acad Sci USA 92:4254–4258

Pusztai L, Clover LM, Cooper K, Starkey PM, Lewis CE, McGee JO (1994) Expression of tumour necrosis factor alpha and its receptors in carcinoma of the breast. Br J Cancer 70:289–292

Mizukami Y, Nonomura A, Yamada T, Kurumaya H, hayashi M, Koyasaki N, Taniya T, Noguchi M, Nakamura S, Matsubara F (1990) Immunohistochemical demonstration of growth factors, TGFα, TGFβ, IGF-I and *neu* oncogene product in benign and malignant human breast tissues. Anticancer Res 10:1115–1126

Moses HL, Yang EY, Pietenpol JA (1990) TGF-beta stimulation and inhibition of cell proliferation: new mechanistic insights. Cell 63:245–247

Mueller H, Flury N, Liu R, Scheidegger S, Eppenberger U (1996) Tumor necrosis factor and interferon are selectively cytostatic in vitro for hormone-dependent and hormone-independent human breast cancer cells. Eur J Cancer 32:2312–2318

Muller T, Kurtz A, Vogel F, Breter H, Schneider F, Angstrom U, Mieth M, Bohmer FD, Grosse R (1989) A mammary-derived growth inhibitor (MDGI) related 70 kDa antigen identified in nuclei of mammary epithelial cells. J Cell Physiol 138:415–423

Muller Fahrnow A, Egner U, Jones TA, Rudel H, Spener F, Saenger W (1991) Three-dimensional structure of fatty-acid-binding protein from bovine heart. Eur J Biochem 199:271–276

Murphy AN, Unsworth EJ, Stetler Stevenson WG (1993) Tissue inhibitor of metalloproteinases-2 inhibits bFGF-induced human microvascular endothelial cell proliferation. J Cell Physiol 157:351–358

Murphy LC, Dotzlaw H (1989) Regulation of transforming growth factor α transforming growth factor β messenger ribonucleic acid abundance in T-47D human breast cancer cells. Mol Endocrinol 3:611–616

Newman MJ (1990) Inhibition of carcinoma and melanoma cell growth by type 1 transforming growth factor beta is dependent on the presence of polyunsaturated fatty acids. Proc Natl Acad Sci USA 87:5543–5547

Ranchalis JE, Gentry LE, Ogawa Y, Seyedin SM, McPherson J, Purchio A, Twardzik DR (1987) Bone-derived and recombinant transforming growth factor βs are potent inhibitors of tumor cell growth. Biochem Biophys Res Commun 148:783–789

Ranges GE, Figari IS, Espevik T, Palladino MA Jr (1987) Inhibition of cytotoxic T cell development by transforming growth factor beta and reversal by recombinant tumor necrosis factor alpha. J Exp Med 166:991–998

Raza H, Pongubala JR, Sorof S (1989) Specific high affinity binding of lipoxygenase metabolites of arachidonic acid by liver fatty acid binding protein. Biochem Biophys Res Commun 161:448–455

Roberts AB, Anzano MA, Wakefield LM, Roche NS, Stern DF, Sporn MB (1985) Type beta transforming growth factor: a bifunctional regulator of cellular growth. Proc Natl Acad Sci USA 82:119–123

Roberts AB, Sporn MB, Assoian RK, Smith JM, Roche NS, Wakefield LM, Heine UI, Liotta LA, Falanga V, Kehrl JH et al. (1986) Transforming growth factor type beta: rapid induction of fibrosis and angiogenesis in vivo and stimulation of collagen formation in vitro. Proc Natl Acad Sci USA 83:4167–4171

Robinson SD, Silberstein GB, Roberts AB, Flanders KC, Daniel CW (1991) Regulated expression and growth inhibitory effects of transforming growth factor-beta isoforms in mouse mammary gland development. Development 113:867–878

Robinson SD, Roberts AB, Daniel CW (1993a) TGF beta suppresses casein synthesis in mouse mammary explants and may play a role in controlling milk levels during pregnancy. J Cell Biol 120:245–251

Robinson SD, Roberts AB, Daniel CW (1993b) TGFβ suppresses casein synthesis in mouse mammary explants and may play a role in controlling milk levels during pregnancy. J Cell Biol 120:245–251

Rocker D, Ravid A, Liberman UA, Garach Jehoshua O, Koren R (1994) 1,25-Dihydroxyvitamin D3 potentiates the cytotoxic effect of TNF on human breast cancer cells. Mol Cell Endocrinol 106:157–162

Rohde Schulz B, Lichtner RB (1995) EGF enhances attachment of metastatic rat mammary adenocarcinoma cell clone MTLn3 to fibronectin and collagen. Invasion Metastasis 15:1–10

Rose DP, Connolly JM (1990) Effects of fatty acids and inhibitors of eicosanoid synthesis on the growth of a human breast cancer cell line in culture. Cancer Res 50:7139–7144

Rothe M, Wong SC, Henzel WJ, Goeddel DV (1994) A novel family of putative signal transducers associated with the cytoplasmic domain of the 75 kDa tumor necrosis factor receptor. Cell 78:681–692

Rothe M, Sarma V, Dixit VM, Goeddel DV (1995) TRAF2-mediated activation of NF-kappa B by TNF receptor 2 and CD40. Science 269:1424–1427

Sacchettini JC, Gordon JI, Banaszak LJ (1988) The structure of crystalline *Escherichia coli*-derived rat intestinal fatty acid-binding protein at 2.5-A resolution. J Biol Chem 263:5815–5819

Sandgren EP, Luetteke NC, Palmiter RD, Brinster RL, Lee DC (1990) Overexpression of TGFα in transgenic mice: induction of epithelial hyperplasia, pancreatic metaplasia and carcinoma of the breast. Cell 61:1121–1135 (Abstr)

Schor SL, Schor AM, Grey AM, Chen J, Rushton G, Grant ME, Ellis I (1989) Mechanism of action of the migration stimulating factor produced by fetal and cancer patient fibroblasts: effect on hyaluronic and synthesis. In Vitro Cell Dev Biol 25:737–746

Schwarz LC, Wright JA, Gingras MC, Kondaiah P, Danielpour D, Pimentel M, Sporn MB, Greenberg AH (1990) Aberrant TGF-beta production and regulation in metastatic malignancy. Growth Factors 3:115–127

Silberstein GB, Daniel CW (1987a) Investigation of mouse mammary ductal growth regulation using slow-release plastic implants. J Dairy Sci 70:1981–1990

Silberstein GB, Daniel CW (1987b) Reversible inhibition of mammary gland growth by transforming growth factor-beta. Science 237:291–293

Silberstein GB, Strickland P, Coleman S, Daniel CW (1990) Epithelium-dependent extracellular matrix synthesis in transforming growth factor-beta 1-growth-inhibited mouse mammary gland. J Cell Biol 110:2209–2219

Silberstein GB, Flanders KC, Roberts AB, Daniel CW (1992) Regulation of mammary morphogenesis: evidence for extracellular matrix-mediated inhibition of ductal budding by transforming growth factor-beta 1. Dev Biol 152:354–362

Specht B, Bartetzko N, Hohoff C, Kuhl H, Franke R, Borchers T, Spener F (1996) Mammary derived growth inhibitor is not a distinct protein but a mix of heart-type and adipocyte-type fatty acid-binding protein. J Biol Chem 271:19943–19949

Spener F, Borchers T, Mukherjea M (1989) On the role of fatty acid binding proteins in fatty acid transport and metabolism. FEBS Lett 244:1–5

Spener F, Unterberg C, Börchers T, Grosse R (1990) Characteristics of fatty acid-binding proteins and their relation to mammary-derived growth inhibitor. Mol Cell Biochem 98:57–68

Sporn MB, Roberts AB (1985) Autocrine growth factors and cancer. Nature 313:745–747

Sporn MB, Roberts AB (1988) Peptide growth factors are multifunctional. Nature 332:217–219

Sporn MB, Roberts AB (1990a) TGF-beta: problems and prospects. Cell Regul 1:875–882

Sporn MB, Roberts AB (1990b) The transforming growth factor-betas: past, present, and future. Ann NY Acad Sci 593:1–6

Stampfer M, Bartley JC (1985) Induction of transformation and continous cell lines from normal human mammary epithelial cells after exposure to benzo-*a*-pyrene. Proc Natl Acad Sci USA 82:2394–2398

Stetler Stevenson WG, Krutzsch HC, Liotta LA (1989) Tissue inhibitor of metalloproteinase (TIMP-2). A new member of the metalloproteinase inhibitor family. J Biol Chem 264:17374–17378

Streuli CH, Schmidhauser C, Kobrin M, Bissell MJ, Derynck R (1993) Extracellular matrix regulates expression of the TGF-β1 gene. J Cell Biol 120:253–260

Suessmilch R, Jaeger J, Behlke J (1990) Nonstandard turn of MDGI protein binding site of fatty acids? Diff J Quantum Chem 38:311–319

Sun L, Wu G, Willson JK, Zborowska E, Yang J, Rajkarunanayake I, Wang J, Gentry LE, Wang XF, Brattain MG (1994) Expression of transforming growth factor beta type II receptor leads to reduced malignancy in human breast cancer MCF-7 cells. J Biol Chem 269:26449–26455

Takahashi K, Suzuki K, Ono T (1990) Loss of growth inhibitory activity of TGF-beta toward normal human mammary epithelial cells grown within collagen gel matrix. Biochem Biophys Res Commun 173:1239–1247

Taketani Y, Oka T (1983) Epidermal growth factor stimulates cell proliferation and inhibits functional diffeentiation of mouse mammary epithelial cell in culture. Endocrinology 113: 871–877

Thanos D, Maniatis T (1995) NF-kappa B: a lesson in family values. Cell 80:529–532

Thompson AM, Steel CM, Chetty U, Hawkins RA, Miller WR, Carter DC, Forrest AP, Evans HJ (1990) p53 gene mRNA expression and chromosome 17p allele loss in breast cancer. Br J Cancer 61(1):74–78

Thompson AM, Kerr DJ, Steel CM (1991) Transforming growth factor beta 1 is implicated in the failure of tamoxifen therapy in human breast cancer. Br J Cancer 63:609–614

Thompson EW, Brunner N, Torri J, Johnson MD, Boulay V, Wright A, Lippman ME, Steeg PS, Clarke R (1993) The invasive and metastatic properties of hormone-independent but hormone-responsive variants of MCF-7 human breast cancer cells. Clin Exp Metastasis 11:15–26

Thompson NL, Flanders KC, Smith JM, Ellingsworth LR, Roberts AB, Sporn MB (1989) Expression of transforming growth factor-beta 1 in specific cells and tissues of adult and neonatal mice. J Cell Biol 108:661–669

Travers MT, Barrett-Lee PJ, Berger U, Lugmani YA, Gazet JC, Powels TJ, Coombes CR (1988) Growth factor expression in normal, benign and malignant breast tissue. Br Med J 296:1621–1625

Tweedie S, Edwards Y (1989) cDNA sequence for mouse heart fatty acid binding protein, H-FABP. Nucleic Acids Res 17:4374

Valverius EM, Walker Jones D, Bates SE, Stampfer MR, Clark R, McCormick F, Dickson RB, Lippman ME (1989) Production of and responsiveness to transforming growth factor-beta in normal and oncogene-transformed human mammary epithelial cells. Cancer Res 49:6269–6274

Veerkamp JH, Vankuppevelt THMSM, Maatman RGHJ, Rrinsen CFM (1993) Review – structural and functional aspects of cytosolic fatty acid-binding proteins. Prostagland Leuk Essent Fatty 49:887–906

Vogel F, Breter H, Erdmann B, M:uller T, Binas B, Grosse R (1990) Characterization and function dependent localization of mammary derived growth inhibitor (MDGI) in mammary glands of bovine and mice. Acta Histochem Suppl 40:77–80

Vonderhaar BK (1987) Local effect of EGF, α-TGF and EGF-like growth factors on lobuloalveolar development of the mouse mammary gland in vivo. J Cell Physiol 132:581–584

Wahl SM, Hunt DA, Wakefield LM, McCartney Francis N, Wahl LM, Roberts AB, Sporn MB (1987) Transforming growth factor type beta induces monocyte chemotaxis and growth factor production. Proc Natl Acad Sci USA 84:5788–5792

Wakefield LM, Letterio JJ, Geiser AG, Flanders KC, O'Shaughnessy J, Roberts AB, Sporn MB (1995) Transforming growth factor-beta s in mammary tumorigenesis: promoters or antipromoters? Prog Clin Biol Res 391:133–148

Wakeling AE (1993) Are breast tumours resistant to tamoxifen also resistant to pure antioestrogens? J Steroid Biochem Mol Biol 47:107–114

Wakeling AE, Valcaccia B, Newboult E, Green LR (1984) Non-steroidal antioestrogens – receptor binding and biological response in rat uterus, rat mammary carcinoma and human breast cancer cells. J Steroid Biochem 20:111–120

Walker RA, Dearing SJ (1992) Transforming growth factor beta 1 in ductal carcinoma in situ and invasive carcinomas of the breast. Eur J Cancer 28:641–644

Walker RA, Dearing SJ, Gallacher B (1994) Relationship of transforming growth factor beta 1 to extracellular matrix and stromal infiltrates in invasive breast carcinoma. Br J Cancer 69:1160–1165

Wallukat G, Boehmer FD, Engstroem U, Langen P, Hollenberg M, Behlke J, Kuehn H, Grosse R (1991a) Modulation of the beta-adrenergic-response in cultured rat heart cells. II. Mammary-derived growth inhibitor (MDGI) blocks induction of beta-adrenergic supersensitivity. Dissociation from lipid-binding activity of MDGI. Mol Cell Biochem 102:49–60

Wallukat G, Kuehn H, Wollenberger A (1991b) Supersensitivity to isoprenaline induced in cultured neonatal rat heart cells by certain eicosatetraenoic acids. Eur Heart J 12 Suppl F:145–148

Wallukat G, Nemecz G, Farkas T, Kuehn H, Wollenberger A (1991c) Modulation of the beta-adrenergic response in cultured rat heart cells. I. Beta-adrenergic supersensitivity is induced by lactate via a phospholipase A2 and 15-lipoxygenase involving pathway. Mol Cell Biochem 102:35–47

Walz DA, Wider MD, Snow JW, Dass C, Desiderio DM (1988) The complete amino acid sequence of porcine gastrotropin, an ileal protein which stimulates gastric acid and pepsinogen secretion. J Biol Chem 263:14189–14195

Welch DR, Fabra A, Nakajima M (1990) Transforming growth factor beta stimulates mammary adenocarcinoma cell invasion and metastatic potential. Proc Natl Acad Sci USA 87:7678–7682

Wiegmann K, Schutze S, Machleidt T, Witte D, Kronke M (1994) Functional dichotomy of neutral and acidic sphingomyelinases in tumor necrosis factor signaling. Cell 78:1005–1015

Wrana JL, Attisano L, Cçrcamo J, Zentella A, Doody J, Laiho M, Wang XF, Massague J (1992) TGFbeta signals through a heteromeric protein kinase receptor complex. Cell 71:1003–1014

Yamamoto M, Maehara Y, Sakaguchi Y, Kusumoto T, Ichiyoshi Y, Sugimachi K (1996) – transforming growth factor-beta-1 induces apoptosis in gastric cancer cells through a p53-independent pathway. Cancer 1628–1633

Yamamoto T, Komura H, Morishige K, Tadokoro C, Sakata M, Kurachi H, Miyake A (1994) Involvement of autocrine mechanism of transforming growth factor-beta in the functional differentiation of pregnant mouse mammary gland. Eur J Endocrinol 130:302–307

Yamasaki N, Shimanaka J, Sonnenberg M (1975) Studies on the common active site of growth hormone. J Biol Chem 250:2510–2514

Zajchowski DA, Band V, Pauzie N, Tager A, Stampfer M, Sager R (1988) Expression of growth factors and oncogenes in normal and tumor-derived human mammary epithelial cells. Cancer Res 48:7041–7047

Zugmaier G, Ennis BW, Deschauer B, Katz D, Knabbe C, Wilding G, Daly P, Lippman ME, Dickson RB (1989) Transforming growth factors type beta 1 and beta 2 are equipotent growth inhibitors of human breast cancer cell lines. J Cell Physiol 141:353–361

Zyad A, Benard J, Tursz T, Clarke R, Chouaib S (1994) Resistance to TNF-alpha and adriamycin in the human breast cancer MCF-7 cell line: relationship to MDR1, MnSOD, and TNF gene expression. Cancer Res 54:825–831

Growth Inhibition of Human Fibroblasts in Vitro

Alvaro Macieira-Coelho[1]

1 Introduction

Growth arrest of human fibroblasts maintained in vitro occurs when the density increases and the cells become confluent; this is the so-called contact inhibition of growth characteristic of all normal cells. Growth arrest also occurs at the end of the cell population's proliferative life span, a state that has been erroneously called senescence but which is irrelevant for aging of the organism since the terminal cell is present in tissues in vivo in amounts unrelated to the age of the donor (Macieira-Coelho 1995). This terminal cell is supposed to be in a differentiated state (Martin et al. 1974), although its putative role in homeostasis has not been identified so far. An increased number of terminal fibroblasts was found in vivo under pathological conditions (Macieira-Coelho 1995).

Human fibroblasts, like many other somatic cells, have a limited potential for cell division (Hayflick and Moorhead 1961). It is generally believed that this is due to an intrinsic property of the cells that are endowed with a fixed potential number of divisions. Other investigators, however, suggested that the finite division potential is the result of genetic damage (Rubin 1997).

Whatever the cause for the finite mitotic life span, during serial proliferation there is a progressive shift in cycling times due mainly to the prolongation of the G1 and G2 periods (Fig. 1).

The cell population becomes increasingly heterogeneous not only in the kinetics of proliferation, but also metabolically and morphologically (for review see Macieira-Coelho 1988). It is this shift in cell behavior through division that is significant for aging of the organism.

Three phases were first proposed to characterize the proliferative life span of human fibroblasts (Hayflick and Moorhead 1961). Later, the study of the kinetics of proliferation allowed definition of four phases (Fig. 2; Macieira-Coelho and Taboury 1982).

[1] INSERM, 73 bis rue Maréchal Foch, 78000 Versailles, France

Progress in Molecular and Subcellular Biology, Vol. 20
A. Macieira-Coelho (Ed.)

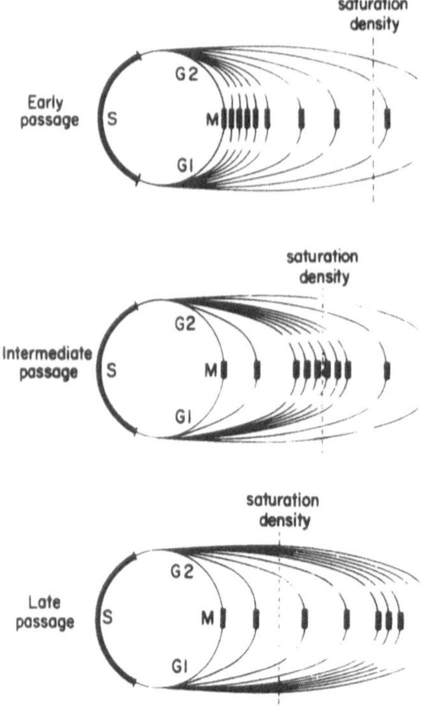

Fig. 1. Schematic representation of the distribution of cell cycles in human fibroblast populations through their proliferative life span before the terminal phase IV. The *circles* represent short cycling times and the *ellipses* progressively longer times spent to finish the division cycle. There is a progressive shift to long cycles, expressing an increased sensitivity to contact inhibition of growth

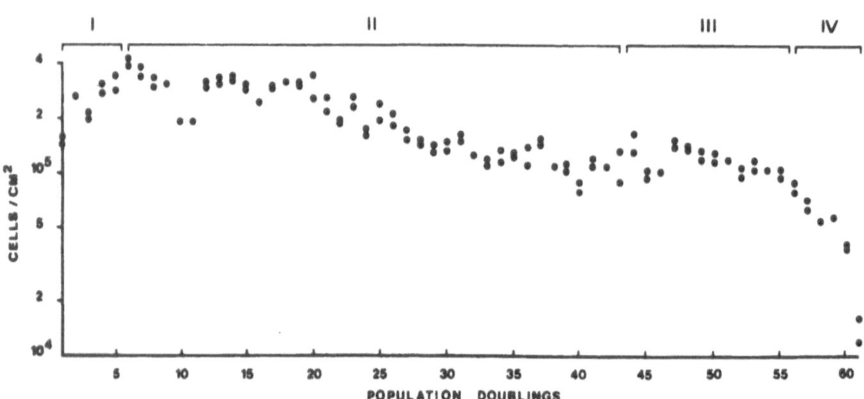

Fig. 2. Proliferative life span of a human embryonic lung fibroblast population expressed as the maximal cell densities reached after each population doubling. The population doubling levels corresponding to the different phases are indicated. For further explanation see text

During phases I–III there is a succession of subtle changes in the division cycle rendering the cell population progressively more heterogeneous (Fig. 1). Finally, during phase IV, abrupt events take place (Macieira-Coelho and Azzarone 1982; Macieira-Coelho 1995) and the cell population enters a state of growth arrest where it can remain metabolically active for an indeterminate period of time. It should be stressed, however, that during this final stage, some cells continue cycling very slowly (Macieira-Coelho 1974; Matsumura et al. 1979). Some terminal cells are produced through phases I–III.

It has been questioned whether there are any similarities between contact inhibition of growth and the terminal postmitotic state. The question is beside the point since the latter is irreversible unless the cells become transformed; moreover, the terminal cells suffer profound structural modifications that must be obviously associated with significantly different, more complex regulations. Thus the terminal inhibition of growth should be examined separately.

2 Mechanisms Responsible for Contact Inhibition of Growth

2.1 Conformational Flexibility

The action of molecular inhibitors of cell division is not the only mechanism responsible for quiescence; loss of conformational flexibility also contributes to the resting stage. In regard to DNA synthesis, the steric position of initiating sites is crucial for entrance into the S period (Macieira-Coelho 1983). DNA conformational flexibility is also necessary for gene expression. During contact inhibition of growth, the cells become confluent, establish stable contacts, are attached to a small area, and become immobilized. This decreases the probability of molecules assuming the favorable conformation to trigger DNA synthesis and contributes to the down-regulation of the different kinases and transcription factors that move the cell along the division cycle and to the upregulation of inhibitors of these kinases from the Cip/Kip family (p21, p27, p57) and INK4 family (p15, p16, p18, p19). At this stage, inhibitors of protein phosphatases can override the block of DNA synthesis in contact-inhibited cells (Afshari and Barrett 1994).

Through the cell proliferative life span there are progressive structural changes associated with a decline in contractility (Macieira-Coelho and Azzarone 1990) which follow a pattern analogous to that of the different phases identifiable in the study of the kinetics of proliferation (Fig. 2). The decreased conformational flexibility becomes apparent when the cells are in the resting phase, and is one of the parameters contributing to the increased sensitivity to contact inhibition of growth that is observed during the serial proliferation of human fibroblasts (Macieira-Coelho et al. 1966).

The numerous metabolic changes that occur during serial proliferation of human fibroblasts, which can contribute to increased contact inhibition of growth, have been reviewed (Macieira-Coelho 1988). Recently, it was reported that the basal level of c-fos and junB mRNAs start to decline towards the middle of the fibroblast proliferative life span (Irving et al. 1992). The products of these two genes form homo- and heterodimeric complexes (AP-1) which act as transcription factors for the activation of genes necessary for the initiation of DNA synthesis. It was suggested that the decline of the expression of c-fos and junB could lead to an increase in jun/jun AP-1 homodimers at the expense of fos/jun heterodimers, in this way contributing to the progressive disregulation of the division cycle (Riabowol et al. 1992).

2.1.1 Inhibitory Glycopeptides

Different growth inhibitors accumulate in a cell culture when cell density increases, contributing to contact inhibition of growth. The first evidence suggesting the presence of negative growth regulators responsible for the resting stage when human fibroblasts reach confluency was of an indirect nature. It was found that cells proliferated further, rinsing the cell monolayer layer with the used medium, that the supernatant of cells that had reached maximal density after several stimulations was inhibitory for sparse cultures, and that the dialyzed supernatant lost its inhibitory activity (Garcia-Giralt et al. 1970).

Subsequently, it was found that rinsing resting-stage monolayers induced a decreased incorporation of glucosamine into glycopeptides obtained from the rinsing fluid (Hiu et al. 1985). The decreased incorporation of glucosamine was associated with the induction of the division cycle by growth factors in a cell cycle-dependent fashion (Fig. 3).

These results led to the search for the molecule associated with decreased sugar incorporation. A glycopeptide could be separated by electric charge, lectin affinity, and molecular mass, which inhibited DNA synthesis when added to sparse proliferating human fibroblasts (Macieira-Coelho and Söderberg 1993). The molecular weight was around 1000 Da. Further purification by capillary electrophoresis showed that the glycopeptide is synthesized in growing- and resting-stage human fibroblasts, but that its inhibitory activity depends upon the modification of the sugar content (Macieira-Coelho 1996).

As cells proliferate to confluency, the relative sugar content of the glycopeptide assumes a composition that is growth-inhibitory.

When resting-stage cells are stimulated by growth factors, the incorporation of glucosamine into the glycopeptide decreases while that of galactose varies little. Simultaneously with these changes in sugar incorporation, the glycopeptide becomes less inhibitory (Macieira-Coelho 1996).

Figure 4 illustrates the decrease in glucosamine incorporated into the glycopeptide when resting-stage human fibroblasts were stimulated by fibroblast growth factor, fetal calf serum, fresh medium, and a growth factor secreted by

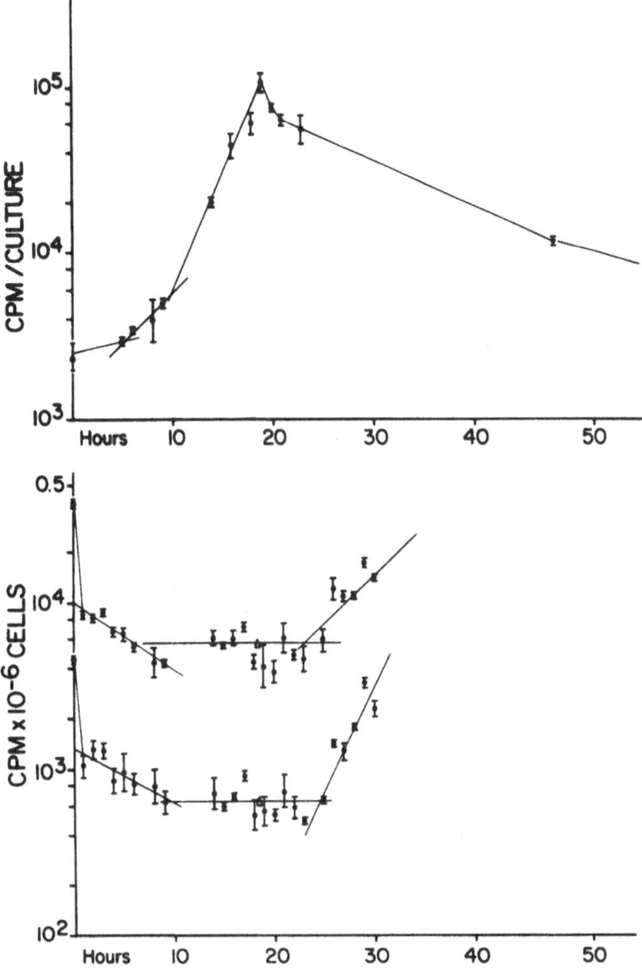

Fig. 3. Kinetics of ^{3}H-thymidine (*upper curve*) and ^{3}H-glucosamine (*lower curves*) incorporation in resting-phase cultures stimulated with a medium change. The *white triangle* in the lower curves corresponds to the middle of the plateau phase of glucosamine incorporation and coincides with the peak of thymidine incorporation. The *bars* represent the extreme values between duplicate samples. (Macieira-Coelho and Söderberg 1993, by permission of Wiley-Liss)

Rous sarcoma virus (RSV)-transformed fibroblasts (Macieira-Coelho et al. 1969).

The change in sugar composition rendering the glycopetide less inhibitory, following renewal of the medium or addition of fresh serum, helps to explain why confluent human fibroblasts reach higher densities when the nutrient medium is renewed daily (Fig. 5). Moreover, the same change induced by the autocrine growth factor of RSV-transformed fibroblasts can explain why trans-

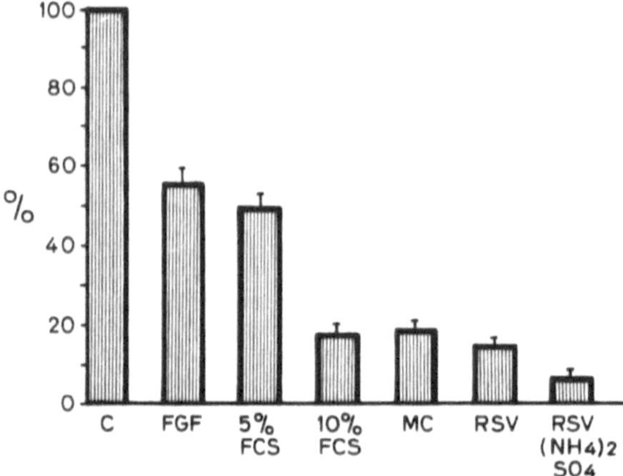

Fig. 4. Change in sugar composition of a growth-regulatory glycopetide, expressed as the percentage of decreased incorporation of glucosamine compared with control cells (*C*), after addition to quiescent cells of fibroblast growth factor (*FGF*), 5 and 10% fetal calf serum (*FCS*), fresh nutrient medium (*MC*), and an autocrine growth factor secreted by RSV-transformed fibroblasts obtained directly from the supranatant or after precipitation with ammonium sulfate

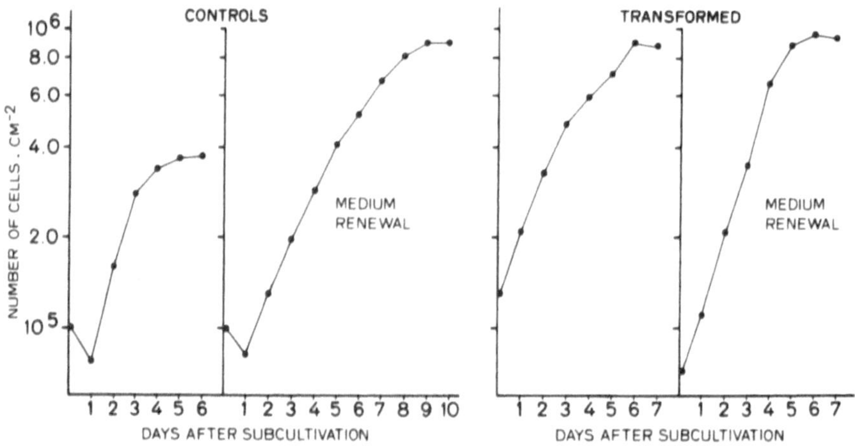

Fig. 5. Growth curves of control and RSV-transformed human fibroblasts maintained without and with daily medium renewals

formed fibroblasts reach identical densities with and without daily medium renewals (Macieira-Coelho 1967).

The glycopeptide was assayed for its potential to influence cell contractility by a test that measures the capacity of cells to retract a plasma clot (Macieira-Coelho and Azzarone 1990). It was found that the decrease in inhibitory activity of the glycopeptide associated with the change in sugar composition is

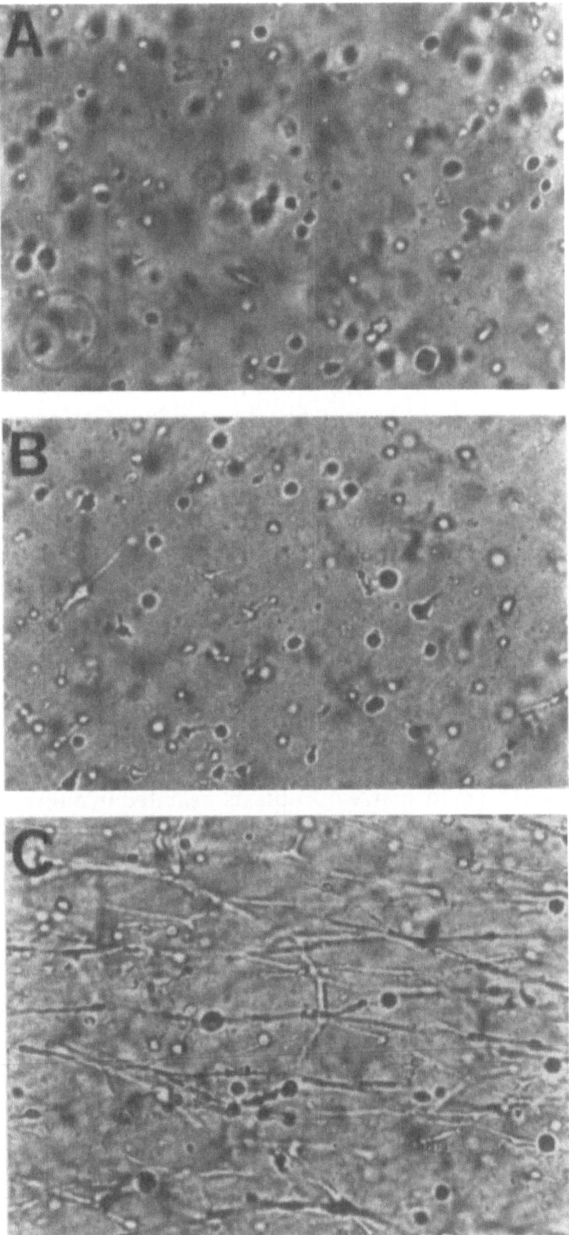

Fig. 6A–C. Morphology of human fibroblasts in a plasma clot without (**A**) and with the glycopetide obtained from resting (**B**) and fresh serum-stimulated (**C**) human fibroblasts. Only the cells suspended in the plasma with the molecule from stimulated cells are stretched with a fiber-like morphology and retracted the plasma clot

accompanied by the acquisition of the competence to potentiate cell contractility (Fig. 6). Hence, this growth inhibitor seems to act via its effect as a regulator of the cell contractile network; for this reason, we propose naming it systolin.

The removal of growth inhibition through the washing of confluent cell monolayers was later also found by other investigators (Froelich and Anastassiades 1975; Stroebel-Stevens and Lacey 1981; Wu et al. 1985). These authors could identify a cationic molecule in the 10000–30000 MW range, that is a protease inhibitor (Stroebel-Stevens and Lacey 1981). This was later identified as a beta-galactoside-binding protein (GBP) obtained from serum-free conditioned medium, which is a substrate for a growth-related protease (Manilal et al. 1993). A murine homolog of the GBP had been previously isolated (Wells and Malluchi 1991). Hence, in this case, growth inhibition is achieved through the binding of the GBP to the growth-promoting protease. Proteases are well known to have this effect through hydrolysis of cell-surface molecules.

Growth-inhibitory activity was also obtained after ultrafiltration of the used medium from human fibroblast cultures (Houck et al. 1973). This inhibitor, which was claimed to be a chalone, could also be found in the aqueous cell extract, had a molecular weight between 10000 and 1000 daltons, was strongly cationically charged, and contained mannose (Houck et al. 1977). It is not known if there is any identity common to this molecule and the ones described above.

Other investigators reported that plasma membrane glycoproteins from human fibroblasts are growth-inhibitory and can induce a state similar to quiescence (Wieser and Oesch 1986). Wieser et al. (1990) could isolate and characterize the plasma membrane glycoprotein, which they named contactinhibin. This is a 60–70-kDa molecule which, when added in immobilized form to sparse fibroblasts, resulted in a reversible 70–80% growth inhibition. In soluble form, the glycoprotein was not inhibitory. Confluent cultures treated with antibodies against contactinhibin were released from contact inhibition of growth. The glycoprotein was associated with vimentin and concentrated at cell-cell contact sites.

The same group reported that immobilized gangliosides inhibited DNA synthesis and that the degree of inhibition was directly related to the density of the cultures from which the gangliosides were isolated (Janik-Schmitt et al. 1987). The inhibitory action of these molecules was also reported by Ohsawa and Senshu (1987); the results suggested that the exogenous ganglioside blocked the cell cycle traverse in the early part of the G1 period.

L-iduronate-rich glycosaminoglycans (GAGs) could also prevent the initiation of the cell cycle by several positive growth factors (Westergren-Thorsson et al. 1993). It was suggested that GAGs may control cell proliferation via direct effects of the intact polymers or via their degradation products on receptor structures linked to the signaling system. The degradation products can also be internalized, be transported to the nucleus, and exert a growth-inhibitory effect there.

These works suggest the presence of different inhibitory molecules at the surface of human fibroblasts that can decrease the probability of initiating the division cycle by different mechanisms when cells become confluent.

In conclusion, the exact mechanism of action of these growth inhibitors has not been ascertained. They must be integrated with the different molecules that constitute the web that controls transit through the division cycle.

2.1.2 Further Evidence for the Presence of Growth Inhibitors

Fusion products between proliferating and quiescent cells also provided indirect evidence in favor of the presence of growth inhibitors. Nuclei from growth-arrested early passage cells were inhibitory for nuclei from replicative early-passage fibroblasts if fusion occurred at least 3 h before the G1/S boundary (Rabinovitch and Norwood 1980; Stein and Yanishevsky 1981). Cells in S phase at the time of fusion completed DNA synthesis.

Furthermore, polyadenylated RNA isolated from early-passage, growth-inhibited fibroblasts had a slight inhibitory effect on sparse proliferating cells (Lumpkin et al. 1986). These experiments eventually led to the cloning of genes expressed in growth-arrested cells and the identification of one gene (SDI1) coding for an important negative growth regulator of the cell cycle, the p21 protein (Noda et al. 1994). High levels of p21 can block the cell cycle in G1 (Harper et al. 1995). The activity of p53, a positive transactivator of p21 gene expression, was found to increase in a stepwise fashion through the different phases of the human fibroblast proliferative life span (Bond et al. 1996). This could be one of several mechanisms at the molecular level underlying the progressive increase in sensitivity to contact inhibition of growth during serial proliferation (Macieira-Coelho et al. 1966). The prolongation of the G2 period that becomes apparent during growth decline (Macieira-Coelho and Berumen 1973) could be the result of DNA damage triggering G2 check points.

Other mRNAs that were inhibitory when injected into human fibroblasts also led to the cloning of a gene supposed to regulate negatively the division cycle; the respective protein, however, could not yet be identified (Dell'Orco et al. 1996).

3 Mechanisms Responsible for the Terminal Postmitotic Stage of Human Fibroblasts

3.1 Structural Modifications

The final growth arrest of the proliferative life cycle of human fibroblasts is a complex event resulting from the interplay of several parameters. Basically, there is a deregulation of the transmission of the different signals that allow a cell to initiate and complete the division cycle. There is not, however, a complete block of the signaling system; some events occur and others do not.

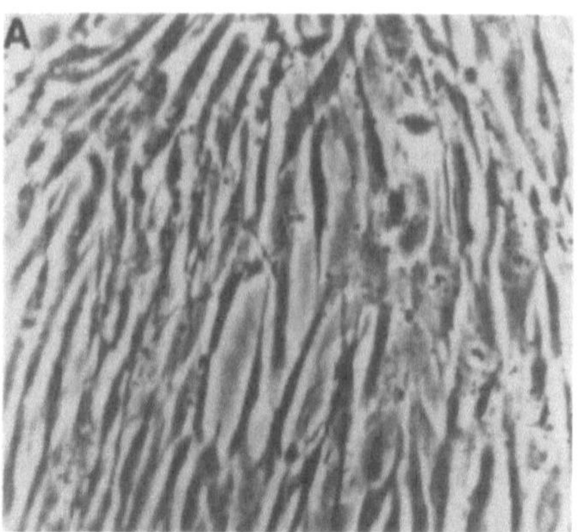

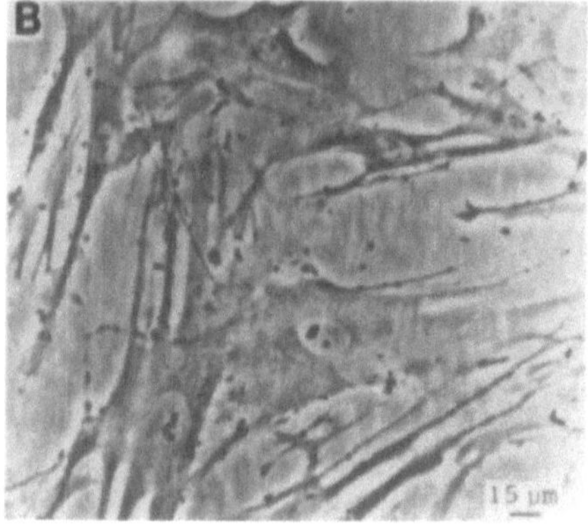

Fig. 7A,B. Morphology of human fibroblasts during their proliferative life span (**A**), and after reaching the terminal phase IV (**B**)

Several changes occur in many respects when the cells enter phase IV, which contribute to the nondividing state. A complete review of the modifications taking place during phase IV has been published (Macieira-Coelho 1988), so here we will give only a brief account of previous data and review the more recent findings.

In the terminal stage of the human fibroblast proliferative life cycle, there are profound structural modifications which are obvious from the observation of the morphology (Fig. 7).

Simultaneously with the sudden changes in the kinetics of proliferation, both the cytoplasm and the nucleus enlarge abruptly and the cells become stretched and less mobile due to modifications of the cells' scaffold. The actin fibers (Kelley et al. 1980) and the microtubules are differently organized, with a loss of polarity of the latter (Raes et al. 1984; VanGansen et al. 1984). This is expressed as a centrifugal depolymerization of the microtubules, contrary to proliferative cells, where depolymerization is centripetal. The structure of the centrosome is also modified with a diminished pericentriolar material.

The evolution of the composition of the cell surface contributes to the loss of response to growth stimuli. Synthesis of glycosaminoglycans decreases (Schachtschabel and Wever 1978) and their molecular size increases (Passi et al. 1993), with a relative increase in the synthesis of heparan sulfate (Wever et al. 1980). The latter has a significant role as a limiting factor for proliferation in the terminal cell (Matuoka and Mitsui 1981).

The composition of membrane glycoproteins is also modified (Milo and Hart 1976; Blondal et al. 1985). Milo (1973) could isolate from postmitotic cells a 50-kDa MW glycoprotein which was inhibitory for proliferating cells. Other investigators have also obtained enriched membrane proteins from terminal cells, with DNA synthesis inhibitory activity when added to proliferating early-passage fibroblasts (Stein and Atkins 1986; Pereira-Smith et al. 1985).

Other molecules of the cell surface that determine cell attachment and thus can contribute to the postmitotic state were shown to be modified in the terminal cells. An increased production of collagen with a change in the ratio alpha-1:alpha-2-chains was reported (Kontermann and Bayreuther 1979). Although less fibronectin is found at the surface of the terminal cell, there is higher production of the molecule (Vogel et al. 1981), which is not as efficient in promoting adhesion (Millis et al. 1977) probably due to structural changes (Sorrentino and Millis 1984); they may be the result of significant shifts in the splicing of fibronectin during transition to the postmitotic state (Magnuson et al. 1991). The deficient adhesion is probably also due to modifications of the alpha-5-beta-1 integrin-fibronectin receptor (Hu et al. 1996). The results suggest a remodeling of the receptor, since there are fewer alpha-5-beta-1 heterodimers with a probable sequestration of this subunit, while the beta-1 subunit has relatively stable levels. Events such as a decline in TGF-beta activity, which has an important role in the remodeling of the extracellular matrix (Zeng et al. 1996), also contribute to the dramatic changes in cell architecture and the decreased probability of initiating the division cycle.

One can infer that the modifications in cell-substratum adhesion contributing to the final growth arrest are most likely caused by a disturbance in the interaction of these different molecules, whose synthesis is differently coordinated and whose structure is altered.

Structural modifications also occur in the genome that render the initiation of DNA synthesis more difficult, inter alia, decreased supercoiling potential of DNA (Almagor and Cole 1989), decline in the hybridization signal of the probe for telomeres suggesting deletions or rearrangements at chromosome ends (Harley et al. 1990), and reorganization of the different hierarchical structures of DNA with decondensation of heterochromatin and accumulation of extra-chromosomal circular DNA (Macieira-Coelho 1990, 1991, 1995).

These structural modifications of DNA must be responsible for the chaotic DNA partition during the last mitoses of the human fibroblast proliferative life span (Macieira-Coelho 1995), which also decreases the probability of reinitiating DNA synthesis.

Structure and function are coupled in the cell, the flow of information depending on structural flexibility; this is what has been called the biology of conformation (Ivanov et al. 1973). The structural reorganization at the supramolecular and cellular levels decreases the conformational flexibility and lowers the probability of activating energy barriers. This is a fundamental feature of the evolution of the fibroblast proliferative life span and in particular of the terminal cell, which has been neglected. It is in itself responsible to a great extent for the deregulation of the different signaling pathways rendering the progression through the division cycle increasingly more difficult. At the very end, only viral gene products have the biological activity necessary to compensate for or bypass the defective mechanisms (Tsuji et al. 1984).

3.1.1 Energy Transduction Mechanisms

There is evidence favoring a disturbance of energy transduction mechanisms all the way from the periphery to the nucleus.

A disturbance seems to be present at the very origin of inorganic phosphate in the terminal cell, since there is a decrease in ATP content after exposure to metabolic poisons (Muggleton-Harris and Defuria 1985). This was considered as an estimate of ATP turnover.

The pathways of the phosphorylation cascade initiated at the cell periphery are impaired. Serum-stimulated phase IV cells demonstrate little activation of phospholipase C and D, failing to generate significant amounts of 1,2-diacylglycerol (DAG), the endogenous activator of protein kinase C (PKC) (Choudhury et al. 1991; Vannini et al. 1994; Venable et al. 1994). An impaired ability to translocate PKC to the cell membrane following mitogenic stimulation is also present (De Tata et al. 1993). Ceramide, which is significantly increased in the terminal cell, has been proposed to be responsible for the inhibition of proliferation by inhibiting this pathway of energy transduction (Venable et al. 1994). After long-term treatment with an agonist, however, there is a significantly higher increase in DAG and higher activation of phospholipase D, the enzyme involved in sustained DAG generation (Meacci et al. 1995).

Another phosphorylation pathway is also disregulated in the terminal fibroblast. There is an enhanced cAMP response to isoproterenol, possibly due to lower cAMP phosphodiesterase activity (Ethier et al. 1992), and in the presence of an agonist there is an increase in the level of both the cAMP-dependent regulatory and catalytic subunits of protein kinase A (PKA) (Liu et al. 1986). The cAMP response element-binding protein (CREB), however, which enhances gene expression when phosphorylated by protein kinase A (PKA), is markedly decreased (Chin et al. 1996).

Thus, the regulation of these pathways seems to be disturbed through a modification of the interaction of the respective components.

The phosphorylation of ribosomal S6 proteins, a prerequisite for the activation of protein synthesis during G1, fails to be enhanced in stimulated Phase IV cells (Kihara et al. 1986). Qualitative modifications of the phosphorylation of proteins have been reported with reinforcement of the phosphorylation in some proteins and a relative decrease in others (Kahn et al. 1982; Atadja et al. 1994). A similar situation was described for the 44- and 42-kDa forms of the mitogen-activated protein kinase (MAP), also called extracellular signal-regulated (ERK) kinase (Afshari et al. 1993). While both forms are phosphorylated in proliferating cells, only the former was phosphorylated in terminal fibroblasts.

Another interesting alteration is the presence of a new 3′:5′-cyclic AMP-independent histone kinase (Kahn et al. 1982).

Postmitotic Phase IV cells are unable to phosphorylate the retinoblastoma gene product (pRb) (Stein et al. 1990), a necessary step to entering into S phase. The phosphorylation of pRb during G1 phase leads to the expression of the transcription factor E2F1, and seems to be mediated by cyclin D/Cdk4 and cyclin E/Cdk2 complexes; in the postmitotic fibroblasts, although the expression of cyclins D1 and E is elevated, associated kinase activity is very low and the cyclin D1/Cdk2 complexes contain exclusively unphosphorylated Cdk2 (Dulic et al. 1993; Lucibello et al. 1993). Another group reported a decreased cyclin D1 mRNA expression (Won et al. 1992). The nonphosphorylation of pRb is one of the causes for the nonexpression of the DNA polymerase-alpha, cdc2, and cyclin A genes (Cristofalo and Pignolo 1996). An additional cause for the deficient expression of E2F1 may be the presence of an inhibitory factor (Good et al. 1996).

Barrett and coworkers pursued the study of the phosphorylation pathways during Phase IV. They found that, although the induction of cdc2 and cdk2 kinases is defective (Richter et al. 1991; Tsuji et al. 1993), a protein phosphatase inhibitor could to a certain extent override the block of DNA synthesis, so that phosphorylation of the retinoblastoma and MAP-kinase proteins took place, as well as the induction of the cdc2 protein (Afshari and Barrett 1994). Hence, protein phosphatases seem to play a role in the negative growth regulation of the contact-inhibited as well as of the postmitotic fibroblast. Morevoer, the cyclin-dependent kinase inhibitor p16 is also implicated in final growth arrest (Alcorta et al. 1996; Hara et al. 1996; Wong and Riabowol 1996); p16 becomes complexed to both cdk4 and cdk6. On the other hand, the p21 inhibitor coded by the SDI1 gene increases before the postmitotic state but decreases when cells are finally arrested (Alcorta et al. 1996). Since it is expressed in postmitotic cells at barely detectable levels, it is not required for the final growth arrest (Medcalf et al. 1996).

Another protein inhibitor that was supposed to play a role in the postmitotic state is p53, since the introduction of a dominant-negative p53 mutant into phase IV cells extended the proliferative life span by approximately 17 doublings (Bond et al. 1994). Although an increased activity of p53 has been reported in terminal fibroblasts (Atadja et al. 1995), cells from individuals with Li-Fraumeni syndrome enter phase IV in spite of the absence of the p53 (Medcalf et al. 1996). In contrast to mitotic cells, induction in the terminal cell of the SDI1 gene that codes for p21 is independent of p53 function (Bond et al. 1995).

Other steps necessary to enter S phase that are missing are the accumulation of cyclin B mRNAs, and $p34^{cdc2}$ protein (Stein et al. 1991). However, not all the events that normally take place in mitotic cells during transit through G1 are blocked. Induction, for instance, of thymidine kinase, c-Ha-ras, c-myc, c-jun, histone H3, and beta-actin occurs in the postmitotic cell (Paulsson et al. 1986; Rittling et al. 1986; Seshadri and Campisi 1989). Actually, the protooncogene c-Ha-ras-1 is amplified and overexpressed (Srivastava et al. 1985) and so is the cyclin D1 (Fukami et al. 1995). Some results are, however, conflicting; the expression of the thymidine kinase gene, for instance, was reported to be depressed (Chang and Chen 1988). It seems that many of the results depended on the way in which the cells were maintained or were considered to be postmitotic, and are probably irrelevant to explain growth arrest.

Other results trying to ascertain which of the G1 events of mitotic cells are missing in the postmitotic fibroblasts are also conflicting. It has been reported that in phase IV cells c-fos expression is not inducible in response to growth factors (Seshadri and Campisi 1989). It can, however, be induced by phorbol esters, which are activators of protein kinase C (PKC) (Riabowol et al. 1992); this could mean that more powerful energy transduction mechanisms are needed to induce this gene at the very end of the fibroblasts' proliferative life span. Other authors, however, found no decline of c-fos expression after stimulating the terminal fibroblasts with serum (Paulsson et al. 1986; Lucibello et al. 1993). Only one of the five fos and jun family members (fosB) showed significantly decreased induction (Lucibello et al. 1993). These authors also came to the conclusion that DNA binding of AP-1 is not reduced to a level that could implicate it in the postmitotic state.

Another source of energy transduction that is modified, contributing to the final growth arrest, is Ca^{2+} mobilization. An elevation of $CaCl_2$ fails to increase the maximal cell density in phase IV cells (Praeger and Gilchrest 1986) maybe because of an uncoupling of the cell-cycle regulation between calmodulin and its respective mRNA (Brooks-Frederich et al. 1993) and the suppression of calcium-dependent membrane currents (Takahashi et al. 1992; Liu et al. 1994). These disturbances in the mobilization of Ca^{2+} may be due to the overexpression of a gene coding for a protein (WS3-10) with a calcium-binding domain (Grigoriev et al. 1996).

An additional finding indicating a general deregulation of energy transduction mechanisms in postmitotic human fibroblasts is the intracellular acidification (Takahashi et al. 1992) and the significant decrease of the NAD/NADH ratio (Passi et al. 1993), possibly due inter alia to the increased expression of the gene for NADH dehydrogenase (Doggett et al. 1992).

3.1.2
Putative Growth Inhibitors and Failure to Induce Positive Growth Factors

Cell-fusion experiments suggested the presence of growth inhibitors responsible for the final postmitotic state. Norwood et al. (1974) measured DNA synthetic activity in heterokaryons of proliferative and postmitotic cells, and found that the postmitotic nucleus inhibited DNA synthesis in the nucleus originated from the proliferating cell. Brief postfusion treatments with cycloheximide or puramycin delayed the inhibition of DNA synthesis in the heterokaryons (Burmer et al. 1982). This suggested that a protein synthesis-dependent growth inhibition is responsible for the suppressive effect on DNA synthesis.

Burmer et al. (1983) and Drescher-Lincoln and Smith (1983) also showed that cytoplasts from postmitotic cells are inhibitory when fused with whole cycling cells. Perfusion treatment of the postmitotic cells with protein synthesis inhibitors eliminated the ability of the postmitotic cytoplast to block DNA synthesis (Drescher-Lincoln and Smith 1984). DNA synthesis-inhibitory activity was associated with the cytoplasmic membrane, and was heat-, trypsin- and periodate-sensitive, suggesting that it is a glycoprotein (Stein and Atkins 1986; Stein et al. 1986).

Furthermore, polyadenylated RNA isolated from terminal human fibroblasts inhibited DNA synthesis in cycling cells after microinjection, whereas the RNA from proliferation-competent cells had no such effect (Lumpkin et al. 1986).

Stein and Yanishevsky (1979, 1981) and Stein et al. (1982) also assayed heterokaryons with postmitotic nuclei and nuclei from different immortalized cell lines. Nuclei from some cell lines could induce DNA synthesis, others were inhibited by the postmitotic nucleus.

Failure to express positive autocrine regulators also contributes to final growth arrest. Ferber et al. (1993) found that terminal cells fail to express insulin-like growth (IGF-1) factor mRNA, which is expressed in moderate amounts in mitotic human fibroblasts. The results suggested that the failure to express IGF-1 may be due to the lack of transcription factors.

A reduction in the levels of polyamines, which are crucial for the progression through G1, has also been reported, due to the reduction of the activity of

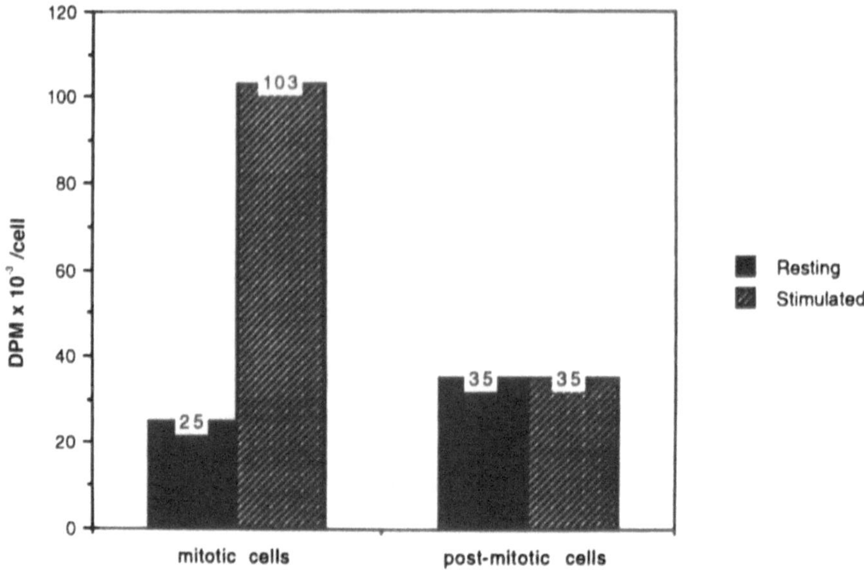

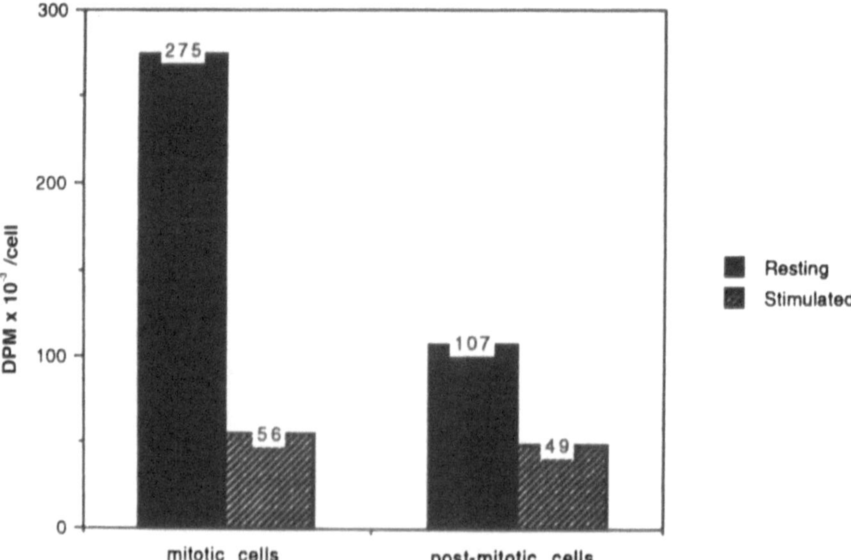

Fig. 8. Thymidine (*above*) and glucosamine (*below*) incorporation in resting phase II cells and in cells at the end of their proliferative life span, before (*black columns*) and after (*gray columns*) stimulation. Cells were stimulated by the addition of fresh medium and harvested 24h later. Identical cultures were exposed for 24h to ^{3}H-thymidine and for 1h to ^{3}H-glucosamine. (Macieira-Coelho and Söderberg 1993, by permission of Wiley-Liss)

ornithine decarboxylase (ODC), the enzyme critical for their synthesis (Duffy and Kremzner 1977; Chen et al. 1986). The reduction in enzyme activity could be due inter alia to the presence of an inhibitor of ODC (Duffy and Kremzner 1977). Another polyamine-dependent mechanism that fails in the terminal fibroblast is the conversion of lysine to deoxyhypusine and to hypusine residue in the initiation factor 5A precursor (Chen and Chen 1997).

Moreover, in the postmitotic cell, growth factors fail to induce the suggar composition (Fig. 8) which renders the glycoprotein described above (systolin) less inhibitory (Macieira-Coelho and Söderberg 1993).

Many investigators engaged in the study of the terminal postmitotic human fibroblast were looking for *the* change that could explain growth arrest. This attitude, although proven elusive, still prevails. Terminal growth arrest, however, appears as a complex interaction between positive and negative growth regulators, decreased molecular and cellular conformational flexibility, coupled with disturbances in energy transduction mechanisms, and reorganizations in the different hierarchical orders of DNA structures. Further studies are needed to ascertain if this is a homeostatic, programmed evolution (Martin et al. 1974) or a degenerative event secondary to genetic damage (Rubin 1997).

References

Afshari CA, Barrett JC (1994) Disruption of G0-G1 arrest in quiescent and senescent cells treated with phosphatase inhibitors. Cancer Res 54:2317–2321

Afshari CA, Vojta PJ, Annab LA, Futreal PA, Willard TB, Barrett JC (1993) Investigation of the role of G1/S cell cycle mediators in celular senescence. Exp Cell Res 209:231–237

Alcorta DA, Xiong Y, Phelps D, Hannon G, Beach D, Barrett JC (1996) Involvement of the cyclin-dependent kinase inhibitor p16 (INK4a) in replicative senescence of normal human fibroblasts. Proc Natl Acad Sci USA 93:13742–13747

Almagor M, Cole RD (1989) Changes in chromatin structure during aging of cell cultures as revealed by differential scanning calorimetry. Biochemistry 28:5688–5693

Atadja PW, Stringer KF, Riabowol KT (1994) Loss of sserum response element-binding activity and hyperphosphorylation of serum response factor during cellular aging. Mol Cell Biol 14:4991–4999

Atadja P, Wong, Garkavetz I, Veillette C, Riabowol K (1995) Increased activity of p53 in senescing fibroblasts. Proc Natl Acad Sci USA 92:8384–8352

Blondal JA, Dick JE, Wright JA (1985) Membrane glycoprotein changes during senescence of normal human diploid fibroblasts in culture. Mech Ageing Dev 30:273–283

Bond JA, Wyllie FS, Wynford-Thomas D (1994) Escape from senescence in human diploid fibroblasts induced directly by mutant p53. Oncogene 9:1885–1889

Bond JA, Blaydes JP, Rowson J, Haughton MF, Smith JR, Wynford-Thomas D, Wyllie FS (1995) Mutant p53 rescues human diploid cells from senescence without the induction of SDI1/WAF1. Cancer Res 55:1404–2409

Bond JA, Haughton M, Blaydes J, Gire V, Wynford-Thomas D, Wyllie F (1996) Evidence that transcriptional activation by p53 plays a direct role in the induction of cellular senescence. Oncogene 13:2097–2104

Brooks-Frederich KM, Cianciarulo FL, Rittling SR, Cristofalo VJ (1993) Cell cycle-dependent regulation of Ca^{2+} in young and senescent WI-38 cells. Exp Cell Res 205:412–415

Burmer GC, Zeigler CJ, Norwood TH (1982) Evidence for endogenous polypeptide-mediated inhibition of cell-cycle transit in human diploid cells. J Cell Biol 94:187–192

Burmer GC, Motulsky H, Zeigler CJ, Norwood TH (1983) Inhibition of DNA synthesis in young cycling human diploid fibroblast-like cell upon fusion to enucleate cytoplasts from senescent cells. Exp Cell Res 145:79–84

Chang ZF, Chen KY (1988) Regulation of ornithine decarboxylase and other cell cycle-dependent genes during senescence of IMR-90 human diploid fibroblasts. J Biol Chem 263:11431–11435

Chen KY, Chang ZF, Liu AYC (1986) Changes of serum-induced ornithine decarboxylase activity and putrescine content during aging of IMR-90 human diploid fibroblasts. J Cell Phys 129: 142–146

Chen PC, Chen KY (1997) Dramatic attenuation of hypusine formation on eukaryotic intiation factor 5A during senescence of IMR-90 human diploid fibroblasts. J Cell Phys 170:248–254

Chin JH, Okazaki M, Frazier JS, Hu ZW, Hoffman BB (1996) Impaired cAMP-mediated gene expression and decreased cAMP response element binding protein in senescent cells. Am J Phys 271:C362–C371

Choudhury GG, Sylvia VL, Sakagushi AY (1991) Decline of signal transduction by phospholipase C in IMR 90 human diploid fibroblasts at high population doubling levels. FEBS Lett 293:211–214

Cristofalo VJ, Pignolo RJ (1996) Molecular markers of senescence in fibroblast-like cultures. Exp Gerontol 31:111–123

Dell'Orco RT, McClung JK, Jupe ER, Liu XT (1996) Prohibitin and the senescent phenotype. Exp Gerontol 31:235–244

De Tata V, Ptasznik A, Cristofalo VJ (1993) Effects of the tumor promoting agent phorbol 12-myristate 13-acetate (PMA) on the proliferation of young and senescent WI-38 human diploid fibroblasts. Exp Cell Res 205:261–269

Doggett DL, Rotenberg MO, Pignolo RJ, Phillips PD, Cristofalo VJ (1992) Differential gene expression between young and senscent, quiescent Wi-38 Cells. Mech Ageing Dev 65:239–255

Drescher-Lincoln CK, Smith JR (1983) Inhibition of DNA synthesis in proliferating human diploid fibroblasts by fusion with senescent cytoplasts. Exp Cell Res 144:455–462

Drescher-Lincoln CK, Smith JR (1984) Inhibition of DNA synthesis in senescent-proliferating cybrids is mediated by endogenous proteins. Exp Cell Res 153:208–217

Duffy PE, Kremzner LT (1977) Ornithine decarboxylase activity and polyamines in relation to aging of human fibroblasts. Exp Cell Res 108:435–440

Dulic V, Drullinger LF, Lees E, Reed SI, Stein GH (1993) Altered regulation of G1 cyclins in senescent human diploid fibroblasts: accumulation of inactive cyclin E-Cdk2 and cyclin D1-Cdk2 complexes. Proc Natl Acad Sci USA 90:11034–11038

Ethier MF, Medeiros M, Romano FD, Dobson JG (1992) Mechanism of enhanced cyclic AMP stimulation by isoproterenol in aged human fibroblasts. Exp Gerontol 27:287–300

Ferber A, Chang CD, Sell C, Ptasznik A, Cristofalo VJ, Hubbard K, Ozer HL, Adamo M, Roberts CT Jr, LeRoith D, Dumenil G, Baserga R (1993) Failure of senescent human fibroblasts to express the insulin-like growth factor-1 gene. J Biol Chem 268:17883–17888

Froehlich JE, Anastassiades TP (1975) Possible limitation of growth in human fibroblast cultures by diffusion. J Cell Phys 86:567–580

Fukami J, Anno K, Ueda K, Takahashi T, Ide T (1995) Enhanced expression of cyclin D1 in senescent human fibroblasts. Mech Ageing Dev 81:139–157

Garcia-Giralt E, Berumen L, Macieira-Coelho A (1970) Growth inhibitory activity in the supernatants of nondividing WI-38 cells. J Natl Cancer Inst 45:649–655

Good L, Dimri GP, Campisi J, Chen KY (1996) Regulation of dihydrofolate reductase gene expression and E2F components in human diploidd fibroblasts during growth and senescence. J Cell Phys 168:580–588

Grigoriev VG, Thweatt R, Moerman EJ, Goldstein S (1996) Expression of senescence-induced protein WS3-10 in vivo and in vitro. Exp Gerontol 31:145–157

Hara E, Smith R, Parry D, Tahara H, Stone S, Peters G (1996) Regulation of p16 cdkn2 expression and its implications for cell immortalization and senescence. Mol Cell Biol 16:859–867

Harley CB, Futcher AB, Greider CW (1990) Telomeres shorten during aging of human fibroblasts. Nature 345:458–460

Harper JW, Elledge J, Keyomarsi K, Dynlacht B, Tsai LH, Zhang P, Dobrowolski S, Bai C, Connell-Crowley L, Swindell, Fox MP, Wei N (1995) Inhibition of cyclin-dependent kinases by p21. Mol Biol Cell 6:387–400

Hayflick L, Moorhead PS (1961) The serial cultivation of human diploiod cell strains. Exp Cell Res 25:585–621

Hiu IJ, Liepkalns V, Azzarone B, Macieira-Coelho A (1985) Metabolic changes in fibroblast adhesion sites induced by growth factors. Cell Biol Int Rep 9:379–387

Houck JC, Sharma VK, Cheng RF (1973) Fibroblast chalone and serum mitogen (anti-chalone). Nat New Biol 246:111–113

Houck JC, Kanagalingam K, Hunt C, Attalah A, Chung A (1977) Lymphocyte and fibroblast chalones: some chemical properties. Science 196:896–897

Hu Q, Moerman EJ, Goldstein S (1996) Altered expression and regulation of the alpha-5-beta-1 integrin-fibronectin receptor lead to reduced amounts of funtional alpha5beta1 heterodimer on the plasma membrane of senescent human diploid fibroblasts. Exp Cell Res 224:251–263

Irving J, Feng J, Wistrom C, Pikaart M, Villeponteau B (1992) An altered repertoire of fos/jun (AP-1) at the onset of replicative senescence. Exp Cell Res 202:161–166

Ivanov VI, Minchenkova LE, Schyolkina AK, Poletayev AI (1973) Different conformations of double-stranded nucleic acid in solution as revealed by circular dichroism. Biopolymers 12:89–110

Janik-Schmitt B, Oesch F, Wieser RJ (1987) Immobilized glycolipids from human diploid fibroblasts inhibit DNA synthesis of cultured human fibroblasts. Exp Cell Res 169:15–24

Kahn A, Guillouzou A, Leibovitch MP, Cottreau D, Bourel M, Dreyfus JC (1982) Modifications of phosphoproteins and protein kinases occurring with in vitro aging of cultured human cells. Gerontology 28:360–370

Kelley RO, Trotter JA, Marek LF, Perdue BD, Taylor CB (1980) Variation in cytoskeletal assembly during spreading of progressively subcultivated human fibroblasts (IMR-90). Mech Ageing Dev 13:127–141

Kihara F, Ninomiya-Tsuji J, Ishibashoi S, Ide T (1986) Failure in S6 protein phosphorylation by serum stimulation of senescent human diploid fibroblasts TIG-1. Mech Ageing Dev 37:27–40

Konterman K, Bayreuther K (1979) The cellular aging of rat fibroblasts in vitro is a differentiation process. Gerontology 25:261–274

Liu AYC, Chang ZF, Chen KY (1986) Increased level of cAMP-dependent protein kinase in aging human lung fibroblasts. J Cell Phys 128:149–154

Liu S, Thweatt R, Lumpkin Jr CK, Goldstein S (1994) Suppression of calcium-dependent membrane currents in human fibroblasts by replicative senescence and forced expression of a gene sequence encoding a putative calcium-binding protein. Proc Natl Acad Sci USA 91: 2186–2190

Lucibello FC, Sewing A, Brüsselbach S, Bürger C, Müller R (1993) Deregulation of cyclins D1 and E and supression of cdk2 and cdk4 in senescent human fibroblasts. J Cell Sci 105:123–133

Lumpkin CK, McClung JK, Pereira-Smith OM, Smith JR (1986) Existence of high abundance antiproliferative mRNAs in senescent human diploid fibroblasts. Science 232:393–394

Macieira-Coelho A (1967) Dissociation between inhibition of movement and inhibition of division in RSV transformed human fibroblasts. Exp Cell Res 47:193–200

Macieira-Coelho A (1974) Are non-dividing cells present in ageing populations of human fibroblasts in vitro? Nature 248:421–422

Macieira-Coelho A (1983) Changes in membrane properties associated with cellular aging. Int Rev Cytol 83:183–220

Macieira-Coelho A (1988) Biology of normal proliferating cells in vitro. Relevance for in vivo aging. Karger, Basel

Macieira-Coelho A (1990) Reorganization in the different hierarchical structures of DNA during cell senescence. In: Finch CE, Johnson CE (eds) Molecular biology of aging, vol 123. Wiley-Liss, New York, pp 351–364

Macieira-Coelho A (1991) Chromatin reorganization during senescence of proliferating cells. Mutat Res 256:81–104

Macieira-Coelho A (1995) The last mitoses of the human fibroblast proliferative life span, physiopathological implications. Mech Ageing Dev 82:91–104

Macieira-Coelho A (1996) A growth inhibitor implicated in the growth arrest of human fibroblasts. FEBS Lett 378:61–63

Macieira-Coelho A, Azzarone B (1982) Aging of human fibroblasts is a succession of subtle changes in the cell cycle and has a final short stage with abrupt events. Exp Cell Res 141:325–332

Macieira-Coelho A, Azzarone B (1990) Correlation between contractility and proliferation in human fibroblasts. J Cell Phys 142:610–614

Macieira-Coelho A, Berumen L (1973) The cell cycle during growth inhibition of human embryonic fibroblasts in vitro. Proc Soc Exp Biol Med 144:43–47

Macieira-Coelho A, Söderberg A (1993) Growth inhibitory activity in extracts from human fibroblasts. J Cell Phys 154:92–100

Macieira-Coelho A, Taboury F (1982) A reevaluation of the changes in proliferation during ageing in vitro. Cell Tissue kinet 15:213–224

Macieira-Coelho A, Pontén J, Philipson L (1966) The division cycle and RNA synthesis in diploid human cells at different passage levels in vitro. Exp Cell Res 42:374–377

Macieira-Coelho A, Hiu IJ, Garcia-Giralt E (1969) Stimulation of DNA synthesis in resting stage human fibroblasts after infection with Rous sarcoma virus. Nature 222:1172

Magnuson VL, Young M, Schattenberg DG, Mancini MA, Chen D, Steffensen B, Klebe RJ (1991) The alternative splicing of fibronectin pre-mRNA is altered during aging and in response to growth factors. J Biol Chem 266:14654–14662

Manilal S, Scott GK, Tse CA (1993) Inhibition of an endogenous growth-related proteinase enhances the recovery of a negative growth regulator from cultured human cells. Cell Biol Int 17:317–323

Martin GM, Sprague CA, Norwood TH, Pendergrass WR (1974) Clonal selection, attenuation and differentiation in an in vitro model of hyperplasia. Am J Pathol 74:137–154

Matsumura T, Zerrudo Z, Hayflick L (1979) Senescent human diploid cells in culture: survival, DNA synthesis and morphology. J Gerontol 34:328–334

Matuoka K, Mitsui Y (1981) Involvement of cell surface heparan sulfate in density-dependent inhibition of cell proliferation. Cell Struct Funct 6:23–34

Meacci E, Vasta V, Faraoni P, Farnararo M, Bruni P (1995) Potentiated bradykinin-induced increase of 1,2-diacylglycerol generation and phospholipase D activity in human senescent fibroblasts. Biochem J 312:799–803

Medcalf ASC, Klein-Szanto AJP, Cristofalo VJ (1996) Expression of p21 is not required for senescence of human fibroblasts. Cancer Res 56:4582–4585

Millis AJT, Hoyle M, Field B (1977) Human fibroblasts conditioned media contains growth promoting activities for low density cells. J Cell Phys 93:17–24

Milo GE (1973) Enhancement of senescence in low passage human embryonic cells by an agent extracted from phase III cells. Exp Cell Res 79:143–151

Milo GE, Hart RW (1976) Age-related alterations in plasma membrane glycoprotein content and scheduled or nonscheduled DNA synthesis. Arch Biochem Biophys 176:324–333

Muggleton-Harris AL, Defuria R (1985) Age-dependent metabolic changes in cultured fibroblasts. In Vitro Cell Dev Biol 21:271–276

Noda A, Ning Y, Venable SF, Pereira-Smith OM, Smith JR (1994) Cloning of senescent cell-derived inhibitors of DNA synthesis using an expression screen. Exp Cell Res 211:90–98

Norwood TH, Pendergrass WR, Martin GM (1974) Dominance of the senescent phenotype in heterokaryons between replicative and post-replicative human fibroblast-like cells. Proc Natl Acad Sci USA 71:2231–2235

Ohsawa T, Senshu T (1987) Exogenous GM1 ganglioside caused G1-arrest of human diploid fibroblasts. Flow cytometric studies. Exp Cell Res 173:49–55

Passi A, Albertini R, Bardoni A, Rindi S, Pallavicini G, De Luca G (1993) Modifications of proteoglycans produced by human skin fibroblasts during replicative senescence. Cell Biochem Funct 11:263–269

Paulsson Y, Bywater M, Pfeifer-Olsson S, Olsson R, Nilsson S, Heldin CH, Westermark B, Betsholtz C (1986) Growth factors induce early pre-replicative changes in senescent human fibroblasts. EMBO J 5:2157–2162

Pereira-Smith OM, Fisher SF, Smith JR (1985) Senescent and quiescent cell inhibitors of DNA synthesis: membrane associated proteins. Exp Cell Res 160:296–306

Praeger FC, Gilchrest BA (1986) Influence of increased extracellular calcium concentration and donor age on density-dependent growth inhibition of human fibroblasts. Proc Soc Exp Biol Med 182:315–321

Rabinovitch PS, Norwood TH (1980) Comparative heterokaryon study of cellular senescence and the serum deprived state. Exp Cell Res 130:101–109

Raes M, Brabander M de, Remacle J (1984) Polyploid cells in ageing hamster fibroblasts in vitro: possible implication of the centrosome. Eur J Cell Biol 35:79–80

Riabowol K, Schiff J, Gilman MZ (1992) Transcription factor AP-1 activity is required for initiation of DNA synthesis and is lost during cellular aging. Proc Natl Acad Sci USA 89:157–161

Richter KH, Afshari CA, Annab LA, Burkhart BA, Owen RD, Boyd J (1991) Down-regulation of cdc2 in senescent human and hamster cells. Cancer Res 51:6010–6013

Rittling SR, Brooks KM, Cristofalo VJ, Baserga R (1986) Expression of cell cycle-dependent genes in young and senescent WI-38 fibroblasts. Proc Natl Acad Sci USA 83:3316–3320

Rubin H (1997) Cell aging in vivo and in vitro. Mech Ageing Dev 98:1–35

Schachtschabel DO, Wever J (1978) Age-related decline in the synthesis of glycosaminoglycans by cultured human fibroblasts (WI-38). Mech Ageing Dev 8:257–264

Seshadri T, Campisi J (1989) Growth-factor-inducible gene expression in senescent human fibroblasts. Exp Gerontol 24:515–522

Sorrentino JA, Millis AJT (1984) Structural comparisons of fibronectins isolated from early and late passage cells. Mech Ageing Dev 28:83–97

Srivastava A, Norris JS, Reis RJS, Goldstein S (1985) c-Ha-ras-1 proto-oncogene amplification and overexpression during the limited replicative life span of normal human fibroblasts. J Biol Chem 260:6404–6409

Stein GH, Atkins L (1986) Membrane-associated inhibitor of DNA synthesis in senescent human diploid fibroblasts. Characterization and comparison to quiescent cell inhibitor. Proc Natl Acad Sci USA 79:5287–5292

Stein GH, Yanishevsky RM (1979) Entry into S phase is inhibited in two immortal cell lines fused to senescent human diploid cells. Exp Cell Res 120:155–165

Stein GH, Yanishevsky RM (1981) Quiescent human diploid cells can inhibit entry into S phase in replicative nuclei in heterodikaryons. Proc Natl Acad Sci USA 79:9030–9034

Stein GH, Yanishevsky RM, Gordon L, Beeson M (1982) Carcinogen-transformed human cells are inhibited from entry into S phase by fusion to senescent cells but cells transformed by DNA tumor viruses overcome the inhibition. Proc Natl Acad Sci USA 79:5287–5292

Stein GH, Atkins L, Beeson M, Gordon L (1986) Quiescent human diploid fibroblasts. Common mechanism for inhibition of DNA replication in density-inhibited and serum-deprived cells. Exp Cell Res 162:255–260

Stein GH, Beeson M, Gordon L (1990) Failure to phosphorylate the retinoblastoma gene product in senescent human fibroblasts. Science 249:666–669

Stein GH, Drullinger LF, Robetorye RS, Pereira-Smith OM, Smith JR (1991) Senescent cells fail to express cdc2, cycA, and cycB in response to mitogen stimulation. Proc Natl Acad Sci USA 88:11012–11016

Stroebel-Stevens JD, Lacey JC (1981) Further evidence for an inhibitor of proliferation elaborated by normal human fibroblasts in culture: partial characterization of the inhibitor. J Cell Phys 106:201–207

Takahashi Y, Yoshida T, Takshima S (1992) The regulation of intracellular calcium ion and pH in young and old fibroblast cells (WI-38). J Gerontol Biol Sci 47:B65–B70

Tsuji Y, Ninomiya-Tsuji J, Torti SV, Torti FM (1993) Selective loss of cdc2 and cdk2 induction by tumor necrosis factor-alpha in sensecent human diploid fibroblasts. Exp Cell Res 209:175–182

Tsuji Y, Ide T, Ishibashi S, Nishikawa K (1984) Loss of responsiveness in senescent human TIG-1-cells to the DNA synthesis-inducing effect as a function of maturation and aging. Mech Ageing Dev 27:219–232

VanGansen P, Siebertz B, Capone B, Malherbe L (1984) Relationship between cytoplasmic microtubular complex, DNA synthesis and cell morphology in mouse embryonic fibroblasts (effects of age, serum deprivation, aphidicolin, cyochalasin B and colchicine). Biol Cell 52:161–174

Vannini F, Meacci E, Vasta V, Farnaro M, Bruni P (1994) Involvement of protein kinase C and arachidonate signaling pathways in the alteration of proliferative response of senescent IMR690 human fibroblasts. Mech Ageing Dev 76:101–111

Venable ME, Blobe GC, Obeid LM (1994) Identification of a defect in the phospholipase D/diacylglycerol pathway in cellular senescence. J Biol Chem 269:26040–26044

Vogel KG, Kelley RO, Steward C (1981) Loss of organized fibronectin matrix from the surface of aging diploid fibroblasts. Mech Ageing Dev 16:295–303

Wells V, Malluchi L (1991) Identification of an autocrine negative growth factor: mouse beta-galactoside-binding protein is a cytostatic factor and cell growth regulator. Cell 64:91–97

Westergren-Thorsson G, Persson S, Isaksson A, Önnervik PO, Malmström A, Fransson LA Fransson (1993) 1-irudonate-rich glycosaminoglycans inhibit growth of normal fibroblasts independently of serum or added growth factors. Exp Cell Res 206:93–99

Wever J, Schachtachabel DO, Sluke G, Wever G (1980) Effect of short or long term treatment with exogenous glycosaminoglycans on growth and glycosaminoglycan synthesis of human fibroblasts (WI-38). Mech Ageing Dev 14:89–99

Wieser RJ, Oesch F (1986) Contact inhibition of growth of human diploid fibroblasts by immobilized plasma membrane glycoproteins. J Cell Biol 103:361–367

Wieser RJ, Schütz S, Tschank G, Thomas H, Dienes HP, Oesch F (1990) Isolation and characterization of a 60–70-kD plasma membrane glycoprotein involved in the contact-dependent inhibition of growth. J Cell Biol 111:2681–2692

Won KA, Xiong Y, Beach D, Gilman MZ (1992) Growth-regulated expression of D-type cyclins in human diploid fibroblasts. Proc Natl Acad Sci 89:9910–9914

Wong H, Riabowol K (1996) Differential CDK-inhibitor gene expression in aging human diploid fibroblasts. Exp Gerontol 31:311–325

Wu KF, Pope JH, Ellem KAO (1985) Inhibition of growth of certain human tumour cell lines by a factor derived from human fibroblast-like cell lines. I. Demonstration by mixed culture and by use of cell washings. Int J Cancer 35:477–482

Zeng G, McCue HM, Mastrangelo L, Millis AJ (1996) Endogenous TGF-beta activity is modified during cellular aging: effects on metalloproteinase and TIMP-1 expression. Exp Cell Res 228:271–276

Subject Index